GENETICS AND TUBERCULOSIS

The Novartis Foundation is an international scientific and educational charity (UK Registered Charity No. 313574). Known until September 1997 as the Ciba Foundation, it was established in 1947 by the CIBA company of Basle, which merged with Sandoz in 1996, to form Novartis. The Foundation operates independently in London under English trust law. It was formally opened on 22 June 1949.

The Foundation promotes the study and general knowledge of science and in particular encourages international co-operation in scientific research. To this end, it organizes internationally acclaimed meetings (typically eight symposia and allied open meetings, 15–20 discussion meetings, a public lecture and a public debate each year) and publishes eight books per year featuring the presented papers and discussions from the symposia. Although primarily an operational rather than a grant-making foundation, it awards bursaries to young scientists to attend the symposia and afterwards work for up to three months with one of the other participants.

The Foundation's headquarters at 41 Portland Place, London W1N 4BN, provide library facilities, open every weekday, to graduates in science and allied disciplines. The library is home to the Media Resource Service which offers journalists access to expertise on any scientific topic. Media relations are also strengthened by regular press conferences and book launches, and by articles prepared by the Foundation's Science Writer in Residence. The Foundation offers accommodation and meeting facilities to visiting scientists and their societies.

Information on all Foundation activities can be found at http://www.novartisfound.demon.co.uk

Novartis Foundation Symposium 217

GENETICS AND TUBERCULOSIS

1998

JOHN WILEY & SONS

Chichester · New York · Weinheim · Brisbane · Singapore · Toronto

Published in 1998 by John Wiley & Sons Ltd,
Baffins Lane, Chichester,
West Sussex PO19 1UD, England

National 01243 779777
International (+44) 1243 779777
e-mail (for orders and customer service enquiries): cs-books@wiley.co.uk
Visit our Home Page on http://www.wiley.co.uk
or http://www.wiley.com

Other Wiley Editorial Offices

John Wiley & Sons, Inc., 605 Third Avenue,
New York, NY 10158-0012, USA

WILEY-VCH Verlag GmbH, Pappelallee 3,
D-69469 Weinheim, Germany

Jacaranda Wiley Ltd, 33 Park Road, Milton,
Queensland 4064, Australia

John Wiley & Sons (Asia) Pte Ltd, 2 Clementi Loop #02-01,
Jin Xing Distripark, Singapore 129809

John Wiley & Sons (Canada) Ltd, 22 Worcester Road,
Rexdale, Ontario M9W 1L1, Canada

Novartis Foundation Symposium 217
ix+269 pages, 45 figures, 14 tables

Library of Congress Cataloging-in-Publication Data

Genetics and tuberculosis.
p. cm. – (Novartis Foundation symposium : 217)
'Symposium on Genetics and Tuberculosis, held at the Sports Science Institute, Cape Town, South Africa on 18–20 November 1997'--Contents p.
Editors: Derek J. Chadwick (Organizer) and Gail Cardew.
Includes bibliographical references and index.
ISBN 0-471-98261-X (hbk. : alk. paper)
1. Tuberculosis–Genetic aspects–Congresses. 2. Mycobacterium tuberculosis–Congresses. I. Chadwick, Derek. II. Cardew, Gail. III. Symposium on Genetics and Tuberculosis (1997 : Cape Town, South Africa) IV. Series.
[DNLM: 1. Tuberculosis–genetics congresses. 2. Hereditary Diseases–genetics congresses. WF 200 G3275 1998]
RC307.G46 1998
616.9′95042–dc21
DNLM/DLC
for Library of Congress 98-23836
CIP

British Library Cataloguing in Publication Data

A catalogue record for this book is available from the British Library

ISBN 0 471 98261 X

Typeset in 10½ on 12½ pt Garamond by Dobbie Typesetting Limited, Tavistock, Devon.
Printed and bound in Great Britain by Biddles Ltd, Guildford and King's Lynn.
This book is printed on acid-free paper responsibly manufactured from sustainable forestry, in which at least two trees are planted for each one used for paper production.

Contents

Symposium on Genetics and tuberculosis, held at the Sports Science Institute, Cape Town, South Africa on 18–20 November 1997

Editors: Derek J. Chadwick (Organizer) and Gail Cardew

Participants

R. M. Anderson Department of Zoology, University of Oxford, South Parks Road, Oxford OX1 3PS, UK

E. D. Bateman Department of Medicine, Medical School, University of Cape Town, Observatory 7925, Cape Town, South Africa

R. J. Bellamy Wellcome Trust Centre for Human Genetics, Windmill Road, Oxford University, Oxford OX3 7BN, UK

S. R. Benatar Department of Medicine, Medical School, University of Cape Town, Observatory 7925, Cape Town, South Africa

A. D. Beyers MRC Centre for Molecular and Cellular Biology, Department of Medical Biochemistry, Faculty of Medicine, University of Stellenbosch, PO Box 19063, Tygerberg 7505, South Africa

J. M. Blackwell University of Cambridge Clinical School, Department of Medicine, Box 157, Addenbrooke's Hospital, Hills Road, Cambridge CB2 2QQ, UK

P. J. Brennan Colorado State University, Department of Microbiology, Fort Collins, CO 80523, USA

S. T. Cole Unité de Génétique Moléculaire Bactérienne, Institut Pasteur, 28 Rue du Dr Roux, 75724 Paris Cedex 15, France

H. Collins (*Bursar*) Department of Immunology, Max-Planck Institute for Infection Biology, Monbijiustrasse 2, 10117 Berlin, Germany

J. Colston Mycobacteria Research Unit, National Institute for Medical Research, The Ridgeway, Mill Hill, London NW7 1AA, UK

J. Davies Department of Microbiology and Immunology, University of British Columbia, #300-6170 University Boulevard, Vancouver BC, Canada V6T 1Z3

P. R. Donald Department of Paediatrics and Child Health, Faculty of Medicine, University of Stellenbosch, PO Box 19063, Tygerberg 7505, South Africa

K. Duncan Glaxo Wellcome Research and Development, Medicines Research Centre, Gunnels Wood Road, Stevenage, Hertfordshire SG1 2NY, UK

M. R.W. Ehlers Department of Medical Biochemistry, Medical School, University of Cape Town, Observatory 7925, Cape Town, South Africa

P. E. M. Fine Infectious Disease Epidemiology Unit, Department of Infectious and Tropical Diseases, London School of Hygiene and Tropical Medicine, Keppel Street, London WC1E 7HT, UK

B. Fourie Medical Research Council National Tuberculosis Research Programme, Private Bag X385, 0001 Pretoria, South Africa

P. C. Hopewell Division of Pulmonary and Critical Care Medicine, San Francisco General Hospital, 1001 Potrero Avenue, Room 5K1, San Francisco, CA 94110, USA

G. Kaplan Laboratory of Cellular Physiology and Immunology, The Rockefeller University, 1230 York Avenue, New York, NY 10021-6399, USA

I. Kramnik Department of Microbiology and Immunology, Howard Hughes Medical Institute, Albert Einstein College of Medicine of Yeshiva University, 1300 Morris Park Avenue, Bronx, NY 10461, USA

L. Miesel Department of Microbiology and Immunology, Howard Hughes Medical Institute, Albert Einstein College of Medicine of Yeshiva University, 1300 Morris Park Avenue, Bronx, NY 10461, USA

V. Mizrahi Molecular Biology Unit, The South African Institute for Medical Research, Hospital Street, PO Box 1038, Johannesburg 2000, South Africa

I. Orme Mycobacteria Research Laboratories, Department of Microbiology, Colorado State University, Fort Collins, CO 80523, USA

V. Quesniaux Novartis Pharma, CH-4002, Basel, Switzerland

S. Ress Clinical Immunology Laboratory, Department of Medicine, Medical School, University of Cape Town, Observatory 7925, Cape Town, South Africa

G. A.W. Rook Department of Bacteriology, University College London Medical School, Windeyer Building, 46 Cleveland Street, London W1P 6DB, UK

D. Russell Washington University School of Medicine, Department of Molecular Microbiology, 660 South Euclid Avenue, St Louis, MO 63110, USA

B. Ryffel Department of Immunology, Medical School, Old Main Building, Groote Schur Hospital, Observatory 7925, Cape Town, South Africa

L. Steyn Department of Medical Microbiology, Medical School, University of Cape Town, Observatory 7925, Cape Town, South Africa

J. B. Ulmer Vaccines Research, Chiron Corporation, 4560 Horton Street, Emeryville, CA 94608, USA

P. D. van Helden MRC Centre for Molecular and Cellular Biology, Department of Medical Biochemistry, Faculty of Medicine, University of Stellenbosch, PO Box 19063, Tygerberg 7505, South Africa

A. A. Wadee Department of Immunology, The South African Institute for Medical Research, Hospital Street, PO Box 1038, Johannesburg 2000, South Africa

D. Young (*Chair*) Department of Medical Microbiology, Imperial College School of Medicine at St Mary's Hospital, Norfolk Place, London W2 1PG, UK

Introduction

Douglas Young

Department of Medical Microbiology, Imperial College School of Medicine at St Mary's Hospital, Norfolk Place, London W2 1PG, UK

This symposium comes at a critical and exciting stage in tuberculosis research. Over the last decade, a renewed appreciation of the sheer magnitude of the global burden of tuberculosis, together with the emergence of untreatable drug-resistant strains of *Mycobacterium tuberculosis*, have combined to raise public awareness of the pressing need for improved tools to control the disease. The re-emergence of tuberculosis as a public health problem has coincided with a renaissance in tuberculosis research. Optimism engendered by dramatic successes in control of tuberculosis in developed countries prompted the withdrawal of funding and enthusiasm for tuberculosis research in the 1960s, reducing it to a relative backwater inhabited by a handful of specialized and committed enthusiasts. This process has now been reversed. Once again, tuberculosis research features prominently in top quality scientific journals, attracting the attention of a new generation of researchers. However, although we undoubtedly have powerful new tools at our disposal, it is increasingly clear that the central problems we have to confront in tuberculosis research are exactly those that were left unsolved by the previous generation of outstanding scientists who tackled this complex disease. The new era in tuberculosis research is now coming of age. Can we move from having tools that allow us to manipulate mycobacteria in the test tube, to tools that allow us to manipulate the global tuberculosis epidemic?

Genetics has been chosen as the theme for this symposium. Analysis of the mycobacterial genome has provided a central focus for recent tuberculosis research, generating an unprecedented explosion in information about the bacterium. The rapid advances in the field of human genome research have as yet had less direct application to tuberculosis, but host genetics clearly have a profound influence on the course of infection. Tuberculosis represents a complex interplay between human and mycobacterial genomes, and the concept underlying this symposium is that an understanding of tuberculosis will require a complex interplay between scientists providing insights into both the host and the pathogen. The format of this symposium differs from many recent tuberculosis meetings in bringing together a relatively small group of researchers in a forum designed to maximize discussion and interaction. This is a format reminiscent of the 'Scientific Working Groups' organized by the World Health Organization in the 1980s. Much of the current tuberculosis research agenda was set out at those meetings, and I would envisage the present symposium as an opportunity to begin to chart out a new agenda for the next decade.

In taking a long-term view of tuberculosis research, it is worthwhile looking back at the experience of our predecessors. In the 1950s René Dubos predicted 'The final step in the conquest of tuberculosis may well depend upon knowledge of the factors that prevent silent infection from manifesting itself in the form of overt disease' (Dubos & Dubos 1952). In spite of recent progress in the understanding of molecular aspects of infection, our concept of latent tuberculosis remains poorly defined. We are unsure of the extent to which clinical disease is due to reactivation or to reinfection, and we lack firm evidence as to whether bacteria persist in some dormant form, or in a steady state of division and immune killing. The success of our current research will be determined by our ability to exploit understanding at a molecular level in order to address such key physiological aspects of tuberculosis. A second quote gives further pause for thought. Dubos wrote 'We need to develop a new science of social engineering that will incorporate physiological principles in the complex fabric of industrial society' (Dubos & Dubos 1952). By this stage, it seems that Dubos had lost faith in the ability of experimental science to address the real problems of tuberculosis, looking instead towards the social sciences. Assessing the epidemiology of tuberculosis, he was clearly struck by the immense influence of socioeconomic factors—increased incidence during wartime, for example—and of age, i.e. the lack of disease during the 'golden age' of adolescence. Then, as now, it is hard to account for these influences within our understanding of disease mechanisms.

We should keep this broader physiological picture in mind as we go on to discuss recent progress in the predominantly molecular framework of our current research efforts. Developments in genomic research—attempts to understand the expression of complex phenotypes derived from the interactions of multiple genes—may help us to think of ways in which we can integrate our molecular models with the physiological and environmental factors that dominate the epidemiology of tuberculosis. While celebrating recent successes, we should remember that the 'captain of all the men of death' still presents a formidable array of conceptual and technical challenges.

Reference

Dubos R, Dubos J 1952 The white plague: tuberculosis, man and society. Little, Brown & Co., Boston

Host genetic susceptibility to human tuberculosis

R. J. Bellamy and A. V. S. Hill

Wellcome Trust Centre for Human Genetics, Windmill Road, Oxford University, Oxford OX3 7BN, UK

Abstract. Convincing evidence exists that host genes influence the outcome of infection in human tuberculosis. We are employing two complementary strategies to find the genes involved: a linkage-based, comprehensive genome screen and an association-based candidate gene study. In a genome screen of 282 markers on 92 affected sib pairs we have found evidence of co-segregation of disease with five markers, but further studies are required to replicate these results. The absence of a single strongly linked marker demonstrates that susceptibility to human tuberculosis is not controlled by a single major gene. Using a candidate gene approach investigating over 400 tuberculosis cases and 400 ethnically matched healthy controls we have found evidence that *NRAMP1* and vitamin D receptor gene (*VDR*) polymorphisms are associated with tuberculosis. It is hoped that by identifying the genes that account for why only a minority of those exposed to tuberculosis develop disease, we will develop new insights into potential therapeutic and preventative strategies.

1998 Genetics and tuberculosis. Wiley, Chichester (Novartis Foundation Symposium 217) p 3–23

Mycobacterium tuberculosis kills more people than any other single pathogen, and it is estimated that one-third of the World's population is infected with this micro-organism (Murray et al 1990). However, among those infected, only 10% will ever develop clinical disease. Even when a large dose of live, virulent *M. tuberculosis* was inadvertently administered to 249 babies in 1926 in Lubeck, Germany, 173 survived the infection, demonstrating that the majority of the population are resistant to tuberculosis (Dubos & Dubos 1952). So what is different about those who succumb to tuberculosis? Clearly, although *M. tuberculosis* is necessary for the development of tuberculosis, it is not sufficient. In 1882, Robert Koch discovered *M. tuberculosis* as the cause of tuberculosis (Koch 1882), and since that time our interest has focused on the pathogen, whereas the importance of the host's genetic make-up has been largely ignored. In this chapter we summarize the evidence that host genes are important in determining the outcome of infection in tuberculosis and describe the work we are doing to find these genes.

Evidence linking tuberculosis and host genetics

Twin studies provide the most convincing evidence that host genes determine the outcome of infection in tuberculosis. Diehl & von Verscheur (1936) found that concordance among 80 monozygous twin pairs was 65%, compared with only 25% in 125 dizygous twin pairs. Kallman & Reisner (1942) similarly found higher concordance for tuberculosis among monozygous compared to dizygous twins (88% among 78 monozygous compared to 28% among 230 dizygous). In Comstock's reanalysis of the Prophit study the overall concordances were lower but again the rate was higher among monozygous twins (33% in 54 monozygous twin pairs and 14% in 148 dizygous; Comstock 1978).

Evidence that genetic factors are important in the resistance and susceptibility of animals to mycobacteria has accumulated from studies on rabbits and mice. The classic studies of Lurie showed that inbred families of rabbits can be divided into two distinct groups with markedly different susceptibilities to mycobacteria. Following infection with a virulent strain of *Mycobacterium bovis* the resistant rabbits developed cavitary pulmonary disease and the susceptible rabbits developed widespread haematogenously disseminated disease (Lurie 1941). When infected with human-type *M. tuberculosis* the resistant rabbits were able to inactivate more tubercle bacilli than the susceptible rabbits (Lurie et al 1952). The segregation pattern of the almost all-or-none nature of the observed resistance led Lurie to conclude that resistance was under genetic control.

Studies on inbred strains of mice have identified two distinct phenotypes in terms of resistance to *Leishmania*, *Salmonella* and mycobacteria, designated Bcg^{r} and Bcg^{s} for resistant and susceptible, respectively (reviewed in Blackwell 1989). A candidate gene for *Bcg* has been isolated by positional cloning and designated *Nramp1* (natural resistance-associated macrophage protein 1; Vidal et al 1993). A single non-conservative amino acid substitution of glycine by aspartic acid at position 169 has been shown to correlate with the Bcg^{s} phenotype in 27 inbred mouse strains (Malo et al 1994). That *Nramp1* and not a closely linked gene is responsible for the *Bcg* phenotype has been proven by the production of a knockout mouse that is phenotypically identical to the homozygous $Nramp1^{D169}$ mouse (Vidal et al 1995) and by restoration of the resistance phenotype in transgenic mice in which the $Nramp1^{G169}$ allele was transferred onto the background of the homozygous $Nramp1^{D169}$ genotype (Govoni et al 1996). The human homologue of the *Nramp1* gene, designated *NRAMP1*, has been cloned and mapped to human chromosome 2q35 (Cellier et al 1994). Several polymorphisms have been described in *NRAMP1* (Liu et al 1995), and in this chapter we summarize data showing that *NRAMP1* gene variants are associated with human tuberculosis.

Mice with targeted gene disruptions of the genes encoding γ-interferon (IFN-γ; Cooper et al 1993), the IFN-γ receptor (Kamijo et al 1993) and β_2 microglobulin (Flynn et al 1992) have also been found to be highly susceptible to mycobacteria. These genes represent important candidate genes for tuberculosis studies in the human population.

Investigating tuberculosis susceptibility in human populations

Family-based linkage and association-based case-control studies are both required to identify the genes involved in a disease such as tuberculosis, where several genes are likely to be involved in the aetiology of the disease. The two study designs are complementary because they each have their advantages and limitations. Linkage studies can now be used to screen the entire human genome to locate regions containing disease susceptibility genes (Davies et al 1994). This comprehensive and systematic approach should identify any gene that exerts a major effect on disease susceptibility, but because it has relatively low power it will fail to detect genes that exert only a moderate effect on disease risk. Risch & Merikangas (1996) discuss the difference in power between linkage and association, and the effects which allele frequency and additional disease risk attributable to possession of that allele have upon the sample size required. For example, they show that if a disease susceptibility allele has a frequency of 0.5 and exerts a twofold risk of disease, then to have an 80% chance of detecting this effect one would require 2498 sib pairs by linkage but only 340 cases by association. Association studies are therefore useful to detect genes that exert a moderate disease susceptibility effect and that would be missed by linkage. However, as association is only detectable over short genetic distances (generally <1 cM) it is not possible to conduct a whole genome screen using this approach, and association studies are currently limited to the screening of candidate regions. The limited regions over which association is detectable can also be used to advantage. Once a linkage has identified a region segregating with disease, association can be used in the fine mapping of the region to localize the disease susceptibility gene, as was recently carried out for haemochromatosis (Feder et al 1996). We will first discuss the methodology of a genome screen and the progress that has been made in tuberculosis, and then describe association-based candidate gene studies.

Linkage-based genome screens

The affected sib pair method, a variation on that originally described by Penrose (1953), is most frequently used for linkage analysis of multifactorial diseases such as tuberculosis. The families required include two full siblings who have both had disease and preferably also their parents (who are preferably unaffected). As shown in Fig. 1, the inheritance of a highly polymorphic marker can be traced

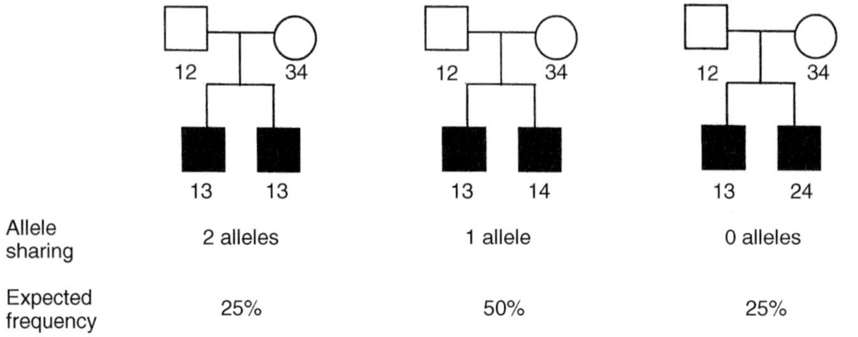

FIG. 1. Affected sib pair identical by descent analysis. The numbers below each family member represent alleles at a marker locus. In the first family, both offspring inherit allele 1 from the father and allele 3 from the mother and thus share two alleles identical by descent. In the second family they have inherited the same paternal allele but different maternal alleles and so share one allele identical by descent. In the third family they have inherited different alleles from both parents and thus share zero alleles identical by descent. If a large number of families are typed, the expected 2 : 1 : 0 sharing frequencies are 25% : 50% : 25%, assuming no linkage between the disease and marker.

from parents to offspring and the number of alleles that the sibs share (i.e. zero, one or two), identical by descent, can be determined. If the marker is not linked to a tuberculosis susceptibility locus then the sibs should share two alleles in 25% of families, one allele in 50% of families and zero alleles in 25%. If the marker locus is linked to a tuberculosis susceptibility locus then there should be a significant excess of families where sibs share two alleles identical by descent and a reduction in the number sharing zero alleles compared to the expected frequencies. This methodology does not make any assumptions about penetrance of genes, mode of inheritance or number of genes involved.

Approximately 100 sib pair families are generally used to carry out a genome screen on a multifactorial disease, and around 300 microsatellite markers, spaced at approximately 10 cM intervals must be typed. Microsatellites are dinucleotide repeats of a $(CA)_n$ sequence. They are highly polymorphic because at a single locus the size of n varies throughout the population. Many thousands of microsatellites have now been isolated, producing a comprehensive map of the entire human genome (Dib et al 1996). Advances in utilizing fluorescence-based typing systems and advanced computer software have allowed semi-automated microsatellite genotyping systems to develop, revolutionizing linkage studies (Reed et al 1994). PCR primers are labelled with one of three fluorescent dyes, FAM™ (blue), HEX™ (yellow) or TET™ (green), and microsatellite PCR products are separated on the basis of size by polyacrylamide gel electrophoresis. A laser is used to detect dye fluorescence as the PCR products migrate through the gel. By utilizing a TAMRA™ (red)-labelled internal size standard the size of the microsatellite PCR

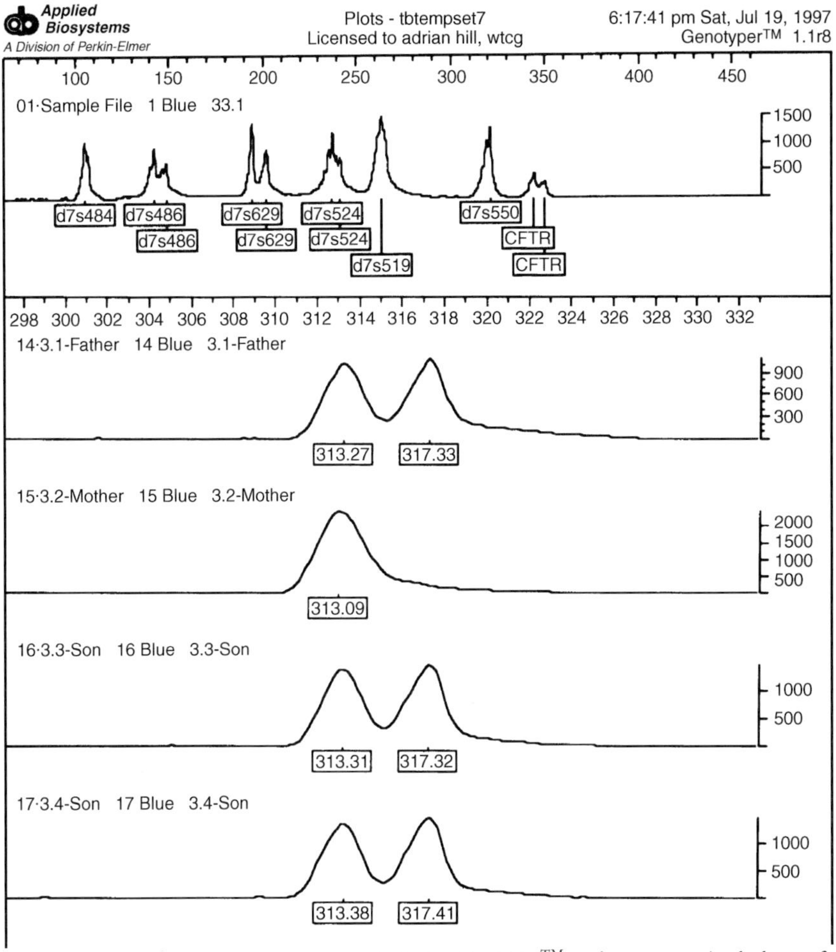

FIG. 2. Seven microsatellite loci each labelled with FAM™ and run on a single lane of an ABI373 sequencer™ (Perkin-Elmer). The Genotyper™ (Perkin-Elmer) software is used to focus on the inheritance of alleles from a single microsatellite locus in one sib pair family. The offspring have inherited the same paternal allele but as the mother is a homozygote it cannot be determined whether or not they have inherited the same maternal allele.

products can be accurately calculated to less than 0.5 bp. Using sets of microsatellite PCR products that do not overlap in size, it is possible to run eight loci labelled with each colour in a single gel lane, i.e. 24 loci can be simultaneously analysed per lane, producing up to 864 genotypes per gel (Reed et al 1994). Figure 2 shows all the microsatellites, labelled with a single fluorescent dye run in one lane and how the Genotyper™ (Perkin-Elmer) computer software can be used to focus on a single locus and compare genotypes in several family members.

We have used this technology to carry out a tuberculosis genome screen on 282 microsatellite markers on 92 sib pair families from The Gambia and South Africa. Five of these markers showed significant evidence of co-segregation with tuberculosis (Table 1). One of these markers, d6s276, is located at the major histocompatibility complex on chromosome 6p. The absence of strong linkage to a single marker demonstrates that tuberculosis susceptibility is not a monogenic trait and that a large number of genes are probably involved. Due to the large number of microsatellites typed, false positive linkages to regions that do not include a tuberculosis susceptibility gene could have arisen. To overcome this problem we are conducting this genome screen in two stages. Having screened the entire genome in our first 92 sibling pairs we now intend to genotype the five markers that demonstrate evidence of linkage in a second set of families, which we are currently recruiting from Africa. This approach minimizes the risk of false positive linkages.

Strong evidence of linkage between microsatellite markers and disease susceptibility is only the first step in identifying the gene(s) of interest. Even when association is present so that the gene can be mapped to a small chromosomal region, the task can still be considerable. This will become easier as the Human Genome Project produces more high resolution genetic and physical maps. A genome-wide map of expressed sequence tags (ESTs) should facilitate the isolation of genes from the region(s) of interest. Hopefully, this approach will eventually lead to the identification of the genes that exert the largest effects on host variability in tuberculosis susceptibility. However, this approach will be likely to miss genes that exert a moderate effect on risk of tuberculosis, and therefore any serious attempt to identify tuberculosis susceptibility genes should also utilize both a linkage-based genome screen and a candidate gene approach.

TABLE 1 Genome screen for tuberculosis

Marker	*2 : 1 : 0*	*1 : 0*	*P value*
d3s1262	24.1/24.4/9.5	84.9/52.8	0.0024
d15s128	21.0/28.9/8.6	87.3/59.3	0.0052
d6s276	17.7/25.2/8.5	79.5/55.9	0.0098
d8s272	19.2/29.6/10	83.9/65.6	0.0160
dxs984	—	—	0.0025

The five markers that show significant co-segregation with tuberculosis are shown. 2 : 1 : 0 sharing refers to the number of sib pairs sharing two, one or zero alleles identical by descent. 1 : 0 refers to whether each allele was co-inherited or not, and therefore includes families where one parent is a homozygote. The values shown are calculated by the program 'sib pair' (Delepine et al 1997) and are not whole integers, as probabilities are inferred when a parental genotype is not available. Genotype sharing frequencies are not comparable for the X chromosome and are therefore not stated.

Candidate gene studies

Association studies require the recruitment of a large number of unrelated cases who are affected by the disease of interest. One advantage over multicase family studies is that because single cases are much easier to recruit one can be more specific in the clinical criteria for study entry. For example, in our tuberculosis case-control study we have only recruited HIV-negative adults (over 16 years), with proven smear-positive pulmonary tuberculosis, from The Gambia. However, in the family-based genome screen we needed to recruit families from South Africa, smear-negative cases, children and a small number of HIV-positive cases in order to obtain sufficient families. By having a more diverse group of patients, heterogeneity could reduce the power of the study.

Candidate gene studies involve comparing the frequency of genetic polymorphisms between tuberculosis cases and ethnically matched controls. Significant differences between the two groups can be due to: (a) chance; (b) a confounding factor; (c) linkage disequilibrium between the polymorphism and the disease susceptibility allele; or (d) the variant itself being a disease susceptibility allele. In order to have adequate power to detect moderate genetic risk factors it is necessary to have several hundred individuals, and many case-control studies examine inadequate numbers. In our study we routinely type over 800 individuals. It is important to minimize the risks of confounding factors, which give rise to a significant difference between cases and controls, by ensuring that cases and controls are adequately matched for ethnic group. Simply designating subjects as Caucasian, African or Asian is probably not adequate, as within each of these categories there is potential for substantial population stratification (i.e. subtle genetic differences between the groups studied). In our study in The Gambia we classify our subjects into Mandinka, Wolof, Fula, Jola, Manjago, Serahule and Gambian other, and these ethnic groups have previously been shown to be closely genetically related (Allsopp et al 1992). It is usually impossible to define ethnic origin so rigorously in western populations and this has the potential to allow population stratification to pass undetected.

Previous studies on genetic susceptibility to tuberculosis have largely focused on the human leukocyte antigen (HLA) system. Associations have been described between tuberculosis and class I HLA antigens, most frequently with HLA DR2. However, these associations have not been consistently demonstrated, and they are unlikely to account for more than a relatively small percentage of the total genetic influence in tuberculosis susceptibility. No non-MHC genes have previously been shown to influence the outcome of infection in tuberculosis; however, some rare immunodeficiency states have been shown to predispose to other mycobacterial infections. Adenosine deaminase deficiency, chronic granulomatous disease and IFN-γ receptor deficiency all predispose to disseminated Bacillus Calmette–Guérin infection (Casanova et al 1995, Jouanguy et al 1996). IFN-γ receptor deficiency is particularly interesting because it has been found to predispose to disease, due to several atypical mycobacteria (*Mycobacterium fortuitum*, *Mycobacterium chelonei* and *Mycobacterium avium*) and to *Salmonella* infections (Newport et al 1996). These disorders provide interesting candidate genes, although it remains to be determined whether common polymorphisms in the genes might affect gene function and explain susceptibility to *M. tuberculosis* itself.

We have been studying many candidate genes in our case-control study of tuberculosis in The Gambia (*NRAMP1,* the HLA genes, and the genes encoding tumour necrosis factor, mannose-binding protein, the vitamin D receptor [VDR], IFN-γ, interleukin-1α, -1β, -RA [receptor antagonist], -4, -9 and -10, complement receptor 1, intercellular adhesion molecule 1, fucosyltransferase-2, inducible nitric oxide synthase, chemokine receptors, and the T helper 2 [Th2] gene cluster). The candidate genes we will focus on here are *VDR* and *NRAMP1*. The active metabolite of vitamin D, 1,25 dihydroxyvitamin D_3 (1,25D_3), regulates calcium metabolism but is also an important immunoregulatory hormone that activates monocytes and stimulates cell-mediated immunity, but suppresses lymphocyte proliferation and cytokine synthesis (Tsoukas et al 1984, Rook et al 1986). These effects are exerted via the VDR, which is present on monocytes and activated B and T lymphocytes (Provvedini et al 1983).

Epidemiological evidence suggests that there is a link between vitamin D deficiency and tuberculosis susceptibility, and prior to the availability of anti-tuberculous chemotherapy treatment of patients with vitamin D suggested it had beneficial effects, particularly in cutaneous tuberculosis (reviewed in Davies 1985). *In vitro* studies have shown that 1,25D_3 can enhance the ability of human monocytes to restrict the growth of intracellular *M. tuberculosis* (Rook et al 1986, Denis 1991). Genetic variation in bone mineral density and susceptibility to osteoporosis have been shown to be associated with single-base change polymorphisms in the VDR gene in many, but not all, populations studied (Morrison et al 1994, Eisman 1996). The genotype most at risk of osteoporosis, designated 'tt', was associated with increased reporter gene mRNA levels *in vitro* (Morrison et al 1994), and individuals with this genotype have the highest levels of circulating 1,25D_3 (Howard et al 1995). We have found that tt homozygotes are significantly under-represented among tuberculosis cases compared to controls (Table 2). This suggests that people with this genotype may be protected against tuberculosis, and supports the suggestion that vitamin D may protect against tuberculosis. This result is particularly interesting because it provides insight into how better understanding of the nature of genetic susceptibility to tuberculosis might provide ideas for new treatment or prophylactic strategies.

We have found that four *NRAMP1* gene variants — a microsatellite 5′ to the transcription start site, a TGTG deletion in the 3′ untranslated region (designated

TABLE 2 The distribution of vitamin D receptor (VDR) genotypes between tuberculosis cases and controls

Genotype	*Tuberculosis cases (%)*	*Controls (%)*
TT	204 (50)	188 (45)
Tt	177 (43)	177 (43)
tt	27 (6.6)	49 (12)
Total	408	414

The overall distribution of VDR genotypes is significantly different between cases and controls (Mantel–Haenszel, stratifying for ethnic group $3 \times 2\chi^2 = 4.93$, $P = 0.026$). The tt genotype was reduced in tuberculosis cases compared to controls (Mantel–Haenszel $\chi^2 = 6.22$, $P = 0.01$, odds ratio 0.53 [95% confidence interval 0.31–0.88]).

1729+55del4), a G to C transversion in intron 4 (469+14G/C) and a non-conservative amino acid substitution at codon 543 (D543N)—are all strongly associated with tuberculosis in this African population. The microsatellite and 469+14G/C polymorphisms are in linkage disequilibrium, as are the 1729+55del4 and D543N variants, and these results are therefore not independent (data not shown). As shown in Table 3 individuals who are heterozygous carriers of both *NRAMP1* variant alleles are significantly over-represented among the tuberculosis cases compared to those who do not possess either variant allele (odds ratio = 4.07; confidence interval = 1.86–9.12). This demonstrates that *NRAMP1* is an important tuberculosis susceptibility gene in human as well as murine populations.

Although susceptibility to tuberculosis in human populations is likely to be due to a complex interaction between a large number of genetic and environmental factors, we have shown that it is possible to identify host genetic factors leading to increased risk of disease. These loci will require further study to assess how

TABLE 3 Combined effects of *NRAMP1* variants and tuberculosis

NRAMP1 variant[a]	*Tuberculosis cases*	*Controls*	*Odds ratio (95% C.I.)*	χ^2*-test*	*P-value*
++/GG	191	251	1.0		
++/GC	45	33	1.79 (1.07–3.00)	5.04	0.02
+del/GG	118	84	1.85 (1.30–2.62)	12.24	0.0005
+del/GC	31	10	4.07 (1.86–9.12)	14.58	0.0001

[a]Variants are a TGTG deletion in the 3′ untranslated region (designated 1729+55del4) followed by a G to C transversion in intron 4(469+14G/C).

There is a highly significant association between heterozygosity for these *NRAMP1* polymorphisms and tuberculosis (overall comparing the four genotypes $\chi^2 = 26.41$, three degrees of freedom, $P = 0.000008$). Odds ratios (95% confidence interval [C.I.]) and χ^2 (Yates', one degree of freedom) values shown are for comparisons with the ++/GG genotype.

important these genetic variants are in different populations and different environments. It is hoped that they will lead to a greater understanding of the key factors in host immunity to tuberculosis, and thus offer new insights into tuberculosis treatment and prophylaxis.

Acknowledgements

We are grateful to the patients for consenting to enter these studies and for the help we received from the Gambian National TB Control Programme Directors, V. Bouchier and K. Manneh, to the field workers Y. Sowe and M. Jawo, and to many other Medical Research Council staff, especially H. Whittle, K. P. W. J. McAdam and T. Corrah. This work is being funded by the Wellcome Trust (Grant ref. 044418/Z/95/139).

References

Allsopp CEM, Harding RM, Taylor C et al 1992 Interethnic genetic differentiation in Africa: HLA class I antigens in the Gambia. Am J Hum Genet 50:411–421

Blackwell J 1989 The macrophage resistance gene *Lsh/Ity/Bcg*. Res Immunol 140:767–828

Casanova J-L, Jouanguy E, Lamhamedi S, Blanche S, Fischer A 1995 Immunological conditions of children with BCG disseminated infection. Lancet 346:581

Cellier M, Govoni G, Vidal S et al 1994 Human natural resistance-associated macrophage protein: cDNA cloning, chromosomal mapping, genomic organization, and tissue-specific expression. J Exp Med 180:1741–1752

Comstock GW 1978 Tuberculosis in twins: a re-analysis of the Prophit study. Am Rev Respir Dis 117:621–624

Cooper AM, Dalton DK, Stewart TA, Griffin JP, Russell DG, Orme IM 1993 Disseminated tuberculosis in interferon-γ gene-disrupted mice. J Exp Med 178:2243–2247

Davies PDO 1985 A possible link between vitamin D deficiency and impaired host defence to *Mycobacterium tuberculosis*. Tubercle 66:301–306

Davies JL, Kawaguchi Y, Bennett ST et al 1994 A genome-wide search for human type 1 diabetes susceptibility genes. Nature 371:130–136

Delepine M, Pociot F, Habita C et al 1997 Evidence of a non-MHC susceptibility locus in type I diabetes linked to HLA on chromosome 6. Am J Hum Genet 60:174–187

Denis M 1991 Killing of *Mycobacterium tuberculosis* within human monocytes: activation by cytokines and calcitriol. Clin Exp Immunol 84:200–206

Dib C, Faure S, Fizames C et al 1996 A comprehensive map of the human genome based on 5264 microsatellites. Nature 380:152–154

Diehl K, von Verscheur O 1936 Der Erbeinfluss bei den Tuberkulose. Gustav Fischer, Jena

Dubos R, Dubos J 1952 The white plague: tuberculosis, man and society. Little, Brown & Co., Boston

Eisman JA 1996 Vitamin D receptor gene variants: implications for therapy. Curr Opin Genet Dev 6:361–365

Feder JN, Gnirke A, Thomas W et al 1996 A novel MHC class I-like gene is mutated in patients with hereditary haemochromatosis. Nat Genet 13:399–408

Flynn JL, Goldstein MM, Triebold KJ, Koller B, Bloom BR 1992 Major histocompatibility complex class-I restricted T cells are required for resistance to *Mycobacterium tuberculosis* infection. Proc Natl Acad Sci USA 89:12013–12017

Govoni G, Vidal S, Gauthier S, Skamene E, Malo D, Gros P 1996 The *Bcg/Ity/Lsh* locus: genetic transfer of resistance to infections in C57BL/6J mice transgenic for the *Nramp1*^{G169} allele. Infect Immun 64:2923–2929

Howard G, Nguyen T, Morrison N et al 1995 Genetic influences on bone density: physiological correlates of vitamin D receptor gene alleles in premenopausal women. J Clin Endocrinol Metab 80:2800–2805

Jouanguy E, Altare F, Lamhamedi S et al 1996 Interferon-γ-receptor deficiency in an infant with fatal Bacille Calmette–Guérin infection. N Engl J Med 335:1956–1961

Kallmann FJ, Reisner D 1942 Twin studies on the significance of genetic factors in tuberculosis. Am Rev Tuberc 47:549–574

Kamijo R, Le J, Shapiro D et al 1993 Mice that lack the interferon-γ receptor have profoundly altered responses to infection with Bacillus Calmette–Guérin and subsequent challenge with lipopolysaccharide. J Exp Med 178:1435–1440

Koch R 1882 Die atiologie der tuberkulose. Berliner Klin Wochenschr 19:221–230

Liu J, Fujiwara M, Buu NT et al 1995 Identification of polymorphisms and sequence variants in the human homologue of the mouse natural resistance-associated macrophage protein gene. Am J Hum Genet 56:845–853

Lurie MB 1941 Hereditary, constitution and tuberculosis. An experimental study. Am Rev Tuberc 44 (suppl):1–125

Lurie MB, Abramson S, Heppleston AG 1952 On the response of genetically resistant and susceptible rabbits to the quantitative inhalation of human-type tubercle bacilli and the nature of resistance to tuberculosis. J Exp Med 95:119–134

Malo D, Vidal S, Skamene E et al 1994 Haplotype mapping and sequence analysis of the mouse *Nramp* gene predict susceptibility to infection with intracellular parasites. Genomics 23:51–61

Morrison NA, Qi JC, Tokita A et al 1994 Prediction of bone density from vitamin D receptor alleles. Nature 367:284–287

Murray CJL, Styblo K, Rouillon A 1990 Tuberculosis in developing countries: burden, intervention and cost. Bull Int Union Tuberc Lung Dis 65:6–24

Newport MJ, Huxley CM, Huston S et al 1996 A mutation in the interferon-γ-receptor gene and susceptibility to mycobacterial infection. N Engl J Med 335:1941–1949

Penrose LS 1953 The general purpose sib-pair linkage test. Ann Eugenics 18:120–124

Provvedini DM, Tsoukas CD, Deftos LJ, Manolagas SC 1983 1,25 dihydroxyvitamin D_3 receptors in human leukocytes. Science 221:1181–1183

Reed PW, Davies JL, Copeman JB et al 1994 Chromosome-specific microsatellite sets for fluorescence-based, semi-automated genome mapping. Nat Genet 7:390–395

Risch N, Merikangas K 1996 The future of genetic studies of complex human diseases. Science 273:1516–1517

Rook GAW, Steele J, Fraher L, Barker S, Karmali R, O'Riordan J 1986 Vitamin D_3, gamma interferon, and control of *Mycobacterium tuberculosis* by human monocytes. Immunology 57:159–163

Tsoukas CD, Provvedini DM, Manolagas SC 1984 1,25-dihydroxyvitamin D_3: a novel immunoregulatory hormone. Science 224:1438–1440

Vidal SM, Malo D, Vogan K, Skamene E, Gros P 1993 Natural resistance to infection with intracellular parasites: isolation of a candidate gene for *Bcg*. Cell 73:469–485

Vidal S, Tremblay ML, Govoni G et al 1995 The *Ity/Lsh/Bcg* locus: natural resistance to infection with intracellular parasites is abrogated by disruption of the *Nramp1* gene. J Exp Med 182:655–666

DISCUSSION

Rook: I would like to ask a question about the vitamin D receptor (VDR). We showed that vitamin D partially activates macrophages and has modest effects on the growth, but the effects were not that impressive (Rook et al 1986). Vitamin D switches off interleukin (IL)-12 and T helper 1 (Th1) lymphocyte functions, so one would expect it to have detrimental effects (Lemire et al 1994). Does anyone have any up-to-date comments on what its involvement might be?

Bellamy: The field is increasingly confusing, certainly in the osteoporosis field, and there seems to be marked variation between one study and another. Around 50 studies have been conducted on the association between *VDR* polymorphisms and bone mineral density. Only half of the studies have confirmed the original association and this heterogeneity is probably due to genuine environmental differences (Eisman 1996). There is the same potential for population heterogeneity in tuberculosis. The original *in vitro* studies which investigated the function of *VDR* polymorphisms suggested that the 't' allele was the most transcriptionally active, but there is some recently published evidence (Verbeek et al 1997) which suggests that the reverse may be the case in certain cell lines. The 'tt' homozygote may not be the allele most responsive to the effects of vitamin D and, therefore, I am now less certain whether vitamin D is good or bad in terms of risk of tuberculosis, but it is certainly doing something.

Kramnik: Do association studies allow the cloning of specific genes, or do you have to rely on candidate gene approaches?

Bellamy: It is possible to use an association study to localize precisely a disease susceptibility gene, and this was recently done for haemochromatosis without any information about a candidate gene (Feder et al 1996). By typing many closely linked markers, a region where there is a peak of association with disease can be defined and this should represent the location of the putative disease susceptibility gene. The problem is that when you think you have found the gene you are looking for it is difficult to be sure that the true disease susceptibility gene is not another gene in strong linkage disequilibrium with your gene of interest. It is important to establish biological plausibility to confirm you have identified the correct gene.

Orme: Some mice strains express one form of *Nramp1* (gene encoding natural resistance-associated macrophage protein 1) and are susceptible to Bacillus Calmette–Guérin (BCG) Montreal, in that it grows a bit faster, whereas other mice strains have the resistance allele and BCG Montreal grows a bit slower. Even more dramatic differences are observed with *Mycobacterium avium*. If I remember correctly, there is one amino acid difference between the resistance and susceptible alleles. When you see *NRAMP1* associated with tuberculosis in humans, do you mean one allelic form of *NRAMP1* or just that the gene is expressed?

Bellamy: That's a difficult question. Part of the problem is that none of these polymorphisms have been shown to be functional, in that they will affect either the function of the gene product or the amount of gene product that is being produced. There are several possibilities: (1) that the mutations affect the amount of product produced; (2) that the mutations are in linkage disequilibrium with another gene variant which either affects the amount produced or affects the structure of *NRAMP1*; or (3) that they are in close linkage disequilibrium with another gene in that region. This is less likely even though the IL-8 receptors are close by because *NRAMP1* is a strong candidate gene for tuberculosis susceptibility.

Kramnik: You mentioned that there may be another gene closely linked to *NRAMP1*, but surely the numbers suggest that such linkage could occur over a much larger distance.

Bellamy: No. This is a case of population association, and associations only act over short genetic distances, probably much less than 1 cM. The region has been fairly extensively studied by Jenny Blackwell's group and the IL-8 receptor gene is a reasonable candidate (White et al 1994), but there isn't that much more space for there to be too many other candidates. *NRAMP1* is a much more likely possibility.

Orme: Two or three labs, including my own, have been looking at this in terms of aerosol tuberculosis in mice, and have found the reverse situation, i.e. the resistant *Bcg*^{r} mouse strains die of tuberculosis, whereas the susceptible *Bcg*^{s} strains have chronic tuberculosis in the lungs. We found that when C57 and DBA mice were infected with even a small amount of tuberculosis aerosol, e.g. 10–50 organisms, where supposedly this effect is strongest, the two mouse strains grow identically for up to 100 days, so one might conclude that there are no differences. However, all the *Bcg*^{r} mice subsequently re-activate and die.

Bellamy: It's worth pointing out that the situation in humans is not that dissimilar, in that no matter what *NRAMP1* genotype individuals possess they can still get tuberculosis, regardless of whether they are in a low or high risk group, because we have detected only a fourfold difference between these two groups. If we had looked at a small number of people we would have missed this difference. The differences are subtle and may only be important at particular key points during infection.

Kaplan: Are we assuming that the *NRAMP1* phenotype is expressed in terms of macrophage function or is there anything new to suggest another level of phenotypic expression?

Blackwell: Members of the NRAMP gene family are polytopic integral membrane proteins. NRAMP2 has recently been shown to be a metal ion transporter (Gunshin et al 1997), and in particular an iron transporter, although this hasn't been definitively proven for NRAMP1. The *NRAMP2* mutation that influences ion transport also causes anaemia in the mouse, providing good evidence for a link between this gene and its function as an iron transporter. The function of the NRAMP gene family as metal ion transporters has been conserved in evolution from yeast through to mammals, so we can be pretty certain that NRAMP1 is also a metal ion transporter. We know that NRAMP1 is in the late endosomal phagosomal compartment and it lines the membrane around the parasite. In *Leishmania* NRAMP1 is localized around the parasitophorous vacuole, and there is a slight difference in the way the mycobacteria enter the macrophage. Prior to macrophage activation mycobacteria remain in early phagosomes, and the activation signal is required for the fusion of the late endosomal lysosomal compartment. We are interested in when NRAMP1 is delivered to the mycobacterial phagosome. From David Russell's work it is clear that some molecules are delivered to the early phagosome and others are delivered later to the late endosomal lysosomal compartment (reviewed in Russell et al 1997). On functional grounds there are various reasons why NRAMP1 might influence the mycobacterium. In terms of an innate resistance mechanism, i.e. before macrophages are activated, if NRAMP1 is around the phagosome then in the mouse the resistant NRAMP1 is the normal functional NRAMP1 molecule, which transports iron out of the vacuole, in contrast to the susceptible mouse strain. Therefore, one of the reasons why bacteria may multiply faster is that they have more iron and other metal ions to work with, whereas the resistant mouse strain flushes it out into the cytoplasm. It is thought, both in the anaemia model and in the macrophage NRAMP1 model, that this may be a major mechanism for how metal ions actually get into the cytoplasm, i.e. the ion is picked up, for example, by a transferrin receptor on the surface of the cell, transported into a phagosome vacuole and then the metal ions are delivered to the cytoplasm. We know that metal ions do many things, so all the pleiotropic effects we see as a result of NRAMP1 action on macrophage activation, for example, may be related to the requirement for metal ions in the cell. We are now in a position to explain the various observations that have been made in mouse systems where NRAMP1 effects are observed before macrophage activation. This process is iNOS (inducible form of nitric oxide synthase) independent, and it seems to be a mechanism that slows down bacterial replication, and then later a killing mechanism occurs which is probably due to the effect of NRAMP1 on the macrophage activation process. We have always thought that it would be difficult to find this murine functional mutation in humans, because it is such a dramatic phenotype that it would probably be selected against. However, we may find functional mutations in the same regions of the molecule, so it is interesting that the functional mutation in *NRAMP1*, which causes infectious disease, and the mutation in *NRAMP2*, which causes anaemia, are in exactly the same locations within the molecule, suggesting that this region is important.

I would like to make one more point. One of the polymorphisms that Richard Bellamy looked at is a type of microsatellite polymorphism in the promoter region of *NRAMP1*. However, this region does not have a straight CA repeat, it has a few CA repeats, followed by a GT, more CA repeats, then a GT and a few CA repeats, so it has a Z DNA structure. We've carried out some reporter gene studies where we have taken the four different types of alleles found in human populations, and shown that they drive different levels of *NRAMP1* expression. The most common allele, which we call allele 3, drives a high level of expression. The next most common allele, which we call allele 2, drives a lower level of *NRAMP1* expression. When we looked at rheumatoid arthritis (Shaw et al 1996) and juvenile rheumatoid arthritis (Dabadghao et al 1998) we found allelic association with allele 3, which drives a high level of expression, but with infectious diseases it seems that the allele driving the lowest level of expression is associated with disease. This provides some evidence that we have one functional polymorphism which regulates the amount of expression of the molecule, but this doesn't preclude the fact that there may be other functional polymorphisms in the coding region of the gene.

Bellamy: Our results are consistent with your *in vitro* data in that what you call allele 3 (and what we call the 199 bp allele) is associated with increased *NRAMP1* expression and individuals homozygous for this allele are the most resistant to the development of tuberculosis.

Young: The *NRAMP1* study is fascinating because it was identified purely from genetics. I'm sure it would have never been discovered by those previously looking at macrophage biology.

Blackwell: This gene was originally cloned by Vidal et al (1993). Not everyone is aware that this gene was worked on for about 20 years as a Mendelian segregating gene. We tried to identify it, and we had markers close to it, but the approach of working back to the gene from the protein product was not successful. It is only through the techniques of positional cloning that such candidates can be identified.

van Helden: We have also found an association between the so-called CA repeat of *NRAMP1* and tuberculosis. However, in the last few weeks we have not found an association with intron 4. Admittedly, our numbers are not as high as yours, but I would like to ask you to comment on this.

Bellamy: There may be several possible reasons why your results differ from ours. First, we find that there is strong linkage disequilibrium between the microsatellite and the intron 4 polymorphism, so it may be that in the Gambian population the intron 4 association is purely a result of the microsatellite association, or it may be due to linkage disequilibrium with another as yet undiscovered functional mutation. You are looking at a rather different population that contains a substantial ethnic admixture. The Cape Coloureds have inherited genes from Khoi San, Caucasian and Asian ancestors, and this can produce less strong linkage disequilibrium compared to that found in an ethnically homogeneous population. It would be useful to know whether there was strong linkage disequilibrium between the microsatellite polymorphism and the intron 4 polymorphism among Cape Coloureds. The difference between the two populations may not represent any difference in terms of the effect of *NRAMP1* on tuberculosis susceptibility. However, there's also the possibility that different environmental effects are operating, or there may be different effects of different modifying genes, given that there may be different tuberculosis susceptibility genes in your population.

van Helden: We have also found an association between some alleles of the mannose-binding protein and tuberculosis. Have you looked at this?

Bellamy: I did not present our results on the mannose-binding protein gene, partly because they are statistically weaker and partly because of the time limitations. We have found that within The Gambia there is an under-representation of carriers of the mutant allele among patients with tuberculosis compared to controls. This is consistent with the suggestion of Garred et al (1994) that mannose-binding protein mutations may have arisen to such a high frequency because they confer some degree of protection against mycobacterial infections.

Ehlers: I have a comment on the mannose-binding protein. Our *in vitro* data (Hoppe et al 1997) suggest that the opsonizing effect is more important for the invasion of non-phagocytic cells than macrophages and monocytes. A component of this may be more important in certain types of tuberculosis or in different stages of the disease. It would be interesting if you were to find that low levels of mannose-binding protein were protective against childhood tuberculosis, disseminated tuberculosis and tuberculosis meningitis.

Bellamy: Unfortunately, in The Gambia childhood tuberculosis is under-diagnosed due to limited resources, so it would be difficult to do that study.

Ehlers: Given that NRAMP1 has now been shown to be an iron transporter, it occurred to me that this may explain the relative resistance of females to tuberculosis. Do females have higher levels of NRAMP1?

Blackwell: No one has looked at such differences in the levels of *NRAMP1* expression. Before we published the functional *NRAMP1* promoter region reporter construct story we wanted to go back to the families that were segregating for the two alleles and see whether we could demonstrate a functional link between these studies and real macrophages in real people, but we haven't yet managed to do this. I know from similar work by Dominic Kwiatkowski (personal communication 1997) on the tumour necrosis factor α (TNF-α) promoter polymorphism that it is difficult to get reporter gene studies to link up with real macrophages in real people because the genetic background varies from person to person, and it's difficult to control for this. This should be looked at in more detail now that we have identified people with different genotypes.

Hopewell: One of the things that makes the epidemiology of tuberculosis difficult is that there are two components: one related to the acquisition of infection; and the other related to the development of disease once infected. It sounds to me that these two components are being pooled together in terms of genetics. In your presentation you commented on the Arkansas data (Stead et al 1990), but in this case the risk that was identified in Blacks was for the acquisition of infection. Once they were infected they had no greater likelihood of developing disease than infected Caucasians.

Bellamy: They had relatively low power to detect a difference in disease incidence following infection because relatively few patients developed tuberculosis. The study had much greater power to detect differences between races in acquisition of infection than for development of disease. It is therefore difficult to draw conclusions about whether Blacks are only more susceptible to infection or whether they have increased susceptibility to disease as well. In addition, they showed that a significantly larger percentage of Blacks developed smear-positive tuberculosis, indicating they had more severe infection.

The problem is in identifying a control group that has been exposed to *Mycobacterium tuberculosis* but that did not develop disease, and a control group that became infected but did not and will never develop clinical tuberculosis. When you're looking at whether or not people develop disease, you can't be certain whether or not the controls have been exposed. In our study we're relying upon the fact that 90% of people infected by *M. tuberculosis* would not develop disease. Patients who develop tuberculosis are therefore an unusual group, and there is substantial evidence to suggest that genetic factors contribute to determining who enters this group. There is no evidence that genetic factors in humans determine who becomes infected following exposure, and environmental factors probably play a part. If you wanted to establish a control group of individuals who were infected by tuberculosis but who were resistant to disease, you would need to establish that they will never develop tuberculosis during their lifetime. Due to the potentially long incubation period of this disease this is not possible. Furthermore, you could not just use a tuberculin skin test to confirm *M. tuberculosis* infection because environmental mycobacteria would produce false positive results.

Ress: I have a similar comment. It is possible that genetic influences could be missed if studies are not designed along those lines. A real challenge is to reconcile the identification of a gene with a biologically plausible mechanism that explains the epidemiological data. If you believe that *NRAMP1* plays a role, presumably it will target innate immunity and would operate at the level of the intrinsic mycobactericidal activity of alveolar macrophages, i.e. it would prevent acquisition of infection. This may be missed if you confine your studies to infected, resistant people. Have you looked for *NRAMP1* influences in people exposed to *M. tuberculosis* who do not skin test convert? Because such individuals may be innately resistant to acquisition of infection.

Bellamy: It's difficult to prove that people cannot be infected. The fact that some people remain tuberculin negative following exposure does not guarantee that they will not become infected, just that they have not developed a particular type of immune response that allows them to respond to purified protein derivative of tuberculin (PPD).

Bateman: Are there other ways of designing the experiment? For example, rather than performing a cross-sectional study, could you have a prolonged follow-up period of, say, 10 or 15 years?

Bellamy: That could be done only in a few populations in the world. In South Africa it may be possible because the incidence of tuberculosis is so high. One could design a prospective study to identify new cases of tuberculosis among individuals who had previously been typed for *NRAMP1* polymorphisms. Due to the low number of people who would develop tuberculosis it would be necessary to genotype a large number of study subjects. This would involve considerable expense and effort because you would have to follow-up a large number of people for a long period of time, so it would be much more difficult than doing a retrospective case-control study.

Hopewell: A study along those lines is being conducted in Uganda. The investigators are looking at household contacts and following those who are infected and uninfected, although I'm not sure what the follow-up period is.

Bellamy: In terms of looking at genetic factors the problem is that a relatively small percentage of the population may have some of these variant alleles. For instance, the *VDR* 'tt' homozygote genotype is present in only 10% of the population, so you would have to screen a lot of people to obtain large enough numbers in each genotype group.

Young: You estimated that something in the order of 50 or 70 genes may be involved. It will be a horrendous feat to identify not only this large number of genes, each of which have a small effect, but also which combinations of alleles are important. In addition, it will be difficult to work out how all the epidemiological factors fit together into the genetic system. Are there any ways of defining algorithms for such complex multivariant analyses?

Bellamy: It's possible, and Roy Anderson is probably the best person to answer this, but if you were to set up a large prospective study and decide in advance that you were going to look at specific environmental and genetic risk factors, it would be possible to work out whether things might be interacting in a particular manner. It all depends on how many genes and how many environmental factors are involved.

Young: But even if you had information about all the factors involved would it then be possible to do mathematical analyses?

Anderson: It depends on whether the effects of these genes are qualitative or quantitative. If they are quantitative, for example, in terms of the quantity of NRAMP1 produced, then analysis is going to be much more difficult. It's a depressing thought, and I hope that at least some of these genes will instead have a simple plus/minus effect.

Kaplan: It's inconceivable that there will be a plus/minus effect of something presented phenotypically in macrophage function because of the complexity of the interaction between the macrophage in the organism.

Anderson: But in the case of CCR5 and HIV, I was surprised that a 32 kDa deletion had a plus/minus effect, such that for homozygotes you either progress to AIDS or you don't. So you could be lucky.

Wadee: I would like to ask Richard Bellamy to comment on the association of human leukocyte antigen (HLA) with tuberculosis. Some time ago you suggested that Class II antigens other than DR4 are associated with tuberculosis, whereas others have suggested an association with DR4 only. Are you suggesting, therefore, that there may be more than one Class II antigen associated with susceptibility to mycobacterial infection?

Bellamy: There are several different possible explanations for HLA associations: they may be due to a specific effect of the HLA molecule; they may be due to linkage disequilibrium because that region contains many genes; or they may be due to chance. There have been many studies on HLA and tuberculosis in different populations, and these are difficult to compare because HLA frequencies vary among different ethnic groups. It is possible that the structure of the HLA molecule is important for epitope recognition, so it may therefore be responsible for the development of immunity in certain individuals. This is probably less likely for *M. tuberculosis* than for other micro-organisms with smaller genomes, because *M. tuberculosis* has many different epitopes that it can potentially present to lymphocytes, and therefore it is unlikely that one or two particular HLA types will be better at recognizing epitopes.

Donald: We know there's a subgroup of people who never become tuberculin positive, no matter how heavy the exposure. Even in societies where tuberculosis is universal, tuberculin positivity never reaches more than about 80%. Therefore, 20% of people are exposed but do not become tuberculin positive. You could hypothesize that you have to have the capacity to become tuberculin positive to become sick, in the sense that you need a fairly aggressive immune response to make holes in your lungs. A group which might be worth looking at are those who remain tuberculin negative despite heavy exposure to tuberculosis, because it would be easier to study their genetics as opposed to those who do become tuberculin skin test positive.

Orme: Is this based on the assumption that delayed-type hypersensitivity (DTH) is bad?

Donald: Yes, and it may be protective, or indeed it may be the degree of DTH that is important. But there is a group that doesn't respond, no matter how exposed they are. It would be interesting to know more about their genetics and what their experience of tuberculosis is, as opposed to those who do respond.

Orme: It depends on how you look at the mechanism. It is time to bury the concept that DTH forms a cavity. My interpretation of your statement that 80% are skin test positive is that they have a recirculating memory T cell pool, because the DTH reactions are mediated by memory T cells, and that the process then relies on an interferon/TNF mechanism in the skin, which creates chemokines, resulting in the recruitment of macrophages and a swelling. It is possible that 20% of people may not react because they're missing some part of this mechanism. The idea that the breakdown of a lesion in caseous necrosis is due to DTH is absolute nonsense.

Donald: Aside from that controversy, it may be that the people who don't develop a positive tuberculin test have some defensive mechanism which enables them to keep the antigen at bay so that their immune system is never exposed to antigens.

Orme: In the mice model studies of about 10 years ago, mice given a high dose, of say 10^8 BCG, were completely anergic to tuberculin. This was the golden age of suppressor T cells, so of course this was easily explained. But if you transfer the T cells from the lesion, where the immunity is expressed, to another mouse that is not immune, you find that the PPD reaction is transferred easily. This could simply be because T cells are sequestered in organs and there is not a large enough pool of T cells in the blood to give rise to a positive reaction. There has been a lot of emphasis so far on innate immunity, in terms of NRAMP1 controlling certain factors, but as far as I'm concerned it doesn't control anything. In my opinion, the generation of acquired immunity determines the level of resistance. For instance, if you take a mouse in a memory-immune state and re-challenge it in the lungs by aerosol, you observe a considerable lag phase before immunity is expressed in the lungs. It's a 'needle in the haystack' situation, so by the time the T cells find the bacteria, the bacteria have grown by 100-fold. An extrapolation of this into a clinical situation would be that if someone lives in an area where there is a lot of *M. tuberculosis*, even if they are immune they will still carry the risk of developing an active lesion in the lungs simply because the T cell response can't find a few bacteria that quickly. The bacteria will increase in number until the T cells find the site of infection.

References

Dabadghao P, Rumba P, Shtuvere AI et al 1998 Natural resistance-associated macrophage protein 1 (NRAMP1) locus polymorphism in juvenile rheumatoid arthritis, in press

Eisman JA 1996 Vitamin D receptor gene variants: implications for therapy. Curr Opin Genet Dev 6:361–365

Feder JN, Gnirke A, Thomas W et al 1996 A novel MHC class I-like gene is mutated in patients with hereditary haemochromatosis. Nat Genet 13:399–408

Garred P, Harboe M, Oettinger T, Koch C, Svejgaard A 1994 Dual role of mannan-binding protein in infections: another case of heterosis? Eur J Immunogenet 21:125–131

Gunshin H, MacKenzie B, Berger UV et al 1997 Cloning and characterization of a mammalian proton-coupled metal ion transporter. Nature 388:482–488

Hoppe HC, de Wet BJ, Cywes C, Daffé M, Ehlers MR 1997 Identification of phosphatidylinositol mannoside as a mycobacterial adhesin mediating both direct and opsonic binding to nonphagocytic mammalian cells. Infect Immun 65:3896–3905

Lemire J, Beck L, Faherty D, Gately MK, Spiegelberg HL 1994 1,25-dihydroxyvitamin D_3 inhibits the production of IL-12 by human monocytes and B cells. Proc 9th Workshop on vitamin D, Orlando (abstr 48)

Rook GAW, Steele J, Fraher L et al 1986 Vitamin D_3, gamma interferon, and controls of proliferation of *Mycobacterium tuberculosis* by human monocytes. Immunology 57:159–163

Russell DG, Sturgill-Koszycki S, VanHeyningen T, Collins H, Schaible U 1997 Why intracellular parasitism need not be a degrading experience for *Mycobacterium*. Philos Trans R Soc Lond B Biol Sci 352:1303–1310

Shaw M-A, Clayton D, Atkinson SE et al 1996 Linkage of rheumatoid arthritis to the candidate gene *NRAMP1* on 2q35. J Med Genet 33:672–677

Stead WW, Senner JW, Reddich WT et al 1990 Racial differences in susceptibility to infection by *Mycobacterium tuberculosis*. N Engl J Med 322:422–427

Verbeek W, Gombart AF, Shiohara M, Campbell M, Koeffler HP 1997 Vitamin D receptor: no evidence for allele-specific mRNA stability in cells which are heterozygous for the Taq1 restriction enzyme polymorphism. Biochem Biophys Res Comm 238:77–80

Vidal SM, Malo D, Vogan K, Skamene E, Gros P 1993 Natural resistance to infection with intracellular parasites: isolation of a candidate for *Bcg*. Cell 73:469–485

White JK, Shaw M-A, Barton H et al 1994 Genetic and physical mapping of 2q35 in the region of the *NRAMP* and *IL8R* genes: identification of a polymorphic repeat in exon 2 of *NRAMP*. Genomics 24:295–302

The epidemiology of tuberculosis in South Africa

P. R. Donald

Department of Paediatrics and Child Health, Faculty of Medicine, University of Stellenbosch, PO Box 19063, Tygerberg 7505, South Africa

Abstract. Reports by reliable observers indicate that tuberculous disease did not occur to any great extent amongst South Africa's indigenous peoples prior to European colonization. Colonization introduced sources of infection and caused rapid urbanization for purposes of commerce and trade. By the start of the 20th century tuberculosis was recognized as a common health problem amongst the Black and Coloured peoples of South Africa. National notification commenced in 1921 and an incidence of 43 per 100 000 rose to 365 per 100 000 in 1958 and declined to 162 per 100 000 in 1986 before rising again to 221 per 100 000 in 1993. High incidences have been consistently recorded amongst the Coloured population of the Western Cape Province: in 1993 713 per 100 000 compared to the national incidence of 225 per 100 000. Using a computerized geographical information system the precise distribution of tuberculosis in two adjacent underprivileged, mainly Coloured communities, with a combined population of 34 000, is being studied. From 1985 to 1994 4044 notified tuberculosis cases gave an incidence of about 1200 per 100 000, varying from 78 to 3150 per 100 000 for the 39 enumerator subdistricts used for census purposes, and was highest in those with the lowest income. Of 5345 housing units 1835 (34%) housed at least one case of tuberculosis and 483 (9%) three or more cases. IS6110 DNA fingerprinting of strains from this community has shown a high degree of strain diversity (209 out of 334 strains evaluated). Clustering, indicative of recent transmission, was found in only 30% of isolates in this high tuberculosis incidence community.

1998 Genetics and tuberculosis. Wiley, Chichester (Novartis Foundation Symposium 217) p 24–41

I tell you naught for your comfort,
Yea, naught for your desire,
Save that the sky grows darker yet,
And the sea rises higher.
(G. K. Chesterton 1911 *Book I. The vision of the king. The Ballad of the White Horse*)

South African tourism has been promoted by the slogan 'see the world in one country'. Similarly, virtually every aspect of the epidemiology of tuberculosis could be studied within South Africa's borders. Tuberculosis has existed in Africa for at least 3000 years and evidence from Egypt in skeletal remains and pictorial representations confirms this fact (Morse et al 1964). It is significant that this evidence comes from a relatively advanced civilization undergoing urbanization. Jumping several millennia we, perhaps, need reminding of the centuries-old interaction between the African east coast and Arabia and the countries of Asia, where tuberculosis had long been present and was well known to indigenous physicians. Via well-established fortified coastal towns trade in slaves and other commodities was carried on long before Vasco da Gama entered the Indian Ocean almost precisely 500 years ago. When da Gama sailed from Malindi across the Indian Ocean he was guided by a pilot from one of several Indian ships in Malindi harbour (Theal 1907).

This intercourse with Arabia and Asia extended at least as far south as Mozambique Island and substantial coastal trading posts existed from where commerce was conducted, amongst others, with the legendary Kingdom of Zimbabwe.

Ample opportunity therefore existed for the introduction of tuberculosis into the Black communities of central and southern Africa. Nevertheless, capable 19th century observers considered that tuberculosis did not exist or was rare amongst the indigenous people of southern Africa (McVicar 1908). European colonization not only introduced new potential sources of tuberculous infection, but also rapid urbanization occurred for purposes of trade, commerce and industry, in particular mining, and the indigenous peoples were willingly or unwillingly drawn into this process, which, perhaps more than any other factor, contributed to the explosive spread of tuberculosis throughout southern Africa.

In South Africa further Asian connections were established when Malay and Indonesian slaves and political malcontents were brought to the Cape by the Dutch East India Company. The Indonesian connection was strengthened indirectly when slaves were also brought from Madagascar, where a large proportion of the population are of Indonesian origin. In 1863 indentured Indian labourers were brought to Natal by the colonial administration to work the sugar plantations. It is likely that many of these immigrants brought tuberculous infection with them.

The epidemiology of tuberculosis in South Africa

Early clinical descriptions of tuberculosis amongst the Black and Khoi San peoples emphasized the acute nature of the disease in those exposed to tuberculosis for the first time; there was much discussion of 'virgin populations' and their implications for the spread and clinical features of tuberculosis (Cummins 1929). By 1920 observers in Black rural areas considered that tuberculosis had now begun to show certain 'endemic' characteristics and was no longer so acute or extensive in its features. Tuberculin testing in Black rural 'homelands' now found approximately 70% of the population to be infected, with 25–60% of children giving a positive reaction by age 10 years (Tuberculosis Research Committee of the Transvaal Chamber of Mines 1932). While this process unfolded, the incidence of tuberculosis amongst the southern African Caucasian population declined as in their countries of origin. Despite continual (and excessive) introduction of new infectious sources in the form of tuberculosis sufferers in search of 'heliotherapy' and 'aerotherapy' this decline accelerated. Much of the above rests on anecdotal evidence. In 1921, however, official notification of infectious diseases, with its acknowledged deficiencies, became country wide and from 1921 three periods can be identified:

(1) From a rate of 43 per 100 000 the incidence rose gradually to a peak of 365 per 100 000 in 1958, the latter stages of this increase being probably stimulated by the availability of chemotherapy.
(2) A period of declining incidence to a low of 162 per 100 000 in 1986. Part of this decline can, no doubt, be ascribed to the artificial exclusion of Black 'homelands' from the figures.
(3) A period of rising incidence reaching 226 per 100 000 in 1993; and it is on this rising arm which we now find ourselves.

This single notification figure obscures regional and racial differences and other anomalies such as sex- and age-related variations in rates.

Figure 1 shows the age- and gender-related notification rates in 1993 (Küstner 1995a). The features are not unexpected in a country with a high incidence of tuberculosis. Thus, high rates with a slight male predominance are experienced in early childhood, relatively low rates during ages 5–15 years ('the golden age of tuberculosis'), a sharp increase in adolescence with a female predominance and from approximately age 25–30 years considerably higher rates in males (Styblo 1991).

Table 1 summarizes the 1993 tuberculosis incidence rates for the four major South African racial groups (Küstner 1995a). The notified incidence of 20 and 11 for Caucasian males and females, respectively, stands in stark contrast to rates of 680 and 500 for Coloured males and females. The majority of the Coloured population reside in the Western Cape Province, and the Western Cape tuberculosis rates are a major point of discussion in South Africa. It should also be noted that a high tuberculosis prevalence (>3000 per 100 000) was recently documented amongst the Khoi San of Namibia (Nel & de Villiers 1993).

In 1989 the tuberculosis incidence in the Western Cape Province was 491 per 100 000 and 141 per 100 000 in the rest of South Africa (Küstner 1991). Not only

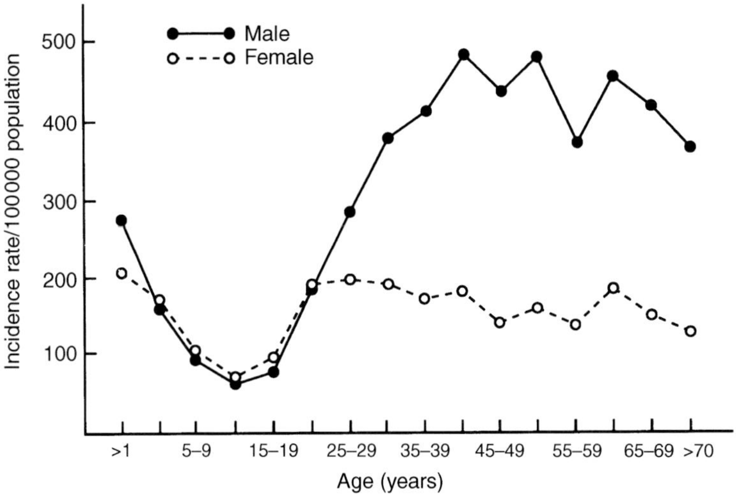

FIG. 1. Age- and gender-specific incidence of tuberculosis for South Africa in 1993. Data from Küstner (1995a).

were incidence rates approximately threefold higher than those in the rest of South Africa, but they also rose more sharply during adolescence. Thus, the incidence rates in the Western Cape Province in the age groups 10–14, 15–19 and 20–24 years were, respectively, 77, 273 and 585 per 100 000; whereas those in the rest of South Africa were, respectively, 42, 69 and 132 per 100 000. Evidence of the serious nature of the tuberculosis situation is also provided by the high rates of tuberculous meningitis in the Western Cape (Berman et al 1992). In 1985–1987 the age-specific incidence rates of tuberculous meningitis for the age groups <1 year, 1–4 years, 5–9 years and 10–14 years were, respectively, 31.5, 17.1, 4.8 and 0.7 per 100 000. Furthermore, age-specific incidence rates of pulmonary tuberculosis for the Coloured population, calculated by five-year birth cohorts, predict a persistent rise, despite a low incidence of HIV seropositivity at the time of the calculation:

TABLE 1 Race and sex-specific tuberculosis incidence rates (per 100 000 people) in South Africa in 1993

	Black	*Coloured*	*Asian*	*Caucasian*
Male	242	680	58	20
Female	137	500	34	11

Data from Küstner (1995a).

<1% seropositivity amongst pregnant women anonymously screened (Department of National Health and Population Development 1992).

The epidemiology of tuberculosis in an urban community in the Western Cape Province with a high incidence of tuberculosis

Against the above background we have, for more than a decade, studied the tuberculosis situation in a small area of the Western Cape Province, i.e. the Cape Town suburbs of Ravensmead and Uitsig (Beyers et al 1996). These suburbs, 2.34 km^2 in area, had a population of 34 294 during the 1991 national census. The tools involved in this evaluation include:

(1) Access to routine tuberculosis notifications.
(2) A long-standing personal relationship with clinic personnel at the suburbs' two local health authority clinics.
(3) The creation of a computerized geographical information system (GIS) map showing all of the suburbs' erven.
(4) The collection of all positive cultures which are evaluated for drug sensitivity and DNA fingerprinting (IS6110).
(5) An in-depth evaluation of sociological and anthropological aspects of the community.

This area, virtually uninhabited in 1960, grew within a decade to become a densely populated residential area. This change is testimony of the pace of urbanization in South Africa and the forced movement of people from other areas of Cape Town under the infamous Group Areas Act. The combined population of these suburbs during the 1991 census was 34 294, of whom 10 775 (31%) were less than age 15 years, and 14 764 (43%) less than 20 years. About 34% of the adult population have an annual income of less than R3000 ($650) and 40% have no income.

In 1985–1994 there were 4044 notified cases of tuberculosis amongst residents of Ravensmead and Uitsig. The distribution of these cases is illustrated in Fig. 2, which is based on the GIS map. Published and unpublished data indicate that approximately 80% of cases are 'sputum' smear positive, 15% are culture positive only and just over 50% of adult patients have cavitation on chest radiography (Beyers et al 1997). During the period under review, 3539 cases of tuberculosis occurred in 1575 single-dwelling units (34%), whereas 505 cases occurred in 260 out of 663 (39%) multiple-dwelling units (flats or apartments).

The 1991 national census subdivided the suburbs into 39 enumerator subdistricts each with a population of approximately 900. Although the overall tuberculosis incidence was 1505 per 100 000 in 1991 it varied from a low of 78 per 100 000 to as high as 3150 per 100 000 in different enumerator subdistricts. The

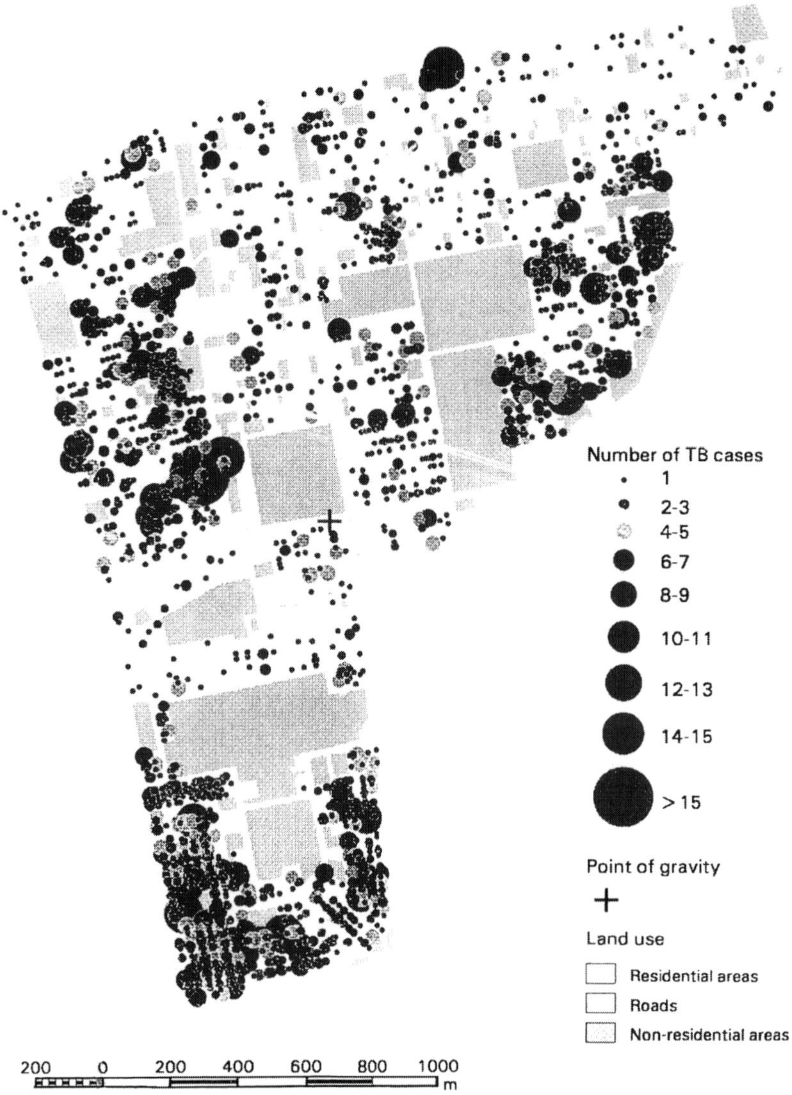

FIG. 2. Distribution of notified cases of tuberculosis in the Cape Town suburbs of Ravensmead and Uitsig 1985–1994.

tuberculosis incidence was also evaluated with regard to various socioeconomic factors, and a significant relationship ($p<0.00004$) was found for mean household income, crowding and maternal education. In Fig. 3 the tuberculosis incidence is illustrated in enumerator subdistricts grouped according to income.

Despite the intensity of the tuberculosis epidemic in this community it was surprising that IS6110-based DNA fingerprinting of 334 *Mycobacterium tuberculosis* isolates found a low rate of clustering (30%) suggesting relatively infrequent recent transmission (Warren et al 1996). Reactivation of latent infection may play a greater role in high incidence areas than was previously suspected.

Drug resistance in South Africa

Resistance to anti-tuberculosis agents is an important factor in the ability of a national programme to manage tuberculosis. In South Africa data from new

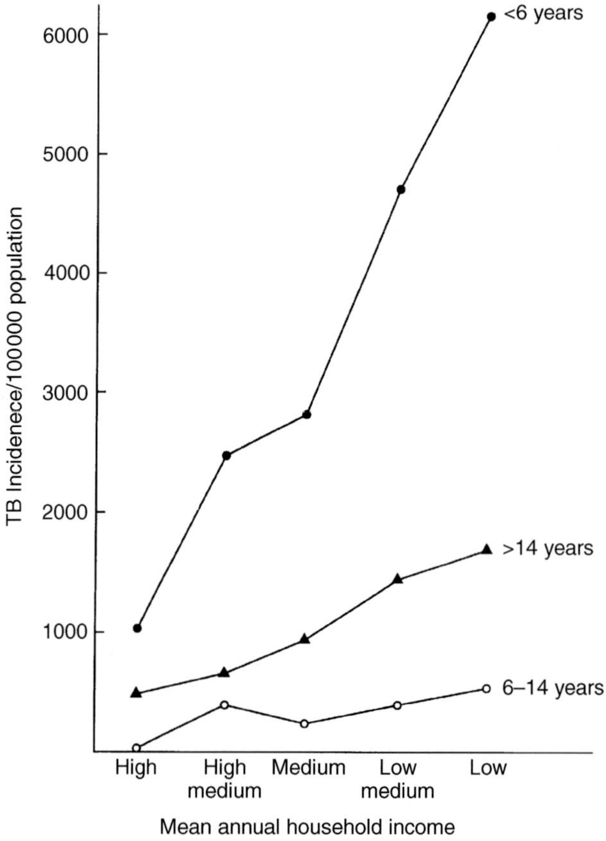

FIG. 3. The notified incidence of tuberculosis (TB) for individuals aged <6 years (●), 6–14 years (▲) and >14 years (○) in enumerator subdistricts of Ravensmead and Uitsig with a mean annual income classified as high (US$6240), medium–high (US$5352), medium (US$4826), medium–low (US$3615) and low (US$1763).

patients at sentinel hospitals have been collected for about 30 years. Primary and acquired isoniazid resistance declined from 14% and 54%, respectively, in 1965–1970 to 9.5% and 15%, respectively, in 1987–1988 (Weyer & Kleeberg 1992). In the Western Cape a recent detailed analysis found initial and acquired isoniazid resistance rates of 3.9% and 10.8%, respectively, and multidrug resistance rates of 1.1% and 4.0%, respectively (Weyer et al 1995).

The dire consequences of multidrug-resistant tuberculosis are highlighted by our experience of following up 443 multidrug-resistant tuberculosis patients five years after diagnosis (Schaaf et al 1996). Amongst those with isoniazid and rifampicin resistance mortality was 48%, compared to 27% in those resistant to either isoniazid or rifampicin and other drugs. Only 33% of patients were cured after five years, 15% were respiratory cripples and 13% remained bacteriologically positive.

Genetic aspects of drug metabolism

Isoniazid remains the single most important antituberculosis agent. Its speed of elimination is genetically determined, and it has recently been shown in a South African population to be trimodal in distribution (Parkin et al 1997). Although it is usually stated not to be a factor in determining the success of an antituberculosis regimen, homozygous fast acetylators of isoniazid have a considerably shorter exposure to therapeutic concentrations of isoniazid than slow or intermediate acetylators. Under operational conditions, or where drug absorption is compromised, this may well be of therapeutic importance.

HIV/AIDS and tuberculosis in South Africa

In common with the rest of sub-Saharan Africa, South Africa is being engulfed by the HIV/AIDS epidemic. This is already having a significant effect upon the epidemiology and clinical presentation of tuberculosis. It has been predicted that compared to an expected incidence of 500 per 100 000 by 2005, the rates may be closer to 2500 per 100 000, and this may be a conservative estimate (Küstner 1995b).

Conclusions

The above characteristics of the epidemiology of tuberculosis are well known, and they highlight a number of issues, many of which are directly or indirectly related to genetics.

Age- and gender-related differences in tuberculosis

The familiar age- and gender-related differences in incidence pose questions, the answers to which might unlock the door to novel means of treatment or better vaccination. During the first year of life a high incidence of tuberculosis is seen and more males than females are affected. This high incidence is even more striking when considered as a proportion of the infected population. Between about five and 15 years of age tuberculous infection rarely proceeds to disease, despite a continuous rise in the proportion of the population infected, as shown by studies of tuberculin hypersensitivity (Hall 1957, Roelsgaard et al 1964). At puberty the very nature of the body's response to tuberculosis alters, and necrosis and pulmonary cavitation become a characteristic feature of disease (Donald et al 1996). This necrotizing response develops earlier in females than males, and it may be related to the earlier onset of female puberty rather than to a specific action of 'female' hormones. It seems likely though that hormones must have a role in this process. After 30 years of age a clear male predominance again emerges. Is this due to some inherent 'male' factor, possibly hormonal, or is it due to greater opportunities for re-infection or suppression of immunity in males who are more likely to be found amongst the neglected destitute in those already deprived communities where tuberculosis flourishes?

Within the boundaries of one country, striking differences exist in the incidence of disease amongst different racial groups. Within the narrow confines of a relatively small suburb, similar differences in the incidence of tuberculosis exist in subgroups of a similar racial background, but of differing socioeconomic status. Despite the existence of multiple cases of tuberculosis within a household, some individuals emerge unscathed.

Are the obvious differences in tuberculosis incidence between the different racial groups in South Africa, and other countries, merely a reflection of undoubted socioeconomic differences or do they reflect an 'inherent' or acquired genetic resistance to tuberculosis based on the elimination of susceptible individuals over centuries? What role does 'tuberculization' play in the apparent greater resistance of one group of people compared to another? If 'tuberculization' exists how does it operate?

The epidemic wave of tuberculosis

> 'The high tide!' Kind Alfred cried.
> 'The high tide and the turn!'
> (G. K. Chesterton 1911 *Book VII. Ethandune. The last charge. The Ballad of the White Horse*)

Tuberculosis is thought to occur in a previously uninfected community in the form of an epidemic wave (Grigg 1958). The initial stage of this wave is marked by rapidly progressive forms of tuberculosis in adults which have been compared to disseminated forms of tuberculosis seen in early childhood (Cummins 1929). Borrel (1920) eloquently described these features amongst Black French colonial troops in France during the First World War and contrasted them with those amongst other colonial troops whose communities had a long experience of 'tuberculization.' In these latter troops a more indolent, slowly progressing form of tuberculosis tended to be seen with a lower mortality. As the epidemic wave proceeds, fewer and fewer children are affected, the adolescent rise flattens and mortality falls. In the final stages of the epidemic mainly older men are affected. The causes of the decline in this wave are speculative, but the elimination of susceptibles by a high mortality in the early stages of the epidemic and the creation of a genetically resistant group has been proposed (Grigg 1958, Bates & Stead 1993).

It has, however, been argued with great vehemence that, even with high mortality rates, the elimination of susceptibles cannot possibly account for the tuberculosis resistance emerging in certain populations, and it has been pointed out that 'epidemic' tuberculosis can again develop should these groups experience adverse socioeconomic conditions (Youmans 1979). In this view the decline in the tuberculosis wave in developed countries should be ascribed to 'suburbanization' and the 'immunizing' effect of an active infection successfully survived.

In presenting our South African data I have emphasized the age-related characteristics of tuberculosis and the fact that children between five and 15 years of age enjoy a degree of protection against tuberculous disease following infection. It has also been shown that those who survive primary infection have 60–80% protection against later disease (Sutherland et al 1982), and that infection in childhood is less likely to be followed by disease in adolescence than an initial infection during adolescence (Barnett & Styblo 1977). In the context of this symposium I suggest that the age-related resistance to disease after infection offers an additional explanation for the phenomenon of the tuberculosis epidemic wave. When tuberculosis enters a community for the first time individuals are infected right across the age spectrum and a high incidence of disease is likely to develop after infection in individuals who have not had the benefit of a 'benign' childhood tuberculous infection to 'immunize' them. A high incidence of disease in adults in turn leads to a high rate of infection, disease and mortality in young children. Only when individuals who have had the dubious 'benefit' of the 'immunizing' effect of a childhood infection enter adult life will rates of disease in adults tend to fall and the nature of adult lung lesions become more indolent and less infectious. In this scenario, subject to socioeconomic influences, the entry of new susceptibles into a particular population and a changing age structure would offer at least a partial explanation of the tuberculosis epidemic wave without necessitating recourse to genetic changes in susceptibility as an explanation.

The role of endogenous reactivation in high incidence communities

It is usually thought that endogenous reactivation plays a relatively minor role in high incidence communities. Thus far, our molecular fingerprinting studies have suggested that a significant proportion of cases arise from endogenous reactivation. If confirmed it should cause a reorientation in our attitude to prophylaxis in high incidence communities where passive case finding is the cornerstone of control policies.

The origin of tuberculosis in South Africa

There is no doubt that European colonization, and subsequent urbanization and industrialization, provided the setting for the explosion of the tuberculosis epidemic in South Africa. There must, however, be doubt as to whether the organisms involved are mainly of European origin. Substantial links exist with Asian countries where tuberculosis was well known to physicians long before it became a significant problem in industrializing Europe. Communities tend to carry their experience of tuberculosis with them, as is the case with the Asian community in the UK (Davies 1994), and it may well be that many of the *M. tuberculosis* strains isolated in South Africa today have their origin in Asia and not in Europe.

Acknowledgements

I am indebted to my collaborators N. Beyers, R. P. Gie, D. P. Parkin, H. S. Schaaf, H. I. Seifart, P. van Helden, R. Warren and H. L. Zietsman for much of the information and data presented, and to the Glaxo Wellcome Action Tuberculosis Research Initiative for their generous funding of our research.

References

Barnett GD, Styblo K 1977 Bacteriological and X-ray status of tuberculosis following primary infection acquired during adolescence or later. Bull Int Union Tuberc 52:5–15

Bates JH, Stead WW 1993 The history of tuberculosis as a global epidemic. Med Clin North Am 77:1205–1217

Berman S, Kibel MA, Fourie PB, Strebel PM 1992 Childhood tuberculosis and tuberculous meningitis: high incidence rates in the Western Cape of South Africa. Tubercle Lung Dis 73:349–355

Beyers N, Gie RP, Zietsman HL et al 1996 The use of a geographical information system (GIS) to evaluate the distribution of tuberculosis in a high-incidence community. S Afr Med J 86:40–44

Beyers N, Gie RP, Schaaf HS et al 1997 A prospective evaluation of children under the age of 5 years living in the same household as adults with recently diagnosed pulmonary tuberculosis. Int J Tuberc Lung Dis 1:38–43

Borrel A 1920 Pneumonie et tuberculose chez les troupes noires. Ann Inst Pasteur 34:105–148
Cummins SL 1929 Virgin 'soil'—and after. A working conception of tuberculosis in children, adolescents and aborigines. Br Med J 2:39–41
Davies PDO 1994 Tuberculosis in immigrants, ethnic minorities and in the homeless. In: Davies PDO (ed) Clinical tuberculosis. Chapman & Hall, London, p 91–209
Department of National Health and Population Development 1992 Tuberculosis through the eyes of cohorts. Epidemiol Comm 19:184–198
Donald PR, Beyers N, Rook GAW 1996 Adolescent tuberculosis. S Afr Med J 86:231–232
Grigg ERN 1958 The arcana of tuberculosis. Am Rev Tuberc Pulm Dis 78:151–172, 426–453, 583–596
Hall S 1957 The prevalence of tuberculosis. In: Heaf FRG (ed) Symposium of tuberculosis. Cassell & Co Ltd, London, p 67–107
Küstner HGV 1991 Tuberculosis in the Cape Province. Epidemiol Comm 18:3–23
Küstner HGV 1995a The scope of the HIV epidemic in South Africa and its impact on tuberculosis. In: Beyers AD, Donald PR, Schaaf HS, van de Wal BW (eds) Tuberculosis and HIV/AIDS. University of Stellenbosch, South Africa, p 1–6
Küstner HGV 1995b Tuberculosis update. Epidemiol Comm 22:13–17
McVicar N 1908 Tuberculosis among the South African natives. S Afr Med Record 6:197–208
Morse D, Brothwell DR, Ucko PJ 1964 Tuberculosis in ancient Egypt. Am Rev Respir Dis 90:524–541
Nel JCWK, de Villiers PJT 1993 Pulmonary tuberculosis amongst the San in Bushmanland, Namibia. S Afr J Epidemiol Infect 8:18–21
Parkin DP, Vandenplas S, Botha FJH et al 1997 Trimodality of isoniazid elimination. Phenotype and genotype in patients with tuberculosis. Am J Respir Crit Care Med 155:1717–1722
Roelsgaard E, Iversen E, Bicher C 1964 Tuberculosis in tropical Africa. Bull WHO 30:459–518
Schaaf HS, Botha P, Beyers N et al 1996 The 5-year outcome of multidrug-resistant tuberculosis patients in the Cape Province of South Africa. Trop Med Int Health 1:718–722
Styblo K 1991 The epidemiology of tuberculosis. The Royal Netherlands Tuberculosis Association Selected Papers 24:17–39
Sutherland I, Svandova E, Radhakrishna S 1982 The development of clinical tuberculosis following infection with tubercle bacilli. Tubercle 63:255–268
Theal GM 1907 History of Africa south of the Zambezi, vol 1. The Portuguese in South Africa from 1505 to 1795. George Allen & Unwin Ltd, London
Tuberculosis Research Committee of the Transvaal Chamber of Mines 1932 Tuberculosis in South African natives with special reference to the disease amongst the mine labourers on the Witwatersrand. Publ S Afr Inst Med Res 30:198–207
Warren R, Hauman J, Beyers N et al 1996 Unexpectedly high strain diversity of *Mycobacterium tuberculosis* in a high incidence community. S Afr Med J 86:45–49
Weyer K, Kleeberg HH 1992 Primary and acquired drug resistance in adult Black patients with tuberculosis in South Africa: results of a continuous national drug resistance surveillance programme. Tubercle Lung Dis 73:106–112
Weyer K, Groenewald P, Zwarenstein M, Lombard CJ 1995 Tuberculosis drug resistance in the Western Cape. S Afr Med J 85:499–504
Youmans GP 1979 Tuberculosis. WB Saunders, Philadelphia, p 356–369

DISCUSSION

Hopewell: I'm struck by how the situation in Ravensmead and Uitsig compares with that in Alaska in the late 1950s and early 1960s, where there was a prevalence of tuberculosis infection in five-year olds of nearly 100%. Mass campaigns promoting chemotherapy for those with disease and preventive therapy for those with tuberculous infection but not the disease resulted in a rapid decrease in prevalence, and by the late 1960s the prevalence of infection was comparable to the low prevalence in the USA. It seems to me that now, although tuberculosis control programmes have remained in place, they are not of the same intensity as those used to bring tuberculosis under control. This suggests that social and biological factors can be overcome by a decent tuberculosis control programme, and that once they are overcome they can be maintained without the same intensity of effort. I realize that the magnitude is much greater in the areas you have studied, but have you considered that they can be used as demonstration areas to test this approach?

Donald: The areas in question have now been identified by the local government as being demonstration areas, and the World Health Organization DOTS (directly observed therapy short-course) approach is being implemented there precisely as a demonstration. Preventive therapy is not a prominent part of it because we haven't been ensuring that the adults who are excreting organisms are actually completing their treatment, but I would like to see a more aggressive chemoprophylactic programme targeted at certain groups. In the Canadian study there was a relatively small group amongst the Inuit who could be identified and intensively treated, but for a whole population chemotherapy would almost be like sprinkling salt over a salad.

Hopewell: You could administer it in drinking water.

Donald: That is probably not a realistic option, but certainly a more explorative approach to chemoprophylaxis may work; for example, a combination of isoniazid and the sterilizing drugs rifampicin and pyrazinamide given to close contacts of sputum smear-positive patients for a period of two to three months may deliver considerable benefits. However, it is unlikely that in our current political climate tuberculosis control programme administration will be considered such an option.

Anderson: The age-specific incidence curves are extremely unusual for any infectious agent that I'm familiar with. Careful interpretation of these curves, which are observed in all regions of the world and are moderately independent of the intensity of transmission, is essential. What are the current hypotheses that explain these curves?

Donald: They may indicate that hormones, and variations of them in different populations, are also involved in modulating the immune response.

Fine: The striking age patterns are identical to those reported in many other populations and over a long period of time; and they have little or nothing to do with local genetics. They reflect how human populations respond to *Mycobacterium tuberculosis*.

van Helden: But I believe that genetics can be used to explain these age-specific curves. Our work with the mannose-binding protein, for example, shows that the risk of tuberculosis meningitis in those with mutant alleles is higher than for pulmonary tuberculosis in adults. I would like to ask Albert Beyers to comment on this with respect to the differences between children and adults.

Beyers: I will discuss our preliminary work on this in my presentation. It is not only the incidence of tuberculosis that changes with age, but also the type of disease because childhood tuberculosis and adulthood tuberculosis differ in terms of pathology. The immunopathology of adulthood pulmonary tuberculosis has previously been ascribed to the Koch phenomenon. Childhood tuberculosis occurs in a person who has not been exposed to the organism before, and adulthood tuberculosis with severe immunopathology is the result of reactivation or reinfection. We do not exclude that the Koch phenomenon contributes to the differences in pathology, but we propose that age-related factors also play a role. There are clear examples of children in high incidence communities who have second bouts of tuberculosis even though they are HIV negative and not immunosuppressed, as far as we can tell. During the second presentation with tuberculosis, children again have the childhood form of the disease. Age-related changes in the type and incidence of disease may be due to alterations in sex steroids, although this is currently purely speculative.

Donald: I have a comment about other infectious diseases that show the same age-related variation in clinical phenomena. For example, when the Epstein–Barr virus is contracted during childhood nothing much happens, but infectious mononucleosis appears during adolescence. The same is true for chickenpox, which amounts to a few vesicles during childhood, but contracted as an adult it becomes a serious disease. It suggests that the immune system undergoes a fundamental change during the process of maturation, which perhaps makes it more aggressive towards infecting agents or intruding antigens.

Fourie: I was struck by the base rate of 3000 tuberculosis cases per 100 000 population in some of the subdistricts, and I thought when has there ever been a rate that high? I recall a report on tuberculosis at the start of the 19th Century in the USA when the mortality rate was around 1600 per 100 000 population in New England (Stead & Dutt 1988). This means that the incidence rate would have been at least twice as high. It might be of interest to note that the highest mortality rate ever recorded was 9000 per 100 000 population in North American Indians in 1886 (Ferguson 1955, see also Daniel et al 1994). Under a good chemotherapy programme, mortality rates drop quickly, thus creating the impression that the tuberculosis epidemic had been brought under control, although transmission is probably still occurring in the survivors. We must surely be seeing a similar pattern in the Western Cape, but the question is whether the exceptionally high incidence rates are true or false. Cummins (1920) reported the tuberculosis rates in South Africa during World War I, and showed that people of mixed race had high rates of tuberculosis morbidity and mortality. This has also been reported in the literature earlier than the 1920s. Therefore, one has to acknowledge that there is something different about this group of people. I believe that the focus should not so much be on the organism, suspecting there might be differences in virulence between strains in different geographical areas, but rather on human genetics.

In addition, one strange phenomenon in the tuberculin skin test response of Coloured groups is that they have two modes above 10 mm following Mantoux testing with 2 tuberculin units of RT23. We would normally expect in areas of high prevalence a mode at 14 or 15 mm. However, in the Western Cape Coloureds, there is a second mode situated between 20 and 22 mm (Fourie 1983). It might be worthwhile investigating this because it suggests that there might be two subgroups, raising the question of differences in susceptibility for disease.

Why do I think the rates in Ravensmead are false? Because the degree of variation is unusual for such a small area. Also, after the introduction of the new tuberculosis registration system by standardized case definition in South Africa at the end of 1995, the number of reported tuberculosis cases in the Western Cape reduced by 30%. We need to look at the statistics of the epidemic in the areas you have studied in terms of the new case definition, and then make statements on the epidemiology.

I would also like to ask, why haven't we seen any impact on the epidemic in that area despite the intensive efforts of a dedicated team over many years? Is it because cases that are being fed back into the community are propagating the transmission of the disease? Is it that the diagnosis is good and the initial treatment phase is appropriate, but the cases are not seen through to the end?

van Helden: If it is true that the majority of cases are not due to recent transmission, then if there is a high infection rate the majority of cases will reactivate. There will be a huge case load despite the treatment of active people, and this will persist for some time because the disease is so endemic in that population. Also, in multiple-case households about half of the multiple cases are infected outside the home, so in other words these people are spending a large amount of exposure time outside the home. We don't know where that exposure time occurs; it could be in the neighbouring suburb because the suburb is continuous with another suburb on at least one side. While a criticism of the work done in this suburb is that it is an isolated component of the whole, the fact of the matter is that we cannot look at a larger area.

Bateman: I would like to comment briefly about the Cape epidemic and the question of over-reporting. My first comment relates to a study of the criteria for diagnosis, used in cases notified to the State authorities, and comparing diagnoses made at a major hospital in the Western Cape with those made in equivalent hospitals in other regions. This study confirmed that the diagnosis was based on sound criteria, usually bacteriological, in the Western Cape, whereas in other regions a large proportion was diagnosed according to X-ray abnormalities alone. This suggests that the high notification figures for the Western Cape are correct. Differences in reporting procedures at local clinics may also play a role. For example, in the Ravensmead/Uitsig area, compared to other regions of the Western Cape, there is a higher paediatric notification rate. This is probably the result of an enthusiastic team in that area that has a special interest in the paediatric population. The ratio of smear-positive to smear-negative diagnosed active disease in this area is considerably higher than it is elsewhere.

My second comment is that we must not expect all the answers concerning differing incidences in different population groups to come from genetics. There has been a dramatic upturn in infection rates in African Blacks in this region. One reason for this is the relatively recent influx of Africans to the region. However, these recent arrivals are displaying the same increase in tuberculosis as is found in the mixed race group that has traditionally populated this region. This suggests that transmission and environment are key factors.

Donald: We can dispute the accuracy of a certain percentage of diagnoses, but the most important points from a genetic point of view are those concerning the varying sex ratios and the age influence upon disease.

Fine: It appears that the incidence of tuberculosis is as high in the Ravensmead/Uitsig area as it is anywhere in the world today, but it does not seem to be as high as the incidence in Europe 100 years ago. Peter Donald talked of mortality rates of 1% per population, and we have been told that the incidence of disease is up to 3000 per 100 000 population—but what is the denominator?

Donald: It is about 900 per enumerator subdistrict.

Fine: The standard error based on that denominator would be rather large. If you look at the national statistics anywhere, for example in the UK, over the last century, you will find some areas that have higher and some areas that have lower disease incidences than the national average. There is always a danger of focusing on the area with the highest disease incidence, particularly if it implies a small denominator.

Bellamy: Worldwide, only 10% of people who are infected with *M. tuberculosis* will develop disease. I'm sure that many of us here have been exposed to the bacterium and are probably infected, but have not developed tuberculosis. In contrast, at least in Ravensmead and Uitsig, the risk of developing disease following infection is probably more like 30% or even 50%. This extremely high rate suggests that there is something different in this region, either a particularly virulent strain of *M. tuberculosis*, an unusually genetically susceptible population or some environmental factor. If it does involve the environment, then it must be specific environmental factors rather than just generally poor living conditions, because there are many populations in the world that live in equally bad conditions but have a much lower rate of tuberculosis.

Fine: I'm surprised by the way the phrase 'the tuberculosis epidemic' is being widely used in the field of tuberculosis nowadays, given that many infectious diseases have declined over the last 200 years in developed countries and we don't talk about '*the* polio epidemic', '*the* diphtheria epidemic' or '*the* typhoid epidemic'. We understand the dynamics of acute infectious disease epidemics, and how they decline because of exhaustion of susceptibles and increase when the susceptibles are replenished. The phrase is simplistic, and I'm convinced that the situation is much more complicated for tuberculosis. It can probably be traced back to Grigg (1958), who implied that when tuberculosis was introduced into populations its incidence first increased and then declined in an almost predictable manner, and he attributed the decline largely to genetic selection. Many people have denied that genetic selection explains the incredible declines of tuberculosis observed all over the world, and there is much evidence that they have been associated with improvements in socioeconomic conditions and treatment, etc. I wonder whether a population geneticist has ever looked at the rates of tuberculosis decline in Europe, for example, and asked 'what would I have to assume of a gene to explain that?'

Donald: I tried to illustrate that there is a debate on this, and I quoted Grigg (1958) as one of the people who has influenced our thinking on this. In his defence, tuberculosis is a different disease from the other infectious diseases in which wave patterns are clearly seen. With tuberculosis one has to stand back much further from the disease than in the case of measles, for example, in which the wave phenomenon can be seen within a matter of months. Youmans (1979) demonstrated mathematically that the elimination of susceptibles cannot be the only factor, and he suggested that surburbanization plays a role.

Bellamy: Whether or not changes in the incidence of tuberculosis can be affected by the weeding out of tuberculosis susceptibility genes may depend on the population being studied. One of the classic examples where genes were used as an excuse to justify appalling social conditions was in South Africa during the time when the Blacks worked in mines, when there were high incidences of tuberculosis. Cummins (1929), for example, said that this was due to their racial susceptibility as opposed to the dire conditions they were working in. However, there is evidence to support his virgin soil hypothesis. For example, if we look at the Qu'Appelle Indians when tuberculosis was introduced in 1890, initially tuberculosis caused an annual mortality rate of 10% of the population, and within 40 years the mortality rate had decreased to 0.2% of the population. Half of the families were wiped out, and it is believed that this dramatic fall in tuberculosis death rates was due to a weaning out of tuberculosis susceptibility genes (Motulsky 1960).

References

Cummins SL 1920 Tuberculosis in primitive tribes and its bearing on tuberculosis in civilised communities. Int J Publ Health 1:10–171

Cummins SL1929 Virgin 'soil'—and after. A working conception of tuberculosis in children, adolescents and aborigines. Br Med J 2:39–41

Daniel TM Bates JH, Downes KA 1994 History of tuberculosis. In: Bloom BR (ed) Tuberculosis: pathogenesis, protection and control. ASM Press, Washington DC, p 13–24

Ferguson RG 1955 Studies in tuberculosis. University of Toronto Press, Toronto

Fourie PB 1983 Patterns of tuberculin hypersensitivity in South Africa. Tubercle 64:167–179

Grigg ERN 1958 The arcana of tuberculosis. Am Rev Tuberc Pulm Dis 78:151–172, 426–453, 583–596

Motulsky AG 1960 Metabolic polymorphisms and the role of infectious diseases in human evolution. Hum Biol 32:28–62

Stead WW, Dutt AK 1988 Epidemiology and host factors. In: Schlossberg D (ed) Tuberculosis. 3rd Edn. Springer Verlag, New York, p 1–15

Youmans GP 1979 Tuberculosis. WB Saunders, Philadelphia, p 356–369

Using conventional and molecular epidemiological analyses to target tuberculosis control interventions in a low incidence area

Philip C. Hopewell

Division of Pulmonary and Critical Care Medicine, San Francisco General Hospital, 1001 Potrero Avenue, Room 5K1, San Francisco, CA 94110, USA

Abstract. To consolidate the gains made in controlling tuberculosis in industrialized countries in the past five years, we must analyse carefully the epidemiology of the disease and the effectiveness of various control interventions. In San Francisco we have performed conventional and molecular epidemiological analyses that have shown that there are in essence two parallel epidemiological patterns, one in the foreign-born population and the second among US-born persons, with little interaction between them. Most tuberculosis in the foreign-born population is a result of endogenous reactivation of latent infection, whereas recent infection with rapid progression to illness is a more frequent course in US-born cases. Among the US-born cases specific risk factors—homelessness, HIV infection, drug abuse—are highly prevalent. Although there has been a progressive reduction in the number of cases and in the proportion resulting from recent infection in San Francisco in the past five years, there continues to be a high proportion of cases that result from recent infection among US-born persons. These findings suggest that existing control interventions should be tailored to specific target groups and that new interventions are needed to provide for increased efficacy.

1998 Genetics and tuberculosis. Wiley, Chichester (Novartis Foundation Symposium 217) p 42–56

The increase in tuberculosis cases and case rates in the US in the late 1980s and early 1990s was a phenomenon previously unseen in an industrialized country (Fig. 1A). With the exception of minor deviations and statistical artefacts, tuberculosis has been decreasing consistently during this century (Fig. 1B), and, between 1953 and 1984, the annual rate of decline averaged at 5–6% (Centers for Disease Control and Prevention 1997). Why should there have been an increase in the incidence of a disease for which both curative and preventive measures are well known and widely available? Moreover, why should the resurgence occur at all

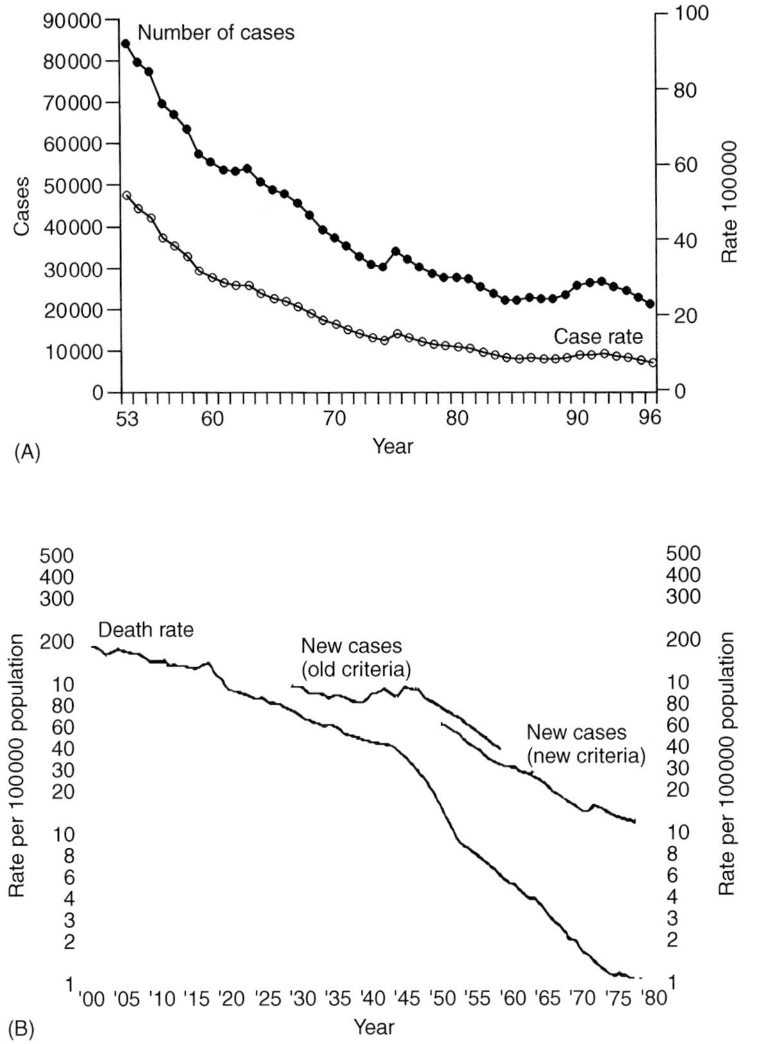

FIG. 1. (A) Reported annual tuberculosis cases and case rates in the US 1953–1996. Data from Centers for Disease Control and Prevention (1997). (B) Reported annual tuberculosis case and death rates in the US 1900–1980. (Modified from Health Education and Welfare Publication No. Centres of Disease Control 78–8360.)

in a country that in 1990 spent more than US$2700 per person on health care (compared with, for example, Tanzania which spent US$4 per person in 1990) (Ad Hoc Committee on Health Research Relating to Future Intervention Options 1996)?

The answers to these questions are not simple and require understanding of a complex interplay of social, biological and political factors that came into unfortunate alignment, most notably in New York City but also in other urban centres, such as San Francisco (Brudney & Dobkin 1991, Cantwell et al 1994). However, the coming together of these factors does not represent a unique occurrence, and the resulting damage to tuberculosis control could occur again in the US or in other countries. Many of the conditions that prevailed in the US in the 1980s and early 1990s are equally prevalent today. The socioeconomic status of the most at-risk segment of the population has not improved, HIV infection is still frequent, and many persons are still in shelters, correctional facilities and hospitals. What has changed is the attention paid to tuberculosis control, and this attention has been translated to substantially increased funding for control interventions (Fig. 2).

Having now re-established control of tuberculosis, it is essential that we evaluate as carefully and precisely as possible the epidemiology of tuberculosis in the US in order to be most efficient in further control interventions, to use the current resources to identify targeted strategies, to indicate where new approaches are needed and to develop new tools to incorporate into the control effort.

The history of tuberculosis in the United States

Although collection of cause-specific mortality data in the US began in 1900, data from all states were not included until 1933. A rate of 'new' cases of tuberculosis

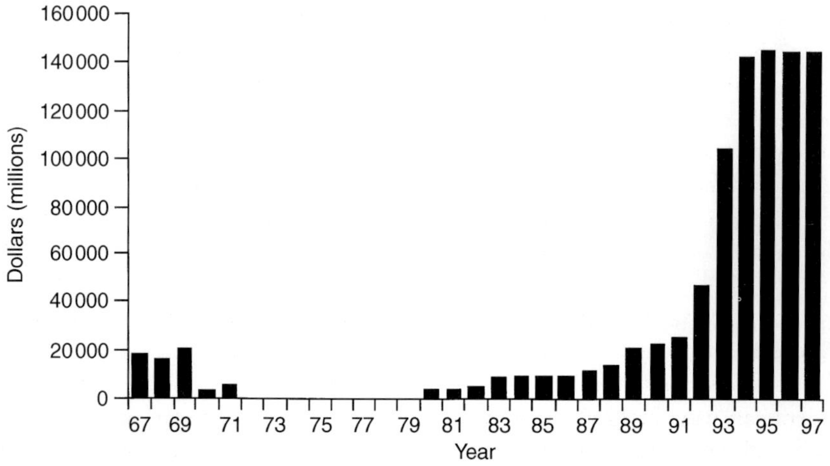

FIG. 2. Funding by the Centers for Disease Control for tuberculosis control.

was determined annually beginning in 1930 but this figure did not become an accurate reflection of disease incidence until 1952 when inactive cases were no longer included. As shown in Fig. 1B, morbidity and mortality rates were consistently decreasing with the exceptions of increased death rates during World War I, a transient levelling of the new case rate in the early 1960s and an apparent increase in new case rates in the mid-1970s due to a change in reporting criteria. Although it is stated frequently that the tuberculosis situation was improving prior to the advent of chemotherapy, it is clear from Fig. 1B that the decline in mortality rates accelerated at the time chemotherapy became available in the late 1940s. Because accurate data on case rates weren't available until after chemotherapy appeared, its impact on morbidity can't be determined.

Although it has been the subject of considerable debate, the decline in tuberculosis mortality prior to the advent of chemotherapy has been generally attributed to improvements in socioeconomic conditions with less crowding and better nutrition. However, there are theoretical reasons, based on mathematical modelling, to think that the decline was in part a function of the natural dynamics of tuberculosis epidemics (Blower et al 1995). Regardless of the cause of the decrease in tuberculosis mortality and morbidity, the common assumptions were that the causative factors, be they biological or sociological, would be more or less constant, and would result in the ultimate disappearance of the disease (Frost 1937). The conditions that came to prevail in the 1970s, 1980s and 1990s in the US were not foreseen.

Not shown in Fig. 1B, but apparent in Fig. 1A, are the two subsequent deviations in the downward trend of morbidity: a levelling of the trend in 1979, 1980 and 1981 and, more notably, a levelling followed by an increase in 1984–1992. The first of these deviations was due to an influx of Southeast Asian refugees and was much more noticeable in data from specific areas, such as San Francisco. This increase in foreign-born cases was a portent of an important trend in tuberculosis epidemiology in the US and in western Europe: the globalization of tuberculosis (McKenna et al 1995, Raviglione et al 1995, Reider et al 1994).

A second factor that is evident is the increase in tuberculosis in persons with HIV infection. Fortunately, in San Francisco the incidence of AIDS has decreased substantially, as has the incidence of tuberculosis and the number of cases with HIV infection. However, HIV infection remains an extremely important risk factor for tuberculosis in most urban centres in the US and in the world (Hopewell 1997).

The second change in direction of the incidence curve of tuberculosis is the subject of this chapter and was a much more serious occurrence that highlighted fundamental flaws in the country's tuberculosis control programme. As can be seen in Fig. 2, targeted funding for tuberculosis at the federal level disappeared in 1972 and didn't reappear until 1978. Even after targeted funds were appropriated, funding levels remained at or below what they had been in the late 1940s. This is

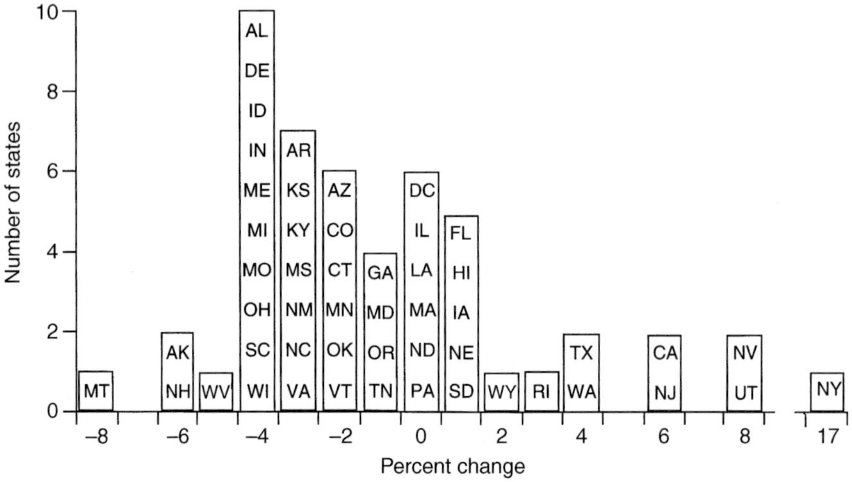

FIG. 3. Per cent change in numbers of reported tuberculosis cases between 1984 and 1992. Initials stand for individual states.

not to say that there was no money for tuberculosis control: local funding continued, although probably at a reduced rate until it was evident that there was a serious problem in the late 1980s.

At least in part the reasons for the lack of broad concern with tuberculosis were related to the fact that the resurgence of the disease was focal. As is shown in Fig. 3, the increase in tuberculosis cases between 1984 and 1992 occurred mainly in the most populous states with New York leading the way. It is worth noting, however, that even among the states that had no increase between 1984 and 1992 the overall percentage decline was less than in previous years.

Perhaps the defining event in attracting public attention to the seriousness of the resurgence of tuberculosis was the death of a prison guard in upstate New York from multidrug-resistant tuberculosis. That this unfortunate occurrence took place against an emerging backdrop of increasing rates of tuberculosis caused by organisms resistant to at least isoniazid and rifampicin succeeded in alerting the public and policy-makers to the need for increased support for tuberculosis control at the national level (Frieden et al 1993). Hence, the Centers for Disease Control funding for tuberculosis increased from approximately US$45 million in 1992 to US$140 million in 1993.

In New York City, the epicentre of both the overall resurgence and the outbreaks of multidrug-resistant tuberculosis, funding for tuberculosis control increased from US$4 million in 1988 to a peak of more than US$40 million in 1994 (Frieden et al 1993, 1995, Fujiwara & Sherman 1997). In all, since the

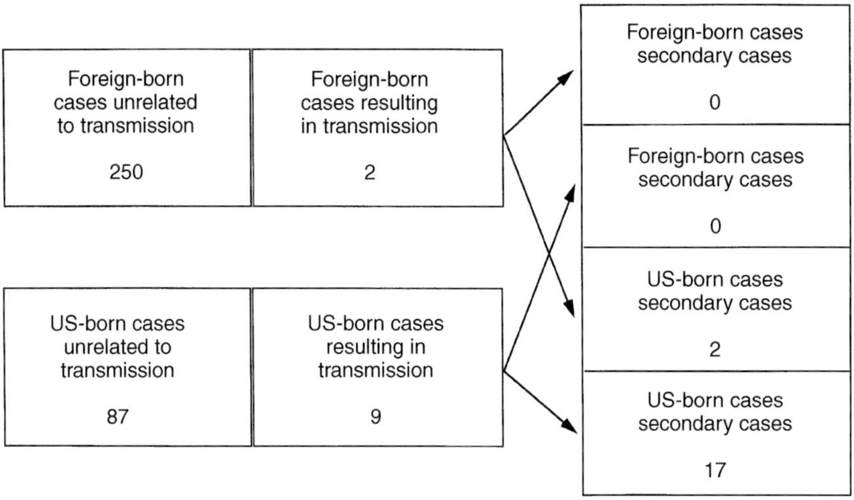

FIG. 4. Interactions between US-born and foreign-born tuberculosis cases in San Francisco 1993–1994. Data from Chin et al (1998).

resurgence began, an estimated total of US$791 million has been spent in New York City to restore control of the disease (Fujiwara & Sherman 1997). Control of tuberculosis has been restored in New York City and in the US as a whole, but it has been an expensive lesson. To avoid repeating the past failures we will have to examine carefully the epidemiology of the disease and to define and focus strategies more precisely and efficiently. In San Francisco we have been using a combination of conventional and molecular epidemiology to examine the dynamics of tuberculosis (Small et al 1994). These investigations have enabled determination of the interactions among risk factors for tuberculosis and the behaviour of the epidemic over time.

Interaction of tuberculosis risk factors in San Francisco

Since January 1991 we have been conducting a population-based study of the molecular epidemiology of tuberculosis in San Francisco, and we have carried out a number of analyses that have served to define the dynamics of the disease in the city. Moreover, we have been able to examine the effects of intensified control measures on the incidence and epidemiological patterns of tuberculosis.

Since the study began, organisms from 86.5% of the culture-positive cases have undergone IS6110-based DNA fingerprinting. In addition, 82.2% of the clustered (identical IS6110 fingerprint) cases that had fewer than six IS6110 bands underwent secondary genotyping using the polymorphic GC-rich sequence (PGRS) approach. The purpose of secondary PGRS typing is to improve on the specificity of matching by IS6110 in low copy number strains for indicating a transmission link.

Because approximately two-thirds of the incident tuberculosis cases in San Francisco occur in persons born outside the US, we were especially interested in the dynamics of the disease within the foreign-born population and the interactions of the US-born and foreign-born populations. In this analysis we sought to determine, in so far as possible, the amount of transmission of *Mycobacterium tuberculosis* that occurred in San Francisco (Fig. 4; Chin et al 1998). To develop the study cohort, we excluded all cases from 1991 and 1992. Without prior data, it could not be determined if they had acquired *M. tuberculosis* during the previous two years. Next, we excluded cases from 1993 and 1994 that had DNA fingerprint matches in 1991 and 1992. Thus, the cohort consisted of persons who had likely developed tuberculosis as a consequence of endogenous reactivation of infection acquired more than two years previously (or possibly acquired from outside San Francisco). Included were all cases from 1995 that had isolates which matched isolates from cases identified from 1993 to 1994, presumably having acquired their infection from the case with the matching isolate. This design, then, enabled us to determine the amount of transmission with progression to disease occurring from endogenously reactivated cases in 1993 to 1995.

This analysis demonstrated that, as we have found in other studies, there is little clustering that involves foreign-born cases, with the possible exception of persons born in Mexico (Small et al 1994, Jasmer et al 1998). This indicates a lack of transmission within the foreign born population and from foreign-born to US-born persons, or vice versa. Thus, presumably, most of the incident cases among the foreign-born persons are the result of reactivation of latent infection.

In the US-born population clustering was more frequent but there was still little evidence of interaction between the two populations. As an indication of the greater frequency of transmission within the US-born population, the ratio of secondary to initial cases in the US-born population was 1 : 5.6 compared with 1 : 121 in the foreign-born population. Of particular note is the finding that all 28 US-born cases in clusters had factors (AIDS, drug use, homelessness) that increased the risk of tuberculosis (Table 1). Neither of the two foreign-born cases in clusters had these risk factors.

This assessment indicates that there are in essence two parallel epidemics of tuberculosis in San Francisco, one (the larger) in the foreign-born population and the second within the indigenous population. These parallel patterns can be characterized in usual sociodemographic terms, but, more importantly, they differ in their pathogenesis. The lack of clustering among the foreign-born population strongly indicates that tuberculosis is resulting from endogenous reactivation of

TABLE 1 Transmission of *Mycobacterium tuberculosis* in San Francisco

A

Population	*Ratio of secondary to initial cases****	*Proportion of source cases****	*Number of clustered cases with risk factors****
US-born	1 : 5.6	9/96 (9.4%)	28
Foreign-born	1 : 121	2/252 (0.8%)	0

***$p < 0.001$

B

Risk factor	n	%
AIDS	23	76.6
Drug use	21	70.0
Homeless	13	43.3
More than one of the above	25	93.3

infection that was likely acquired in their countries of origin. In contrast, among the US-born population tuberculosis is much more likely to be a consequence of infection from an infectious source case in San Francisco. Not surprisingly, all of the clustered US-born cases involved either a source case or secondary cases with one or more risk factors — HIV infection, homelessness or drug abuse.

These findings suggest that different control strategies are necessary to address the indigenous and foreign-born populations in San Francisco. In the foreign-born population, screening for tuberculous infection and wider use of preventive therapy would be predicted to decrease disease resulting from endogenous reactivation. Current indications for the use of isoniazid preventive therapy are conservative, and broadening its use with appropriate monitoring has been shown to be safe (Salpeter et al 1997). Moreover, newer and, potentially, less toxic agents could be more applicable than isoniazid (Grosset & O'Brien 1997).

Investigation of the contacts of newly discovered cases of tuberculosis and isoniazid preventive therapy for those found to be infected is effective in foreign-born populations but less so among the US-born populations. The reasons for this difference are probably predominantly sociological. Foreign-born persons are much more likely to have an easily defined set of contacts — household and family members — and their more casual contacts are not likely to have risk factors for developing tuberculosis. We would assume that one of the reasons for the lack of clustering among these people is the effectiveness of contact investigation and preventive therapy with isoniazid.

In contrast, contact investigation is substantially less effective as currently done among the US-born population. By its nature, a molecular epidemiological approach to the evaluation of contact investigation can identify only the failures; thus, rates of success can't be determined. Nevertheless, in our analysis, nearly 70% of the time the contact investigation associated with the source case failed to identify the eventual secondary case as a contact. Even after extensive record reviews and patient interviews, clear relationships could not be established between putative source and secondary cases. Thus, it isn't clear if the failures of contact investigation were preventable. It is clear, however, that traditional contact investigation that focuses on close household contacts isn't relevant for homeless persons. Considerably more needs to be learned about the social patterns of the risk groups and interviews with index cases reformatted to take these patterns into account.

An additional means of tuberculosis control among US-born persons suggested by this analysis is the screening of persons in identified risk groups for tuberculosis and tuberculous infection. Although screening in general is not a useful means of finding cases in high risk populations, it may be of value (Schluger et al 1997). This is discussed in more detail in the subsequent section.

Tuberculosis in the homeless

Homelessness is a well-recognized risk factor for tuberculosis, although it is difficult to disassociate housing circumstances from a variety of other factors that overlap in this population (Zolopa et al 1994). Nevertheless, homeless persons probably constitute a 'core group' within the population that has epidemiological significance out of proportion to its absolute size (Barnes et al 1997).

In San Francisco there are an estimated 10 000 homeless or unstably housed persons, of whom 2774 were in a cohort that was evaluated prospectively by Moss et al (1998). The rates of tuberculosis were extremely high: 10 persons (0.36%) were found to have active tuberculosis at the time of initial evaluation, and there were an additional 25 incident cases during a median follow-up of 3.2 years (0.27% per person/year).

Molecular epidemiological analysis showed that 15 of the 20 incident cases for whom DNA fingerprinting was available were in 12 different clusters. Three of these 15 cases appeared to be the sources of three clusters.

The reasons for the high incidence of tuberculosis in the homeless and their epidemiological effect in causing new cases are multiple. In spite of a high rate of completion of chemotherapy overall in San Francisco, the highest rates of loss were among the homeless. At least some of these cases continue to be infectious. Many of the persons in this environment are vulnerable to tuberculous infection because of HIV infection or other factors that are more difficult to quantify, such as drug abuse, alcoholism, poor nutrition or generally poor health (Zolopa et al 1994).

In addition to the greater difficulty of completing treatment for tuberculosis among the homeless, traditional approaches to the administration of isoniazid preventive therapy are marginally effective at best. Without incentives, only 20% of homeless persons in the San Francisco cohort who were candidates for isoniazid preventive therapy completed six months of the drug. These findings concerning the homeless suggest that targeted screening programmes designed both to identify persons with active tuberculosis as well as candidates for preventive therapy are useful approaches to tuberculosis control in this high risk population (Schluger et al 1997). Considerable thought and individualization should go into determining the best ways of completing therapy. Directly observed therapy in itself is insufficient but must be combined with comprehensive case management and provision of a range of necessary services. Screening with the goal of identifying candidates for preventive therapy should not be undertaken unless there are provisions to ensure that evaluations are completed and preventive therapy can be given successfully. Given that such persons are at high individual risk and, if disease develops, pose a high community risk, directly observed preventive therapy should be used. As with treating active tuberculosis, individualized and imaginative ways of supervising drug administration must be applied. Because completion of preventive therapy should be the major goal of screening programmes, such programmes should not be undertaken unless the resources are available to follow administration of the preventive therapy to its successful completion.

Molecular epidemiological analyses of tuberculosis trends in San Francisco

In addition to the analyses described previously, we have also examined the trends in tuberculosis in San Francisco between 1991 and 1996 (Jasmer et al 1998). In part, this assessment was intended to provide a global estimate of the effectiveness of a package of tuberculosis control interventions that were implemented mainly in 1992 and 1993.

The number of cases of tuberculosis in San Francisco peaked in 1992 at 370 and declined in each subsequent year. As the total number of cases decreased, the numbers and proportions of cases that were in clusters (using six-month intervals) also declined. Throughout the period, there was a persistent large difference between the proportions clustered in the foreign-born and the US-born populations, as described previously. The proportions of clustered cases did not decline significantly among persons born in the US, persons with HIV infection, those less than 50 years old at diagnosis and African–Americans. These findings indicate that although the new interventions, together with the traditional programme, were effective in decreasing cases overall, there was still an important amount of transmission occurring among persons with risk factors, and that these groups required increased attention and new approaches to control as described above.

Implications for tuberculosis control

The results of several conventional and molecular epidemiological assessments of tuberculosis control highlight the need for careful subgroup analyses and for new targeted approaches to tuberculosis control in San Francisco. Although in some ways tuberculosis epidemiology and control differs between San Francisco and other urban centres in the US, there are many similar features, thus enabling the information from San Francisco to be generalized. Country-wide, tuberculosis occurring in foreign-born persons is an increasing proportion of the total cases reported each year. As described previously, screening for tuberculous infection among foreign-born persons (not just for active tuberculosis as is done currently) and wider use of preventive therapy is a logical means to reduce the likelihood of endogenous reactivation of latent infection. This approach could be greatly facilitated by at least three innovations: (1) more accurate tests to identify tuberculous infection; (2) tests (for genetic susceptibility, for example) that would identify high risk persons so preventive therapy could be targeted more precisely; and (3) better preventive agents that could be administered more easily (for example, once per week) and more safely.

The long-term approach to tuberculosis among foreign-born persons is more effective global tuberculosis control. Such an effort will require a co-ordinated multi-pronged attack guided by a well thought-out plan that specifies roles, responsibilities and resources.

Among persons born in the US, tuberculosis occurs predominantly among those with specific social or biological risk factors and is likely to result from infection with *M. tuberculosis* that has been acquired recently. Among these risk groups, targeted screening and directly observed preventive therapy would be expected to decrease the incidence of tuberculosis. Again, better tests for infection and a safer more efficacious preventive regimen would greatly facilitate this approach.

It is among the indigenous population, given the difficulties with adherence with preventive therapy and the frequency of tuberculosis developing as a result of recent infection, that a vaccine, were an effective compound available, would be particularly useful (Orme 1997).

Acknowledgement

Much of the work cited in this review was supported by a grant from the National Institutes of Health, National Institute of Allergy and Infectious Diseases (AI 34238–04).

References

Ad Hoc Committee on Health Research Relating to Future Intervention Options 1996 Investing in Health Research and Development. World Health Organization, Geneva (Document TDR/Gen/96.1)

Barnes PF, Yang Z, Preston-Martin S et al 1997 Patterns of tuberculosis transmission in Central Los Angeles. JAMA 278:159–163

Blower SM, McLean AR, Porco TC et al 1995 The intrinsic transmission dynamics of tuberculosis epidemics. Nat Med 1:815–821

Brudney K, Dobkin J 1991 Resurgent tuberculosis in New York City. Am Rev Respir Dis 144:745–749

Cantwell MF, Snider DE, Cauthen GM, Onorato IM 1994 Epidemiology of tuberculosis in the United States 1985 through 1992. JAMA 272:535–539

Centers for Disease Control and Prevention 1997 Reported tuberculosis in the United States 1996, p 5

Chin DP, DeRiemer K, Small PM et al 1998 Differences in contributing factors to tuberculosis incidence in the United States-born and foreign-born populations. JAMA, submitted

Frieden TR, Sterling T, Pablos-Mendez A, Kilburn JO, Cauthen GM, Dooley SW 1993 The emergence of drug-resistant tuberculosis in New York City. N Engl J Med 328:521–526

Frieden TR, Fujiwara PI, Wookko RM, Hamburg MA 1995 Tuberculosis in New York City — turning the tide. N Engl J Med 333:229–233

Frost WH 1937 How much control of tuberculosis? Am J Public Health 27:759–763

Fujiwara PI, Sherman LF 1997 Multidrug-resistant tuberculosis: many paths, same truth. Int J Tuberc Lung Dis 1:297–298

Grosset JH, O'Brien RJ 1997 Advances in tuberculosis preventive therapy. Semin Respir Crit Care Med 18:449–457

Jasmer RM, Hahn JA, Small PA et al 1998 A molecular epidemiologic analysis of tuberculosis trends in San Francisco 1991–1996. Ann Intern Med, submitted

Hopewell PC 1997 Tuberculosis in persons with human immunodeficiency virus infection: clinical and public health aspects. Semin Respir Crit Care Med 18:471–484

McKenna MT, McCray E, Onorato I 1995 The epidemiology of tuberculosis among foreign-born persons in the United States 1986 to 1993. N Engl J Med 332:1071–1076

Moss AR, Hahn JA, Tulskey JP et al 1998 Transmission of tuberculosis in the homeless: a prospective study. Lancet, submitted

Orme IM 1997 Progress in the development of new vaccines against tuberculosis. Int J Tuberc Lung Dis 1:95–100

Raviglione MC, Snider DE, Kochi A 1995 Global epidemiology of tuberculosis. Morbidity and mortality of a worldwide epidemic. JAMA 273:220–226

Reider HL, Zellweger JP, Raviglione MC, Keizer ST, Migliori GB 1994 Tuberculosis control in Europe and international migration. Eur Respir J 7:1545–1553

Salpeter SR, Sanders GD, Salpeter EE, Owens DK 1997 Monitored isoniazid prophylaxis for low-risk tuberculin reactors older than 35 years of age: a risk–benefit and cost-effectiveness analysis. Ann Intern Med 127:1051–1061

Schluger NW, Huberman R, Wolinsky M, Dooley R, Rom WN, Holgman RS 1997 Tuberculosis infection and disease among persons seeking social services in New York City. Int J Tuberc Lung Dis 1:31–37

Small PM, Hopewell PC, Singh SP et al 1994 The epidemiology of tuberculosis in San Francisco: a population-based study using conventional and molecular methods. N Engl J Med 330:1703–1709

Zolopa AR, Hahn JA, Gorter R et al 1994 HIV and tuberculosis infection in San Francisco's homeless adults: prevalence and risk factors in a representative sample. JAMA 272:455–461

DISCUSSION

Young: Why is there so much less transmission in foreign-born people than in people who were born in San Francisco?

Hopewell: There are several reasons. One is that in San Francisco, and probably in most of the US and Western Europe, relatively recent arrivals tend to stay in fairly circumscribed communities, rather than travelling around the area, so it's a function of their relatively recent arrival and their behaviour. Because Asians, in particular, in San Francisco tend to live in nuclear families, it's easier to identify their contacts and give them preventive therapy. They also tend to be fairly compliant, which prevents them from developing the disease. Second, because of a much higher prevalence of tuberculous infection among foreign-born people they are protected against new exogenous infection, so they don't progress rapidly to disease. Third, they don't have any of the risk factors that are prevalent in the US-born segment of the population. There is a low rate of HIV infection among foreign-born persons in San Francisco, and not many of these people are homeless.

van Helden: Many of the *Mycobacterium tuberculosis* strains from the foreign-born population matched, even though there was no contact. Is this because the foreign borns largely had the Beijing Asian strain, which dominates in Asia, and that therefore the matching is meaningless in terms of fingerprinting?

Hopewell: It's difficult to know whether it's meaningless or not. We haven't looked at whether these people may have had any contact before they came to the US. Some of them are in the Beijing pattern family, so it's probably a less meaningful than when this pattern is found in the US-born people.

van Helden: If the Beijing Asian strain dominates in Asians, the fact that you couldn't trace recent transmission between them doesn't negate the fact that fingerprinting is useful provided you know in what setting to use it. Has fingerprinting made any impact on tuberculosis control in San Francisco?

Hopewell: It has made some impact. For example, we've been doing screenings in so-called single-room occupancy hotels. We have not only been looking for cases but also for preventive therapy candidates. The people who live in these hotels have many tuberculosis risk factors, so we have been targeting preventive therapy at those patients. In addition, as a result of earlier molecular epidemiological findings we have started to screen people who are admitted into long-term care, particularly those who are HIV positive. We thought that the transmission rates in HIV-positive populations in San Francisco were low because this population tends to be composed of middle-class gay Caucasian men, who have a fairly low incidence of tuberculosis in general, and yet we found a fair amount of transmission occurring in this group. Some of this information could have been discerned by looking at conventional epidemiology, although the outbreak in the hotel could not have been because it's impossible to get detailed information from homeless people who take drugs about where they have been and when.

Fine: Were there any trends in clustering by age in either the foreign-born or the US-born groups? Data from The Netherlands suggest that the proportion clustered decreased with age (van Soolingen 1996). This makes sense, given that the proportion of individuals who are eligible to reactivation of old infections increases with age, whereas younger people are more likely to have been infected recently and thus have isolates that are identifiable as clustered.

Hopewell: We find that clustering is more common in those aged less than 50. We haven't broken it down further than that because the numbers are too small.

van Helden: But you only monitored this for a short period. If different strains are present, such that rapid transmitters are present in the young, but slow-growing strains, which may be quantitatively more dormant and have a longer cycle in the host, are present in the older age group, then you would only see clustering over a much longer time period.

Fine: Do you have evidence of that?

van Helden: No, but can you draw any conclusions from clustering other than that it's a social interaction?

Fine: There are certain observations one would expect from our current understanding of tuberculosis. We haven't talked about variables such as the time window, which will affect the proportion of clustering. There may be certain types of *M. tuberculosis*, identifiable by IS6110 fingerprinting, that tend to lead to late appearance of disease after endogenous reactivation. However, that hypothesis aside, one would expect to see a decrease in clustering by age, as has been observed in Holland by van Soolingen (1996) and in San Francisco by Phil Hopewell, just as one would expect the proportion of disease attributable to reactivation to increase with age.

Hopewell: The situation in San Francisco is confounded by the fact that older people who develop tuberculosis tend to be foreign-born people, so it might not be a function of age per se, rather they were previously infected in a high prevalence area.

Bateman: Why did you only look at the 50-year age demarcation? If you found a small difference at age 50, would you not expect it to be more pronounced at age 65 and above? Also, did you look at the age at which subjects migrated to the US? Because migrants don't tend to represent a normal age spectrum any more than the homeless do, should you not correct for social behaviour before you correct for age?

Hopewell: I agree. We don't know whether the association is age related or if it is because most of the older patients are foreign born and they don't tend to cluster. And in answer to your first question, I don't know why we picked age 50.

Young: Do you think there are certain people who are genetically prone to one type of disease?

Hopewell: Yes, although dissecting the genetic influences out of this morass of social and biological influences is going to be extremely difficult.

Bateman: Does the Beijing Asian strain have identical characteristics to strains from San Francisco?

van Helden: No. There is a large dominance of a certain pattern in Beijing Asian strains, and I suspect the majority of foreign borns in San Francisco are of Asian origin, which would explain why there is a lot of clustering but no contact.

Hopewell: But we only found a small amount of clustering with no contact.

van Helden: This may be because the Beijing Asian strain is so dominant that the clustering is incidental.

Hopewell: We do see a lot of Asian patients, and it is in those patients where, if we find clustering, we often cannot find the epidemiological connection. On the other hand, I should point out that it has been widely accepted for a long time that even in young children who are tuberculin positive it is difficult to find their source of infection, in spite of their having a small circle of contacts. Therefore, our inability to detect the transmission link using epidemiological approaches in either Asians, or for that matter those who are born in the US, does not mean that the connection doesn't exist.

Reference

van Soolingen D 1996 Use of DNA fingerprinting in the epidemiology of tuberculosis. PhD thesis, University of Utrecht, The Netherlands

Vaccines, genes and trials

Paul E. M. Fine

Infectious Disease Epidemiology Unit, Department of Infectious and Tropical Diseases, London School of Hygiene and Tropical Medicine, Keppel Street, London WC1E 7HT, UK

Abstract. This symposium takes place at the intersection of an acute success (the extraordinary recent developments in genetical research relevant to mycobacteria) and a chronic failure (the embarrassing failure of Bacillus Calmette–Guérin [BCG], the world's most widely used vaccine, to control tuberculosis). I will summarize the arguments that the variable efficacy of BCG is attributable, at least in part, to genetic differences in either the vaccine, the host populations or *Mycobacterium tuberculosis*. I will then address the problem of building upon our experience with BCG in order to develop and evaluate an improved vaccine against tuberculosis. The fact that the great burden of tuberculosis is associated with pulmonary disease in adults, much of which is associated either with reactivation of old infection or else with relatively recent reinfection, poses an immense problem for vaccine strategy and evaluation. One approach may be to vaccinate early in life, prior to initial infection with the tubercle bacillus, and then to follow up individuals for many decades. The alternative is to develop vaccines to boost appropriate protective immune mechanisms in individuals who have already been infected, and/or who have already been vaccinated (e.g. with BCG), and/or who may also have been sensitized to various environmental mycobacteria that influence the immune response to tubercle bacilli. The latter approach is novel, and may ultimately prove impossible, but its potential logistic and public health advantages are so great that it warrants serious attention by the research community before being abandoned.

1998 Genetics and tuberculosis. Wiley, Chichester (Novartis Foundation Symposium 217) p 57–72

This symposium is being held at the intersection of an acute success and a chronic failure — the incredible scientific success implicit in the recent breakthroughs in mycobacterial and human genetics, and the embarrassing public health failure explicit in the story of Bacillus Calmette–Guérin (BCG). Given the enormous use of BCG — more people have received BCG over the past 50 years than have received any other vaccine — there would be little need for more meetings on tuberculosis if it had done its job. But it didn't. And we are here, hoping that discoveries inherent in the various genetical advances discussed at this symposium may rescue the situation. It is my role to discuss the development and evaluation of vaccines, with initial emphasis on genetical research. I will do this by building upon our experience with BCG.

BCG: looking for order in chaos

The variable efficacy of BCG vaccines under different conditions has been discussed almost *ad nauseam* (Bloom & Fine 1994, Fine 1995). In summary, we know that at least some BCG vaccines perform well, in terms of protection against pulmonary tuberculosis, under certain conditions. But the same vaccines can perform poorly in different contexts, and we remain ignorant of the determinants of this variance. Among the reasons often discussed are several that do not call immediately for genetical analyses: (1) that the differences reflect methodological differences between studies; (2) that they reflect physiological (e.g. nutritional) differences between the populations: or (3) that they reflect exposure to ultraviolet light, which either inactivates the vaccine or suppresses dermal Langerhans cells. But there are other explanations, which are explicitly genetic in flavour.

Genetics of Mycobacterium tuberculosis

It has been proposed that differences in BCG vaccine efficacy might reflect genetical differences between *Mycobacterium tuberculosis* strains prevalent in different populations or parts of the world. This hypothesis first received attention after the 'failure' of both Paris and Danish strain BCG vaccines in the Chingleput trial in south India, when it was noted that many strains of *M. tuberculosis* which had been isolated in that region of the world were of relatively low virulence for guinea-pigs (Tuberculosis Prevention Trial 1980, Mitchison 1964). The argument ran something like this: that the *M. tuberculosis* strain current in the trial population had peculiar virulence properties, such that it led rarely to primary progressive disease (against which BCG is purported to be most effective), but that it led frequently to reinfection or reactivation of disease, against which BCG is less effective. This explanation was consistent with the observation of low incidence of disease among those considered tuberculin negative at the onset of the Chingleput trial, and with observed high risks of disease among those considered tuberculin positive at the onset of the trial (ten Dam & Pio 1982).

The hypothesis was tested in guinea-pigs by Hank et al (1981), who were unable to note any difference in the effect of either Danish or Prague strain BCGs between supposedly low virulence *M. tuberculosis* isolates from the trial area and recently isolated high virulence strains from North America. Although these animal studies related more to protection against primary disease than against reinfection or reactivation of disease, their publication eclipsed the argument about the so-called 'south India variant', which has since been labelled a 'red herring' by D. A. Mitchison, upon whose work the hypothesis was originally based (personal communication 1996).

That geographic differences in *M. tuberculosis* might have something to do with the variable efficacy of BCG has more recently arisen in the context of discussions of fingerprint analyses of isolates from different populations (Hermans et al 1995). Unfortunately, the studies appropriate to test this hypothesis have yet to be made (e.g. comparison of fingerprint genotypes between cases with and without history of prior BCG vaccination). Such comparisons will soon be made, using IS6110 fingerprints as markers. But we have a logical problem, in that even if the results are negative, this would of course not be sufficient to refute the hypothesis convincingly, given that the fingerprint-dependent strain designation is crude and may be totally irrelevant as far as virulence or antigenicity is concerned. Much more powerful and relevant markers will be revealed in the months to come, as we have a better understanding of the entire genome of *M. tuberculosis*, and these will need to be examined as covariates in the context of appropriate epidemiological studies of BCG. I look forward to such studies, and to the comparisons of field isolates from different cases and populations to the benchmark sequences being derived in the current sequencing of the H37Rv and Oshkosh (CSU isolate 93) 'strains'.

Genetics of Homo sapiens

The story of the influence of host genetics on tuberculous infection and disease is discussed elsewhere in this volume. The classic evidence is based upon anecdotes implying differences in susceptibility between races, the recognition of family clustering, then twin studies, and, more recently, family segregation and case-control studies indicating that specific polymorphisms, e.g. the human leukocyte antigen DR2, influence susceptibility in certain populations (Fine 1981, Brahmajothi et al 1991, Comstock 1978). Inevitably, such studies have led to suggestions that genetical differences between populations might explain the observed differences in the behaviour of BCG. But the direct evidence is thin, at best.

Comstock & Palmer (1966) compared the efficacy of BCG between Caucasians and Blacks in the US Public Health Service trials in Georgia and Alabama. The observed protection was greater in Caucasians (46%; 95% confidence interval: −33% to 78%) than in Blacks (−6%; 95% confidence interval: −102% to 44%) but the numbers of cases were small (21 and 37, respectively), and the difference non-significant statistically. More recently, the finding of low efficacy in the south India Chingleput trial led to two case-control studies of BCG and tuberculosis among Asians in the UK, both of which showed evidence of appreciable protection in the UK (Rodrigues et al 1991, Packe & Innes 1988). Of course one could argue that the Asians included in these studies were not south Indian Dravidians, and hence were not comparable to the population of Chingleput District, and thereby declare the argument to be still open. Several studies are now employing genome scanning techniques to identify new genes associated with tuberculosis, and polymorphisms of candidate genes are being recognized, including those determining the various cytokines involved in cellular responses to mycobacteria. A next step in this evolution will be to look for differences in frequencies of particular polymorphisms between BCG-vaccinated and BCG-unvaccinated cases, and BCG-vaccinated and BCG-unvaccinated individuals similarly exposed but free of disease. The epidemiology implicit in such studies is not simple, and I hope we won't see too many overenthusiastic false positive (or negative) claims flooding our journals before sufficient substantial evidence accumulates allowing some firm conclusions.

Genetics of BCG

BCG has never been cloned. It originated from an isolate of *Mycobacterium bovis* taken from a cow with mastitis, was attenuated over 13 years of passage in media containing ox bile, and has since been maintained in a variety of culture media in various laboratories around the world (Bloom & Fine 1994). As a consequence, there are many different strains of BCG, with different properties; and it has long been wondered whether these differences might not be responsible for some of the observed differences in efficacy. It would be most convenient if this were so, as it could point us towards particular antigens or epitopes that are important in protection.

There are several examples which appear (at least at first) to be consistent with this explanation. Comstock reanalysed case-control study data on the efficacy of BCG in Indonesia and in Colombia, and found evidence for a possible change (decline) in efficacy when the programmes shifted from one (Japanese or British) to another (French or Danish) vaccine (Comstock 1988).

More interestingly, a controlled trial comparing Paris and Glaxo vaccines was carried out in Hong Kong under the auspices of the World Health Organization (WHO) and the Hong Kong tuberculosis services. All infants born between 1978 and 1982 (more than 300 000) were randomized (by alternating vaccine supplies to vaccination centres) to receive one or the other vaccine, by either the intradermal or percutaneous route. They were followed up over the following six years, during which 129 cases were ascertained. Recipients of the Paris-type vaccine were approximately 40% less likely to contract tuberculosis than were those who received the Glaxo vaccine, by either route ($p<0.05$; ten Dam 1993). It is interesting that the differences implied by this trial go in the opposite direction to those suggested by the observational study analyses by Comstock (1988).

Molecular analyses have now identified particular genomic regions found in *M. bovis*, but which appear to be absent from some or all BCGs (Mahairas et al 1996, Philipp et al 1996). In particular, it has been found that the so-called RD-2 region, which contains the *mpt-64* (also called *MPB-64*) gene, is present in the 'primitive' BCG strains (represented by current Brazilian [Moreau], Japanese and Russian substrains), but is absent from those substrains derived from the original BCG Pasteur strain after 1925, represented by today's Pasteur, Copenhagen and Glaxo-Evans substrains (Mahairas et al 1996). The full immunological implications of these deletions, if any, are as yet unknown.

It has recently been suggested that conscious selection by BCG vaccine manufacturers for decreased reactogenicity may have led to decreased immunogenicity of the vaccines (Behr & Small 1997). Despite the absence of human evidence that induction of tuberculin delayed-type hypersensitivity (DTH) was associated with protective immunity (Comstock 1988, Hart et al 1967, Fine 1993), the standard potency assay for BCG vaccines has long been their ability to induce tuberculin hypersensitivity in guinea-pigs. It is thus credible that manufacturers may have selected for strains which maintained (and perhaps maximized?) their ability to induce tuberculin hypersensitivity, but which minimized their induction of regional lymphadenopathy. Ironically, this selection might have favoured a decrease in protective immunizing potential. This is suggested by trends for decreased vaccine protection with increased passage number of several strains of BCG vaccines (Behr & Small 1997). The argument is interesting, perhaps even plausible, but not compelling, for two reasons.

First, there are several examples of identical vaccines providing different levels of protection in different populations. The most obvious example of this is the high efficacy of the Glaxo freeze-dried strain, which continues to provide good efficacy against pulmonary tuberculosis in the UK, but no evidence of any protection against pulmonary tuberculosis in Malawi (Karonga Prevention Trial Group 1996).

Second, it happens that the earlier trials, which used lower passage number vaccines, were carried out in northern Europe and the US, whereas the high passage number strains were evaluated at lower latitudes. A tendency for lower efficacy of BCG vaccines at lower latitudes has been noted by many authors (Palmer & Long 1966, Fine 1995), but has generally been attributed to the interference provided by exposure to environmental mycobacteria or to ultraviolet light. We may thus ask whether this latitude effect is likely to reflect genetical differences between vaccines or vice versa. Given the examples of fixed strains performing differently at different latitudes, and the abundant animal evidence for masking of BCG effects by prior exposure to environmental mycobacteria, it would seem inappropriate to attribute all the latitude effect in efficacy differences to variation between BCG strains.

Genetics of environmental mycobacteria

The implications of environmental mycobacteria have been referred to above. This subject has attracted attention for more than three decades, and much evidence has accumulated indicating that regional differences in exposure to various mycobacteria are responsible for some of the observed differences in observed efficacy of BCG (Palmer & Long 1966, Fine 1995). Unfortunately, the arguments have yet to come down to species or molecular detail, largely because of the bewildering variety and complexity of the environmental mycobacterial flora in different regions of the world. Classic experiments on guinea-pigs suggested that exposure to *Mycobacterium fortuitum*, *Mycobacterium avium* or *Mycobacterium kansasii* imparted 15%, 50% and 85% as much protection, respectively, as did BCG, as measured in terms of survival after challenge with H37Rv (Palmer & Long 1966). It is reasonable to presume that these differences somehow reflect antigenic relatedness to the tubercle bacilli.

On the other hand, the logic may be more perverse. Evidence to date suggests that BCG provides stronger protection against the more distantly related *Mycobacterium leprae* than against its closer relative, *M. tuberculosis* (Fine 1995). This has led some investigators (e.g. Rook & Hernandez-Pando 1998, this volume) to suggest that simple antigenic similarity is not the key to protection, and that it might even be detrimental by eliciting hypersensitivity, which is ultimately harmful. Not all mycobacterial immunologists would agree with this interpretation, but, given our ignorance of the natural history of mycobacterial infections, and of the nature of protective immune responses against them, we would be wise to keep such possibilities on the table.

Bettering BCG, and proving it

There is much talk of developing a new tuberculosis vaccine, and much exciting basic science around that endeavour. Among the approaches being applied are attempts to produce subunit products, attenuated and auxotrophic *M. tuberculosis* strains and DNA vaccines. Some of these new products are now moving towards Phase I and even Phase II human testing. The ultimate evaluation will involve Phase III trials of efficacy.

When it comes to evaluating new antituberculosis vaccines, we are fortunate to have the experience of several well-conducted trials of BCG. But the message of these trials is not a simple one — they have been large, expensive, time-consuming and difficult to interpret. It would be naïve to think that evaluation of new vaccines will be easy (Comstock 1994).

What is the goal?

Before setting up any trial, one needs a clear formulation of a hypothesis of what one is trying to do. In considering an antituberculosis vaccine, there is now increasing appreciation that the real challenge is prevention of pulmonary disease, particularly in adults and in developing country settings. This goal reflects the real burden of tuberculosis. It is (pulmonary) disease in adults which is of greatest epidemiological importance, in that it is responsible for infection transmission, and it is adult disease in developing countries which is of greatest public health importance, making *M. tuberculosis* the number one pathogen in terms of global mortality and is hence responsible for WHO's 1993 declaration of tuberculosis as a global emergency. In addition, one of the ironies of BCG is that it is consistently effective in protecting against childhood forms of tuberculosis (Rodrigues et al 1993).

This is an extremely difficult target, given that, according to our current understanding of tuberculosis, the majority of adult disease in developing countries occurs in individuals who were infected many years previously, who successfully dealt with their primary infection, yet who were unlucky enough to succumb to disease long after, either through reactivation of a latent focus of infection or through reinfection with a similar or new strain of the tubercle bacillus. In either case, it represents the failure or breakdown of an immunity which was developed by the individual in response to the initial homologous exposure. In other words, in order to prevent such disease, we have one of two options: either to get in early, and prevent initial infection; or else to go in after that initial infection and in effect 'improve upon natural homologous immunity'.

The first option, infection prevention, represents an enormous logistic challenge. It leads to proposals to vaccinate early in life and then to follow up individuals for many years until they reach adulthood, when there is a high risk of tuberculous disease. This approach will have to contend with the problem of predicting declines in any vaccine-induced protection, and hence the evaluation of boosters along the way. Only the Gambian trial of hepatitis B vaccine in infancy against adult liver cancer has been this ambitious in design—and that trial has a good surrogate marker for infection, which we lack in tuberculosis (Gambia Hepatitis Study Group 1987). The alternative, i.e. the recruitment of uninfected adolescents or adults, might appear to shorten the duration of the trial, but it raises considerable logistic difficulties of its own, given the weak tools at our disposal for recognition of infection in the first place, and the fact that there will still be a need for long-term follow up, given the age distribution of the target disease.

The second option, post-infection vaccination, represents an enormous immunological challenge, as it implies manipulating the host's immune response so as to improve upon its 'natural' response to infection. It may best be considered as a boosting exercise (this may sound like a more feasible challenge than to 'improve upon nature'), i.e. the enhancement of natural response so as to lead to destruction of latent bacilli in the body and/or to an increased resistance to reinfection. Articulated in this simplistic way, we are requesting two different tasks of the immune system: identifying and destroying dormant bacilli, versus enhancing local defences in the lung. An improved understanding of the pathogenesis of adult tuberculosis would be of immense help in directing our immunological researches along either of these lines (ten Dam & Pio 1982).

Any trial will face many challenges in its design and implementation, as there will be major issues surrounding the identification of infection (still not resolved), the need and appropriate schedule for boosters (which may depend upon the type of vaccine chosen but which adds greatly to the complexity of trial design and to the numbers of subjects required), the handling and implications of HIV infection (with its attendant ethical dimension and particular constraints on live vaccines) and the mere challenge of long-term follow up of large numbers of individuals. What is more, the history of variable efficacy of BCG in different populations indicates that multiple trials may be necessary in order to provide convincing evidence of the superiority of any new vaccine product. In this context it is important to note that populations in which BCG vaccines have been reported to have failed (e.g. Chingleput in south India or Karonga District in northern Malawi) are of particular interest; these are the populations most in need of an improved vaccine, and hence may be preferred sites for the evaluation of a new vaccine.

Correlates of protection

Given the resource implications of a Phase III trial, it would be foolish to embark upon such an enterprise without some prior evidence that the product under evaluation will have a beneficial effect. What sort of evidence might we expect to have? Basically, this can be of two kinds: (immunological) evidence based on humans, e.g. from Phase II trials, that the new vaccine does something to the immune system which we believe to be associated with protection against disease; or (immunological or bacteriological or pathological) evidence from some animal model system that the new vaccine does something which we believe will be associated with protection against human disease.

Human evidence. The search for immunological correlates of protection in humans is an old and thorny problem. Most of us here are aware of the long-standing saga of tuberculin sensitivity: the idea that the ability of a vaccine to induce tuberculin sensitivity was an indicator of its ability to protect against disease. This is now a dead horse, in that several studies have failed to demonstrate any association between the two either in animal models or in human studies (Fine 1993). Rather than go over this ground once again, I would emphasize two lessons from this story.

The first is that dogmatic statements made a quarter of a century ago, by some of the most prominent figures in the immunological world, such as George Mackaness, were wrong on this issue, but their weight set back this aspect of tuberculosis science by many years (Mackaness 1972). It is a classic example of the fallacies of false authority and inappropriate extrapolation. Dogma is dangerous.

The second point reflects another over-simplification. The tuberculin test is surely the most widely used of all immunological tests ever, and there is an enormous literature on the test and on tuberculin DTH (Palmer & Edwards 1967, Rieder 1995). The vast majority of work on tuberculin testing uses a simple criterion to dichotomize individuals or populations into 'positives' versus 'negatives' (infected or not infected, at risk or not at risk, to be vaccinated or given chemoprophylaxis or not, etc.). The criteria change from study to study, depending upon the tuberculin used and the method of administration, but few researchers have appreciated that tuberculin reactivity is not a binary variable. Much evidence indicates that individuals with moderate amounts of sensitivity behave differently to those with strong sensitivity; in particular, they have lower risks of subsequent disease (Fine 1993). Whether this is because these low levels of tuberculin sensitivity indicate that these individuals have already met *M. tuberculosis*, and have responded to that exposure in some efficient way, or whether it means that these individuals have met some other mycobacteria, which has induced some degree of heterologous immunity, is not really known. But we should not pretend that this J shape is not there. Despite such evidence, the history of mycobacterial immunology reads like a litany of dichotomies: positive versus negative DTH; helper versus suppressor cells; CD4 versus CD8; and T helper (Th) 1 versus Th2. My own suspicion is that the ultimate correlate will not appear as a simple binary variable, but will represent a complex function of several pathways involved in the host's responses to these agents

Animal evidence. There is a long history of animal model work on tuberculosis immunology and antituberculosis vaccines, involving mice, guinea-pigs and rabbits. Some of this work is discussed elsewhere in this volume. Despite the fact that this work has contributed immensely to our understanding of immunological processes, it has been based upon systems which are unnatural, not only in host–parasite combination, but also in pathogenesis. Many different models have been employed by different workers, employing different strains of animals, routes and schedules of vaccination, routes and doses of challenge and outcome measures (e.g. host survival, host immune response or bacillary counts in one or another tissue). Wiegeshaus & Smith (1989) have pointed out the difficulty inherent in this plethora of models: the fact that they failed to correlate, for example, in their evaluation of the immunogenicity of different strains of BCG. Given this lack of correspondence, it is by no means clear which model will provide the best evaluation of a new vaccine product, although there is a general intuitive feeling that the model should be based upon low dose aerosol challenge so as to mimic at least that aspect of the human disease. Three other aspects of these animal models may be questioned.

The first is the time scale and outcome measure employed. Most models look for a reduction in numbers of bacilli over relatively short periods of time. But whether reduction of numbers of tubercle bacilli (an unnatural pathogen) over a matter of a few weeks (e.g. as short as two in some current models) is really a credible surrogate of what happens in the human immune response to tubercle infection over months and years is far from clear. My own view is that the pressure on scientists by granting agencies for rapid results has actually worked against the best long-term interests of the scientific community in this regard. There are models of latent tuberculosis—e.g. the so-called Cornell model, which involves low dose challenge leading to a chronic infection that may last for many months until stimulated by corticosteroids or various forms of stress (Brown et al 1995)—which look like more reasonable models of what goes on in the human infected with *M. tuberculosis* than do most of the short-term models. Unfortunately, almost no work has been done on these models in terms of vaccine evaluation. Such experiments are and will be expensive and time consuming, but my hunch is that investment in such work will be worthwhile.

A second point follows directly from this. Virtually all animal studies to date have been based upon a paradigm of vaccination prior to challenge. It is not surprising why this should have been so. However, given the need for protection of adults, and the fact that the vast majority of young people alive today have received BCG, let alone have experience of various environmental mycobacteria, there is a need to develop and evaluate models in which animals are vaccinated after exposure to various mycobacteria. Among the comparisons of interest would be assessments of vaccines given after exposure to *M. tuberculosis* (e.g. in the Cornell model), after various combinations of acute or chronic exposures to environmental mycobacteria (extending the classical studies of Palmer & Long [1966]), or after prior BCG (particularly interesting, as none of the recently derived vaccines has been able to provide greater protection than a primary BCG in animal systems).

A third problem with animal models returns us to the issue of the observed variation in BCG's protection against tuberculosis. Clearly, if a single vaccine product works differently in different populations, its action cannot be mimicked by any single narrowly defined model. Something must be added, e.g. variation in strains of vaccine or of *M. tuberculosis* or in exposure to environmental mycobacteria. Although several authors have provided elegant evidence that prior exposure of mice or guinea-pigs to one or another environmental mycobacteria does influence the measurable protection subsequently observed from BCG (Palmer & Long 1966, Orme & Collins 1984), this observation has not been followed up sufficiently. It provides support for one explanation for the observed behaviour of BCG, but it now needs to be taken to the next step and employed in the basic design of experiments to demonstrate that any new vaccine can do better than BCG.

Conclusion

The development and evaluation of an improved vaccine against tuberculosis represents one of the great challenges facing medical biology today. The history of BCG provides a constant reminder of the difficulties of this challenge. It also shows what a 'good' tuberculosis vaccine can do under appropriate circumstances. There is no doubt that at least certain BCG vaccines do provide considerable protection against adult pulmonary tuberculosis in certain populations, under certain conditions. Surely the exploration of what is involved in these favourable circumstances, and comparison with what happens — or fails to happen — in circumstances where a similar vaccine fails to protect, should provide an important clue to an effective immune response. By building upon appropriate comparisons, we should be able first to understand the reasons for BCG's fickle behaviour and then, hopefully, to learn from its successes — be they based upon the genetics of the pathogen, the genetics of the host, the genetics of the vaccine, the genetics of related organisms in the environment or whatever.

It is possible that the only feasible approach will be to vaccinate prior to natural infection with the tubercle bacillus. If this is so, we will have to face the formidable logistic difficulties of evaluating such a product against adult disease. There are populations in which such evaluations could be carried out in the relatively short term — such as recruits to health care professions and to some mining industries — but the trials would probably be large and need to continue for decades in order to evaluate the duration of protection and any need for boosters.

On the other hand, perhaps fortune will shine upon us, and it will in the end prove feasible to boost the protective machinery in hosts already infected or already sensitized by one or another mycobacterial species. Such a product would have great advantages in being relatively easily evaluated, as it could be tried in the highest risk communities, in which prevalence of infection and tuberculin sensitivity is already high. Furthermore, if successful, such a vaccine would be directly and immediately appropriate for that vast number of adults who are now at highest risk of disease. We yet to have evidence that such an approach is feasible. Maybe it is not. But the prize is too valuable to give up without a try.

References

Behr MA, Small PM 1997 Has BCG attenuated to impotence? Nature 389:133–134

Bloom BR, Fine PEM 1994 The BCG experience: implications for future vaccines against tuberculosis. In: Bloom BR (ed) Tuberculosis: pathogenesis, protection and control. American Society of Microbiology, Washington DC, p 531–557

Brahmajothi V, Pitchappan RM, Kakkanaiah VN et al 1991 Association of pulmonary tuberculosis and HLA in South India. Tubercle 72:123–132

Brown DH, Miles BA, Zwilling BS 1995 Growth of *Mycobacterium tuberculosis* in BCG -resistant and susceptible mice: establishment of latency and reactivation Infect Immun 63:2243–2247

Comstock GW 1978 Tuberculosis in twins: a reanalysis of the Prophit survey. Am Rev Respir Dis 117:621–624

Comstock GW 1988 Identification of an effective vaccine against tuberculosis. Am Rev Respir Dis 138:479–480

Comstock GW 1994 Field trials of tuberculosis vaccines: how could we have done them better? Controlled Clin Trials 15:247–276

Comstock GW, Palmer CE 1966 Long-term results of BCG vaccination in the southern United States. Am Rev Respir Dis 93:171–183

Fine PEM 1981 Immunogenetic aspects of leprosy, tuberculosis and leishmaniasis In: Humber D (ed) Proceedings of AHRI conference on immunology of leprosy, tuberculosis and leishmaniasis. Excerpta Medica, Amsterdam, p 214–222

Fine PEM 1993 Immunities in and to tuberculosis: implications for pathogenesis and vaccination. In: Porter JDH, McAdam KPWJ (eds) Tuberculosis: back to the future. John Wiley & Sons Ltd, Chichester, p 53–74

Fine PEM 1995 Variation in protection by BCG: implications of and for heterologous immunity. Lancet 346:1339–1345

Gambia Hepatitis Study Group 1987 The Gambia hepatitis intervention study. Cancer Res 47:5782–5787

Hank JA, Chan JK, Edwards ML, Muller D, Smith DW 1981 Influence of the virulence of *Mycobacterium tuberculosis* on protection induced by Bacille Calmette–Guérin in guinea-pigs. J Infect Dis 143:734–738

Hart PD'A, Sutherland I, Thomas J 1967 The immunity conferred by effective BCG and vole bacillus vaccines, in relation to individual variations in tuberculin sensitivity and to technical variations in the vaccines. Tubercle 48:201–210

Hermans PWM, Messadi F, Guebrexabher H et al 1995 Analysis of the population structure of *Mycobacterium tuberculosis* in Ethiopia, Tunisia and the Netherlands: usefulness of DNA typing for global tuberculosis epidemiology. J Infect Dis 171:1504–1513

Karonga Prevention Trial Group 1996 Randomized controlled trial of single BCG, repeated BCG, or combined BCG and killed *Mycobacterium leprae* vaccine for prevention of leprosy and tuberculosis in Malawi. Lancet 348:17–24

Mackaness GB 1972 Delayed hypersensitivity and its significance. In: Chamberlayne EC (ed) Fogarty International Proceedings No 14: status of immunization in tuberculosis in 1971. US DHEW, Washington DC, p 69–89
Mahairas GG, Sabo PJ, Hickey MJ, Singh DC, Stover C 1996 Molecular analysis of genetic differences between *Mycobacterium bovis* BCG and virulent *M. bovis*. J Bacteriol 178:1274–1282
Mitchison DA 1964 The virulence of tubercle bacilli from patients with pulmonary tuberculosis in India and other countries. Bull Int Union Tuberc 35:287
Orme IM, Collins FM 1984 Efficacy of *Mycobacterium bovis* BCG vaccination in mice undergoing prior pulmonary infection with atypical mycobacteria. Infect Immun 44:28–32
Packe GE, Innes JA 1988 Protective effect of BCG vaccination in infant Asians: a case-control study. Arch Dis Child 63:277–281
Palmer CE, Edwards LB 1967 Tuberculin test in retrospect and prospect. Arch Environ Health 15:792–806
Palmer CE, Long MW 1966 Effects of infection with atypical mycobacteria on BCG vaccination and tuberculosis. Am Rev Respir Dis 94:553–568
Philipp WJ, Nair S, Guglielmi G, Lagranderie M, Gicquel B, Cole ST 1996 Physical mapping of *Mycobacterium bovis* BCG Pasteur reveals differences from the genome map of *Mycobacterium tuberculosis* H37Rv and from *M. bovis*. Microbiology 142:3135–3145
Rieder HL 1995 Methodological issues in the estimation of the tuberculosis problem from tuberculin surveys. Tubercle Lung Dis 76:114–121
Rodrigues LC, Gill N, Smith PG 1991 BCG vaccination in the first year of life protects children of Indian subcontinent ethnic origin against tuberculosis in England. J Epidemiol Community Health 45:78–80
Rodrigues LC, Diwan VK, Wheeler JG 1993 Protective effect of BCG against tuberculous meningitis and miliary tuberculosis: a meta-analysis. Int J Epidemiol 22:1154–1158
Rook GAW, Hernandez-Pando R 1998 Immunological and endocrinological characteristics of tuberculosis that provide opportunities for immunotherapeutic intervention. In: Genetics and tuberculosis. Wiley, Chichester (Novartis Found Symp 217) p 73–98
ten Dam HG 1993 BCG vaccination. In: Reichman LB, Hershfield ES (eds) Tuberculosis: a comprehensive international approach. Marcel Dekker Inc., New York, p 251–274
ten Dam HG, Pio A 1982 Pathogenesis of tuberculosis and effectiveness of BCG vaccination. Tubercle 63:225–233
Tuberculosis Prevention Trial 1980 Trial of BCG vaccines in South India for tuberculosis prevention. Indian J Med Res 72 (suppl):1–74
Wiegeshaus EH, Smith DW 1989 Review of the protective potency of new tuberculosis vaccines. Rev Infect Dis 11:S484–S490

DISCUSSION

Donald: Sol Rosenthal said a long time ago that one mustn't expect Bacillus Calmette–Guérin (BCG) to protect against 'excessive and indiscriminate exposure' (Rosenthal et al 1961). It seems to be that BCG trials have been most effective when there is a declining epidemic, such as the MRC (1963) trial in the UK when incidences were falling at the time of the trial. The intensity of the tuberculosis exposure to individuals would have been less than in, say, an area experiencing a more intense epidemic. It may be that BCG only works effectively when the intensity of the exposure is low, in contrast to communities where there's a continuous heavy exposure. It is possible, therefore, that in the US trials, where varying results were found by Comstock (1988), epidemics of different intensity were occurring in different communities.

I have a second point about the importance of trials in children. Even with our current concepts of pathogenesis in adults, trials in children might have some relevance and might provide us with a quicker answer. A substantial proportion of adult tuberculosis is derived from lympho-haematogenous dissemination of organisms from the primary focus to Simon foci in the lungs. Therefore if dissemination could be prevented, then endogenous reactivation at a later point in time may also be prevented. One of the most accurate measures of dissemination is the occurrence of tuberculous meningitis and other disseminated forms of tuberculosis in childhood. In a high incidence area, such as the Western Cape, one could measure it within a couple of years. If we could prevent this form of dissemination we would likely prevent the dissemination to the lungs in persons at a later point in life. Even with an annual risk of infection in a high incidence area of 3% by the age of 10 years, when the incidence of disease following infection increases, two-thirds of the population would still be susceptible, so a vaccine acting at that time might have a measurable influence. Therefore, our follow-up period needn't be as long as 30 years.

Fine: There is little evidence for a correlation between risk of infection in the trial population and observed efficacy. The UK data would suggest efficacy actually declines with decreasing risk of infection (Hart & Sutherland 1977), but this correlation is not convincing. This study is often cited, however, because one would expect there to be more reinfection disease in populations that have high infection transmission rates. ten Dam & Pio (1982) attributed the failure of BCG in Chingleput to a high proportion of disease by reinfection.

I take your second point that 30 years is a long time, but we are studying adult disease. The evidence suggests that BCG is effective against tuberculous meningitis, although I understand that some studies in South Africa may contradict this. It is possible that this does reflect protection against dissemination of the bacteria in the body, so that in the long run protection against adult disease will be afforded. However, this is only a hypothesis, and to evaluate it we need to do long-term trials. These may not have to be carried out for 30 years; it may be possible to perform them within 10–15 years. However, we do not know the duration of protection attributable to BCG in any population. For example, in the UK study it seemed as though there was no protection after 20 years, but the number of cases after that time was small. One could imagine that a booster would be required periodically for a tuberculosis vaccine, but this should be evaluated before it is instituted as policy.

Anderson: You mentioned that there has been no demonstration of an association between genetic variability in the pathogen itself and progression to disease later in the life. Are there genetic data on the pathogen that show that reinfection truly is reinfection? Assessment requires a longitudinal cohort-based study, with the assumption that you know the genetic type of the mycobacterium that caused the initial infection and that infection occurs later in life. I imagine that that such data are not available.

Fine: There are some anecdotal data. For example, in the study of a Boston shelter for the homeless in the early 1980s, one individual was identified as having *Mycobacterium tuberculosis* with a peculiar drug resistance pattern, and subsequently this strain spread to other individuals who were known to have been previously tuberculin positive (Nardell et al 1986). Therefore, there's no doubt that reinfection happens, but we need to determine its frequency and the conditions under which it occurs. We hope that the fingerprinting studies will give us some information on this.

Anderson: You commented on the requirement for large-scale vaccine trials to test more subtle questions, such as whether two or three doses should be given and at what age. These problems are similar to those being confronted in other areas, such as in malaria vaccine trials, where vaccines may protect against disease but not against infection. How problematic is the requirement for a large cohort? Because if it is too problematic then we are not going to be able to test any of these vaccines.

Fine: The sample size calculations are straightforward once you have an estimate of incidence in the population you're dealing with. There are certain populations that may have an incidence as high as 1%, but these are unusual. For most populations, sample sizes would be of the order of many thousands, which is not a trivial problem, especially if a long follow-up period is required.

Davies: Given the problems with BCG, why spend money on re-evaluating it when this investment should be spent on designing a new type of DNA vaccine, for example?

Fine: The vast majority of on-going work relating to the development of a new tuberculosis vaccine is not being carried out on BCG. A few groups are using new tools, for example new ways of measuring cellular immune responses, to try to understand why BCG behaves differently in different populations. These approaches are making important contributions, and therefore I support the use of new tools to look at BCG in different ways. But I'm not saying that this is the only approach that should be taken.

Anderson: I have a quick comment about where to spend the money. If you don't understand the correlates of protection then you are taking an enormous gamble with a new product, because if you go into Phase III trials you face the same problems as you do with BCG. Large-scale trials will be required.

Bellamy: Perhaps our thoughts that we are going to develop a new vaccine are premature. We first need to demonstrate some reliable correlates of protective immunity, and this may take 10 or 20 years. We need a prospective trial on BCG in which we measure a whole gamut of cellular and humoral immune responses, and follow people up over a long period of time to see which immune responses correlate well with long-term protection from disease. If we do not have a reliable marker of protection we will have the same problems evaluating the efficacy of any new vaccine that we have had in assessing BCG.

Orme: The correlates of protection are right here under our noses: T cells make interferon-γ see certain proteins.

Kaplan: We can't just ignore BCG because it is being used, and ethically we can't stop using BCG until we know how it compares with any new vaccine and how it interacts with any new vaccine. So, we are going to have to continue working with BCG, whether we like it or not.

References

Comstock GW 1988 Identification of an effective vaccine against tuberculosis. Am Rev Respir Dis 138:479–480

Hart PD'A, Sutherland I 1977 BCG and vole bacillus vaccines in the prevention of tuberculosis in adolescence and early adult life: final report to the Medical Research Council. Brit Med J 2:293–295

MRC 1963 BCG and vole bacillus vaccines in the prevention of tuberculosis in adolescence and early adult life. Third report of the Medical Research Committee by their tuberculosis vaccines clinical trials committee. Br Med J 1:973–978

Nardell E, McInnis B, Thomas B, Weidhaas S 1986 Exogenous reinfection with tuberculosis in a shelter for the homeless. N Engl J Med 315:1570–1575

Rosenthal SR, Loewinsohn E, Graham ML et al 1961 BCG vaccination in tuberculous households. An Rev Respir Dis 84:690–704

ten Dam HG, Pio A 1982 Pathogenesis of tuberculosis and effectiveness of BCG vaccination. Tubercle 63:225–233

Immunological and endocrinological characteristics of tuberculosis that provide opportunities for immunotherapeutic intervention

Graham A. W. Rook and Rogelio Hernandez-Pando†

Department of Bacteriology, University College London Medical School, Windeyer Building, 46 Cleveland Street, London W1P 6DB, UK and †Instituto Nacional de la Nutricion 'Salvador Zubiran', Calle Vasco de Quiroga 15, Delegacion Tlalpan, 14000 Mexico DF

Abstract. Immunity to tuberculosis requires a T helper 1 (Th1) response which can be compromised by excessive release of inflammatory cytokines or Th2 activity. Environmental saprophytes can protect against tuberculosis by inducing Th1 recognition of the common antigens, or make mice more susceptible to tuberculosis than unimmunized controls by evoking a Th2 response. A mixed Th1+Th2 response increases the local toxicity of tumour necrosis factor α (TNF-α). Some saprophytes are potent immunogens. A killed preparation of *Mycobacterium vaccae* can cause systemic activation of spontaneously Th1 cytokine-secreting cells in humans, and can non-specifically suppress pre-existing IgE formation and interleukin 5 (IL-5) production in murine models of allergy. These effects, and the Th2-inducing effects of other species, may explain the variable efficacy of Bacillus Calmette–Guérin, and suggest the need for new approaches to the screening of vaccines before trial in humans. The balance of Th1 to Th2 and the function of inflammatory cytokines are also regulated by cortisol. Glucocorticoid metabolism is abnormal in tuberculosis, suggesting overactivity of 11-β-hydroxysteroid reductase enzymes. The reductase activity of these enzymes is enhanced by TNF-α and IL-1β. The roles of Th2, inflammatory cytokines, common antigens and changes in cortisol metabolism suggest several strategies for immunotherapy, and several sites where genetic polymorphisms may affect susceptibility.

1998 Genetics and tuberculosis. Wiley, Chichester (Novartis Foundation Symposium 217) p 73–98

The only prophylactic vaccine in use for tuberculosis gives a protective efficacy that varies between 80% in some countries, such as the UK and Uganda, to 0%, or even results in a slight increase in disease incidence in other areas such as Georgia and Illinois, and perhaps India (Fine 1995).

Therapy for the disease is equally unsatisfactory because of the need for a minimum of six months of treatment. This need is attributable to two factors. First, a percentage of the bacteria are in a dormant state and are therefore unaffected by the drugs (Grange 1992). Secondly, and at least as important, is the fact that the immune response in tuberculosis patients is manifested by the necrotizing Koch phenomenon, which can wall off lesions but has little bactericidal activity. It has been demonstrated in guinea-pigs (Wilson et al 1940) and mice (Hernandez-Pando et al 1997) that animals manifesting this type of response before infection with virulent *Mycobacterium tuberculosis* are more susceptible to the disease than are non-immunized control animals. Thus, even small numbers of persisters can cause relapse in incompletely cured tuberculosis patients because their pattern of immune response is inefficient, and not corrected by conventional treatments.

The only currently conceivable solution to the problem of the unmanageable six-month treatment regimen is an immunotherapeutic approach to convert the necrotizing Koch phenomenon to protective immunity. This would enable the immune response to contribute to cure, and deal with dormant organisms as they restart their metabolism and release antigen, without the need for continuing chemotherapeutic and antibiotic cover. In this chapter we first consider the difference between the Koch phenomenon and optimal immunity, and then discuss immunological and endocrinological approaches to correcting the defect. In accordance with the theme of this volume, areas where genetic polymorphisms may contribute to variable susceptibility are highlighted.

The immunopathology of tuberculosis

Immunity to *M. tuberculosis* requires a T helper 1 (Th1) pattern of response accompanied by several types of cytotoxic T cell (Flynn et al 1993, 1995a,b, Rook & Hernandez-Pando 1996). A Th2 response is disastrous: even a small Th2 component exacerbates the infection in mice (Lindblad et al 1997, Rook & Hernandez-Pando 1996). Thus, if the animals are pre-immunized so that they have a 'pure' Th1 response to culture filtrate antigens of *M. tuberculosis,* or to the common antigens present in a distantly related environmental saprophyte, they are protected (Figs 1 & 2). However, if they are preimmunized so that they have a mixed Th1+Th2 pattern of response to these common antigens, they have increased susceptibility to pulmonary tuberculosis manifested as accelerated death (Fig. 3a), increased percentage of the lung affected by pneumonia (Fig. 3d) and accelerated appearance of Th2 cells in the lesions (Fig. 3b) (Hernandez-Pando et al 1997, Rook & Hernandez-Pando 1996).

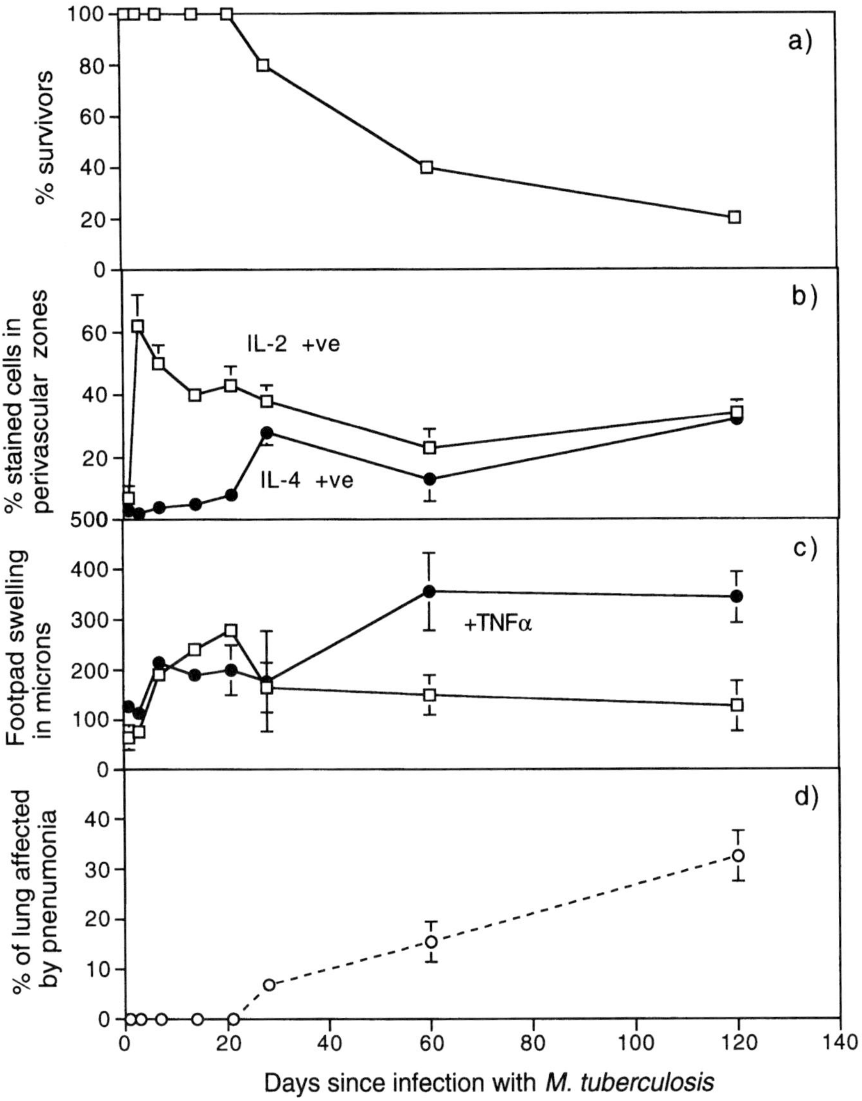

FIG. 1. Unimmunized mice were infected with *Mycobacterium tuberculosis* H37Rv by the intratracheal route, and the following parameters studied at intervals for 120 days. (a) Percentage survival. (b) The percentage of interleukin (IL)-2-positive (□) or IL-4-positive (●) lymphocytes in the zones of perivascular inflammation, as detected by immunohistochemistry. (c) The delayed hypersensitivity response (foot-pad test) to soluble antigen of *M. tuberculosis*, 24 h after challenge (□). After reading the swelling, 1 μg of tumour necrosis factor α (TNF-α) was injected into the same site, and swelling was read again 20 h later (●). (d) Percentage of the lung affected by pneumonia. The pneumonic zones always contained numerous IL-4-positive cells. Data adapted from Hernandez-Pando et al (1997). Error bars represent S.D.

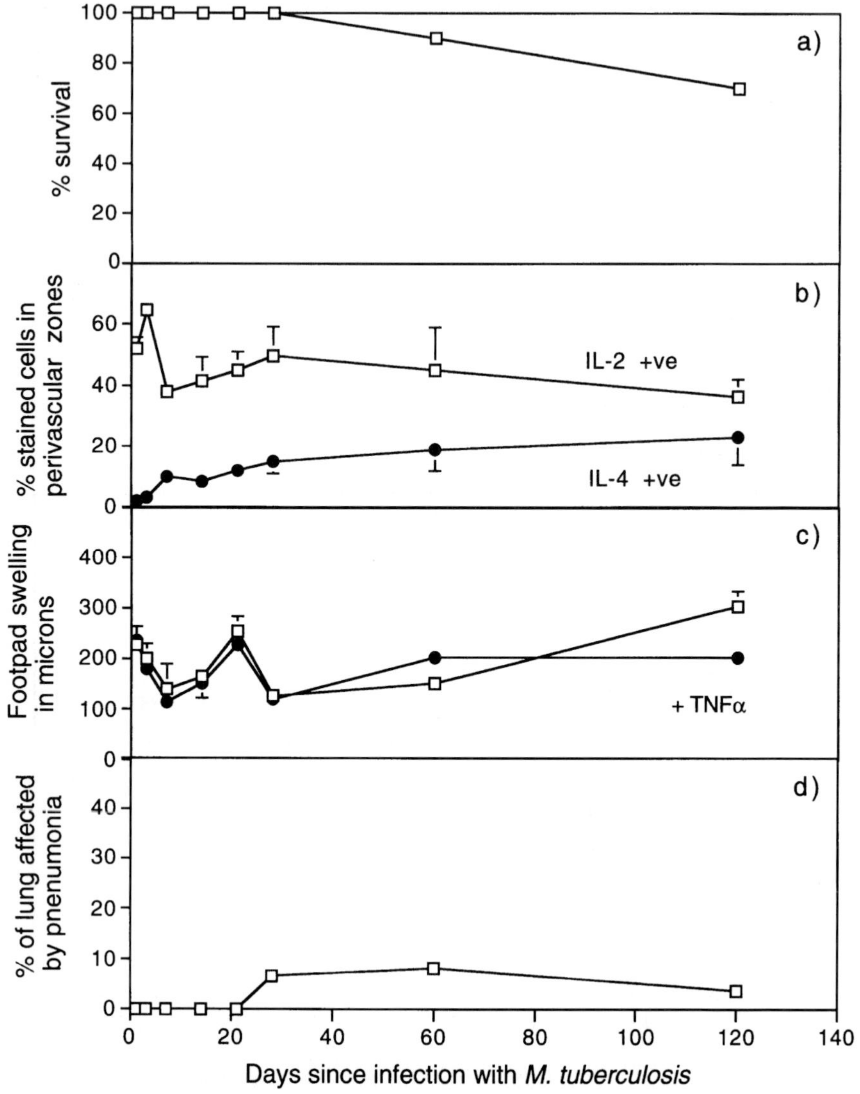

FIG. 2. Mice were immunized once with the optimally T helper 1-inducing dose (10^7 heat-killed bacilli) of the environmental saprophyte *Mycobacterium vaccae*. Two months later they were infected with *Mycobacterium tuberculosis* H37Rv by the intratracheal route. The parameters studied are identical to those described in the legend to Fig. 1, with which these data should be compared. Data adapted from Hernandez-Pando et al (1997). Error bars represent S.D.

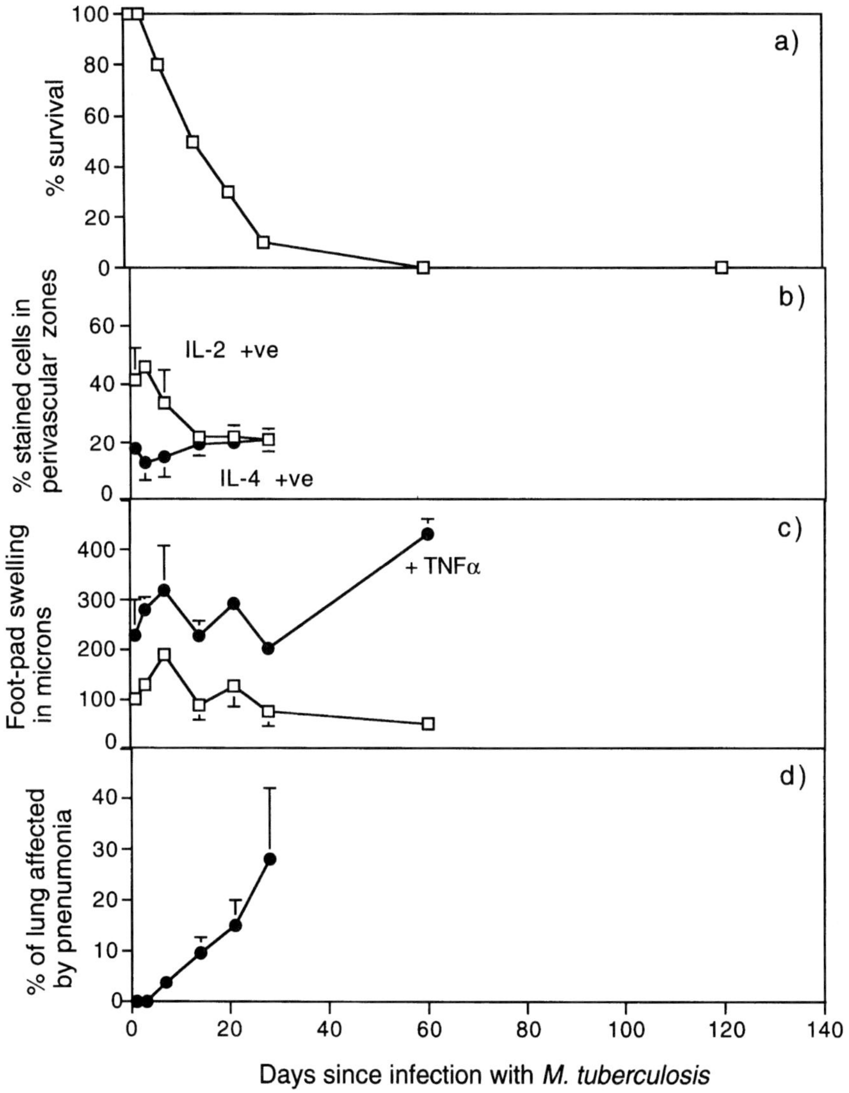

FIG. 3. Mice were immunized once with a 100-fold excessive dose (10^9 heat-killed bacilli) of the environmental saprophyte *Mycobacterium vaccae*. This evokes a mixed T helper 1 (Th1)+Th2 pattern of response. Two months later they were infected with *Mycobacterium tuberculosis* H37Rv by the intratracheal route. The parameters studied are identical to those described in the legend to Fig. 1, with which these data should be compared. Data adapted from Hernandez-Pando et al (1997). Error bars represent S.D.

The Th2 component in human tuberculosis

There is strong evidence that progressive human disease also has an inappropriate Th2 component, though its manifestation is more subtle than in the murine model. The peripheral blood of tuberculosis patients contains Th2 cells that respond to mycobacteria by expressing (Schauf et al 1993) and releasing (Sanchez et al 1994, Surcel et al 1994) interleukin 4 (IL-4). T cells in such blood also express subnormal levels of IL-2 (Schauf et al 1993), and there is free IL-10 (N. Beyers, personal communication 1997) and tuberculosis-specific IgE antibody (Yong et al 1989). It is interesting that blood flow in skin test sites correlates with specific IgE (Gibbs et al 1991) because decreased blood flow in such sites is also a correlate of the Koch phenomenon, the tissue-damaging immunopathology that accompanies progressive disease (Anderson 1891).

Barnes and colleagues have argued that they found little Th2 activity in pleural effusions associated with tuberculosis, or in tuberculous lymphadenitis (Barnes et al 1993, Lin et al 1996, Zhang et al 1995). However, there are several objections to these conclusions. First, these are the high resistance forms of the disease, whereas immunocytochemical analysis of lesions from pulmonary tuberculosis in mice (Hernandez-Pando et al 1997) shows that the abundant Th2 cytokine-secreting cells (Figs 1b, 2b & 3b) appear in progressive disease. The same is likely to be true in humans. Second, Th2 cytokines such as IL-4 tend to be produced at levels that are at least 1000-fold lower that the levels of γ-interferon (IFN-γ), so a low but detectable IL-4 output may be biologically significant. In our own laboratory we have used flow cytometry of peripheral blood T cells activated *in vitro* with TPA (12-*O*-tetradecanoylphorbol 13-acetate) and ionomycin. These studies show an excess of T cells, which can be triggered to secrete IL-4, and reduced secretion of IL-2 (N. Thapa, G. Rook & J. L. Stanford, unpublished work 1997), in agreement with the mRNA studies of Schauf et al (1993). It is important to remember that immunity to *M. tuberculosis* seems to be sensitive to the presence of even a minor Th2 component (Hernandez-Pando et al 1997, Lindblad et al 1997).

The inappropriate Th2 component and the toxicity of tumour necrosis factor α

Tumour necrosis factor α (TNF-α) is usually required for protection against *M. tuberculosis* (Flynn et al 1995a). However, this cytokine becomes toxic when there is a mycobacterial lesion with a mixed Th1+Th2 cytokine profile (Figs 1c, 2c & 3c; Hernandez-Pando et al 1997, Hernandez-Pando & Rook 1994, Rook & Hernandez-Pando 1996). This is similar to the findings in the schistosomiasis model, where immunopathology correlates with the simultaneous presence of Th1 and Th2 cytokines, and TNF-α. This immunopathology, and the permanent fibrosis and tissue damage that result, can be reduced by blocking induction of the

Th2 component (Wynn et al 1995). This phenomenon may partially explain the detrimental effect of a Th2 component in tuberculosis (Hernandez-Pando et al 1997, Lindblad et al 1997). We therefore hypothesize that the Koch phenomenon is a consequence of simultaneous release of Th1 and Th2 cytokines, and TNF-α (Koch 1891).

Environmental saprophytes, the variable efficacy of BCG and the testing of new vaccines

It is important to emphasize that Figs 1–3 demonstrate that heightened susceptibility can be achieved by evoking a Th2 response to an organism that is only distantly related to *M. tuberculosis*, clearly indicating the crucial role of responses to 'common' antigens, probably including the stress proteins (Hernandez-Pando et al 1997). T cells that recognize these proteins constitute a major part of the T cell repertoire at birth (Fischer et al 1992). There is presumably a genetic component to the details of this repertoire. Subsequently, the pattern of response to these antigens, which have a crucial regulatory role (Cohen 1996), is 'educated' by exposure to environmental organisms. Again, there may be a genetic component because surveys of skin test responsiveness to the 'common' antigens revealed HLA associations (van Eden et al 1983).

Mycobacterial antigens are encountered by all humans as a result of contact with environmental mycobacteria, which are ubiquitous in soil and water supplies, but they are not part of the normal commensal flora, so the nature and extent of contact with them depends on where and how you live. Variable contact, priming protective Th1 responses (Fig. 2) or 'anti-protective' Th2 responses (Fig. 3) can explain the variable protective efficacy of Bacillus Calmette–Guérin (BCG). We suggest that in some environments where BCG fails, the recipients already have a Th2 component in their recognition of common antigens. If so, then the data in Fig. 3 suggest that a new vaccine, whether for therapeutic or even for prophylactic use, will need to be able to induce a Th2 to Th1 switch if it is to do better than BCG. In order to test novel prophylactic vaccines in a more realistic way, the scheme outlined in Fig. 4 is suggested. The essence of this scheme is that animals should first be primed so that they have a Th2 response to selected components of *M. tuberculosis*, particularly to shared epitopes. Then the vaccine can be tested in a situation that mimics that seen in humans.

Immunological properties of an environmental saprophyte and systemic activation of Th1

Mycobacterium vaccae is the only environmental saprophyte to have undergone extensive immunological study, and it turns out to have such potent

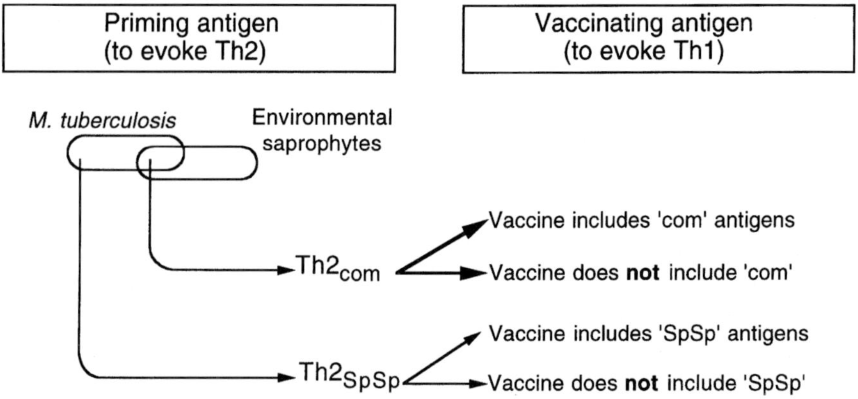

FIG. 4. Suggested schemes for the realistic testing of candidate vaccines before they are tried in humans. If, as suggested in this chapter, pre-existing priming of a T helper 2 (Th2) component by common antigens in the environment is a factor that contributes to the failure of Bacillus Calmette–Guérin vaccination in some countries, candidate vaccines must prove that they can work in mice primed to mimic this situation. It may also be important to consider whether or not the vaccine contains the epitopes to which a Th2 response already exists. com, common antigens; SpSp, species-specific antigens.

immunological effects that the influence of these common organisms on efficacy of BCG (Fine 1995) and on tuberculosis (Figs 1–3) is not surprising. When multiple doses of the *M. vaccae*-derived preparation SRL172 were administered intradermally to 36 patients with late-stage melanoma (in the care of Angus Dalgleish, St. George's Hospital), 41% showed an increase in the peripheral blood of T cells that spontaneously secreted IL-2 *in vitro*. These were detected by flow cytometry (B. Baban, A. Maraveyas, G. A. W. Rook, M. Westby, J. M. Grange, P. Lydyard, J. L. Stanford & A. G. D. Dalgleish, unpublished work 1997) and reached 60–70% of all T cells in some patients. There was a correlation between the appearance of these cells and a good clinical outcome, although the study was not designed to test therapeutic efficacy. The point that is relevant to tuberculosis is that SRL172 can systemically activate Th1 responses and cause the patients to 'dose themselves' with IL-2. This striking release of IL-2 was not accompanied by toxicity, which is the usual constraint upon the use of this cytokine.

M. vaccae also acts as a potent Th1 adjuvant. Mice given a single subcutaneous injection of SRL172 develop powerful Th1 cell-mediated responses to the hsp65 of *M. leprae* in their lungs (C. Wang, R. Hernandez-Pando & G. Rook, unpublished work 1997). The Th1 adjuvanticity can also be demonstrated using recombinant forms of *M. vaccae* expressing foreign proteins (Abou-Zeid C et al 1997), or modified forms to which antigens have been conjugated by Schiff base formation (C. Wang & G. Rook, unpublished work 1997).

Systemic down-regulation of Th2 responses

Before treatment with SRL172, some of the cancer patients discussed above had numerous T cells that secreted IL-4 after stimulation with TPA and calcium ionophore. These cells are rarely present in the blood of normal individuals and they tended to disappear before the spontaneously IL-2-secreting cells increased. This suggested that SRL172 might down-regulate Th2, and this was confirmed experimentally. Balb/c mice were pre-immunized with ovalbumin so that they had high and rising levels of IgE. The IgE production *in vivo* and the potential for IL-5 release by spleen cells cultured with allergen were both shut off by two injections of SRL172 (Wang & Rook 1998). This was a systemic, non-specific effect on Th2 activity, which was not accompanied by induction of an allergen-specific Th1 response and did not require simultaneous injection of the allergen.

Immunotherapy

Immunotherapy with crude extracts of *M. tuberculosis* is not possible, and it causes necrosis and disease exacerbation (Anderson 1891, Koch 1891). In contrast to the necrotic skin test responses to the antigens of tuberculosis itself, tuberculosis patients have defective skin-test responses to the common antigens (Kardjito et al 1986), so injection of these antigens, or of environmental saprophytes that contain them, causes trivial local reactions. These antigens can be protective when they evoke an exclusively Th1 response (Fig. 2), and they are safe to administer. These points, together with the Th2-to-Th1 switching properties of the *M. vaccae*-derived preparation SRL172, justify its investigation as a potential immunotherapeutic in tuberculosis. It has been shown to be effective in mice (without any chemotherapy; Rook & Hernandez-Pando 1996) and in pilot studies in humans (Corlan et al 1997).

The Phase III clinical trial of SRL172 in Durban

Several small pilot studies performed under field conditions had indicated that a single injection of SRL172 could speed up clearance of bacilli from the sputum, and speed up improvement of secondary clinical correlates such as weight and erthyrocyte sedimentation rate (ESR; Corlan et al 1997, Onyebujoh et al 1995). Therefore, a randomized placebo-controlled trial, performed to Good Clinical Practice was performed in Durban, in collaboration with the South African Medical Research Council. There were no significant differences in rate of bacterial clearance, weight gain or ESR at eight weeks. These results may indicate that in spite of its interesting immunological properties, SRL172 is inactive in tuberculosis. However, it is more likely that the speed of cure in hospitalized patients receiving optimal directly observed chemotherapy is already maximal, and cannot be improved with immunotherapy. This view is consistent with the observation that the effect of *M. vaccae* in previous studies was greatest when chemotherapy was most deficient (Onyebujoh et al 1995), or the organisms drug resistant; whereas there were no apparent effects in well-treated, uncomplicated disease (Corlan et al 1997). Thus, *M. vaccae* may be able to compensate for the problems faced in the real world—such as drug resistance, non-compliance, truncated chemotherapy or faulty drugs—but unable to contribute to the rigorous regimen that had to be used in the trial. The format of the trial was dictated by regulatory requirements, but it was clearly unsuitable, because it has not resolved this dilemma.

The endocrine component

There is a further physiological correlate of the events seen in Figs 1–3. During the phase of Th1 dominance, the adrenals increase in weight, presumably because of activation of the hypothalamopituitary-adrenal axis by cytokines (see Fig. 5). However, when the response switches to Th2 the adrenals start to atrophy (Rook & Hernandez-Pando 1996). This is not an acute renal cortical necrosis but rather progressive atrophy with apoptosis. These observations have led to detailed studies of endocrine function in human tuberculosis. The subject is complex, and Fig. 4 shows some of the pathways that control glucocorticoid function in the tissues. The regulatory pathways shown are rich in genetic polymorphisms, and could be a source of new genetic insights into individual susceptibility to tuberculosis.

The cortisone/cortisol shuttle

Local control of glucocorticoid activity is important because T cells that mature in the presence of raised glucocorticoid levels tend to mature into Th2 cells (Ramirez et al 1996). Moreover, simultaneous exposure of T cells to IL-2 and IL-4, as will occur in mixed Th1 + Th2 inflammation, causes reduced affinity of glucocorticoid receptors, and therefore glucocorticoid resistance (Kam et al 1993).

A study of the steroid metabolites in 24 h urine samples from tuberculosis patients has shown that paradoxically the total 24 h output of glucocorticoids is often not raised, and can be low (Rook et al 1996). However, cortisol levels are maintained because there is a striking switch in the balance of metabolites of cortisol and cortisone in favour of cortisol, the active compound. This implies a

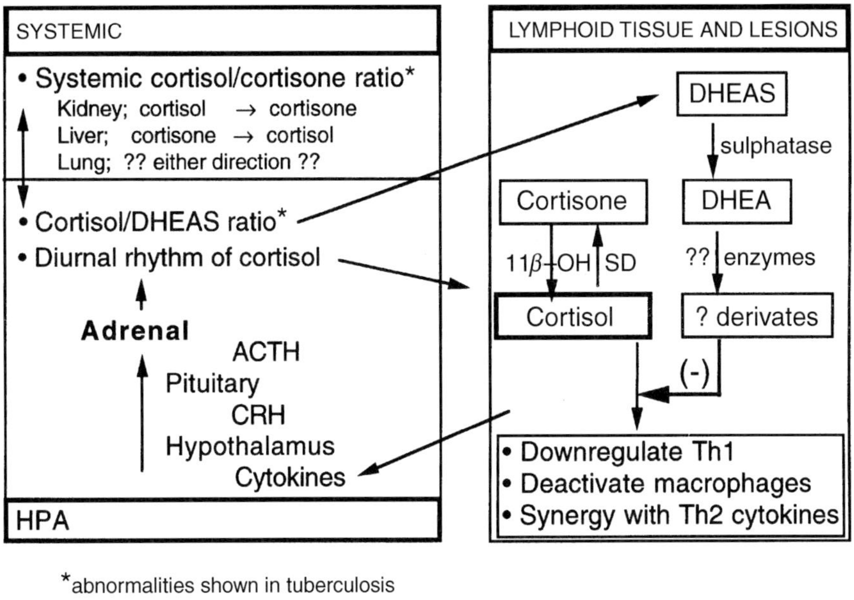

FIG. 5. Factors that determine the effective cortisol function in lymphoid tissue and lesions. Cytokines from the immune response (bottom right) contribute additional drive to the hypothalamopituitary-adrenal (HPA) axis. The eventual output of the adrenal (left) can vary in diurnal rhythm and in cortisol/sulfated dehydroepiandrosterone (DHEAS) ratio. The circulating cortisol level can then be modulated by changes in the cortisol/cortisone ratio, due to the 'cortisol/cortisone shuttle' enzymes in the kidney, liver, lung and other tissues. Within lesions and lymphoid tissues (box on right), both DHEAS and cortisol are subject to regulated local metabolism. Tumour necrosis factor α and interleukin-1β upregulate the reductase activity of 11-β-hydroxysteroid dehydrogenase type 1 (11β-OH SD) activity and so locally increase the proportion of active cortisol (surrounded by heavy rectangle). The pathway of conversion of DHEA to the putative active antiglucocorticoid, indicated as (-), is unknown, because we do not know which are the active metabolites. ACTH, adrenocorticotrophic hormone; CRH, corticotrophin-releasing hormone. Th, T helper.

change in the activity of the enzymes of the cortisol/cortisone shuttle (Fig. 5), which we have confirmed by showing accelerated conversion of an oral cortisone load to active cortisol (B. Baker, B. R. Walker, R. J. Shaw, A. Zumla, J. W. Honour, S. L. Lightman & G. A. W. Rook, unpublished work 1997). This is of particular interest because the appropriate change in the function of 11β-hydroxysteroid dehydrogenase type 1 (11βHSD1) has been shown to be induced by TNF-α and IL-1β (Escher et al 1997). We do not yet know where this is happening, but the tuberculous lung seems a likely site, since it is where cytokine concentrations will be highest. In the normal lung there is also evidence for inactivation of cortisol to cortisone (Escher et al 1994, Hubbard et al 1994).

The antiglucocorticoid role of dehydroepiandrosterone

A second set of the pathways shown in Fig. 5 is also important in tuberculosis. Many of the effects of cortisol on the immune system are opposed by dehydroepiandrosterone (DHEA). In humans, levels of the sulfated form of this hormone (DHEAS) are low up to adrenarche (about five years), intermediate between five and 10 years, and at adult levels by puberty. Levels decline again later in life. These age-related changes correlate with changes in susceptibility to tuberculosis, and changes in the type of disease that results (Table 1; Donald et al 1996). Could these correlations be biologically meaningful? In recent experiments we have manipulated the ratio of glucocorticoid to DHEA (or appropriate DHEA metabolites) in a murine model of pulmonary tuberculosis, and found that too much DHEA relative to glucocorticoid, or too little, both exacerbate disease, but they cause differing pathological changes that are reminiscent of the pathology of tuberculous babies and adults, respectively (Table 1).

Therapeutic possibilities

Endocrine manipulations

Some therapeutic effect can be observed if animals are treated with a combination of a DHEA metabolite and a glucocorticoid (R. Hernandez-Pando, K. Arriaga, L. Pavon & G. A. W. Rook, unpublished work 1997) intended to mimic the ratio seen during the 'safe school age' (5–10 years) when children are resistant to the

TABLE 1 Suggested correlations between dehydroepiandrosterone (DHEA) levels (or DHEA/glucocorticoid ratios) and immunopathology of tuberculosis in humans at different ages. Manipulating the ratios of glucocorticoid to DHEA in tuberculous mice causes similar changes in pathology

Age (human)	*DHEAS level*	*Disease type*	*Ratio of glucocorticoid to DHEA in tuberculous mice*
0–5 years	Low	Consolidation, pneumonia, no cavitation	Normal mouse, i.e. low DHEA and normal corticosterone
5–10 years	$\sim 1\ \mu g/ml$	Resistant (safe school age)	Mouse with late tuberculosis treated with DHEA and corticosterone
Puberty	$\sim 4\ \mu g/ml$	Adult type, with necrosis and cavitation	Mouse with late tuberculosis treated with DHEA but no corticosterone supplement

DHEAS, sulfated form of DHEA.

disease (Donald et al 1996; Table 1). It is possible that this strategy could help human drug-resistant cases. It is also possible that drugs such as derivatives of glycyrrhetinic acid that can block enzymes of the cortisol/cortisone shuttle would have beneficial effects if reductase activity in the lung were limited.

Trial design for testing immunotherapy in tuberculosis

In view of the increasingly widely accepted need for immunotherapy as the only real hope for the improvement of treatment under realistic field conditions, the failure of the Durban trial to yield interpretable data raises important questions about future exploitation of immunotherapy, whether immunological or endocrinological. Is it possible to devise a trial in patients with drug-resistant disease that would conform to regulatory requirements? The sporadic distribution of these patients and their lack of uniformity has frustrated attempts to devise such trials in the past. Secondly, and more importantly, would it be possible to design a study that could be performed under realistic field conditions that would satisfy regulatory bodies? Until these dilemmas are resolved, it may not be possible to test immunotherapy for tuberculosis in a meaningful way.

Acknowledgements

We are grateful to the Wellcome Trust and the Neuroendocrinology Charitable Trust for supporting the investigation of the endocrinology of human tuberculosis. R.H.-P. is grateful to the CONACyT (Consejo Nacional Ciencia y Tecnologia) for supporting laboratory work in Mexico (grant O235P-M9506), and to both the Royal Society and the European Community (INCO-DC contract no. ERBIC18CT960060) for supporting visits to the UK.

References

Abou-Zeid C, Gares M-P, Inwald J et al 1997 Induction of a type 1 immune response to a recombinant antigen from *Mycobacterium tuberculosis* expressed in *Mycobacterium vaccae*. Infect Immun 65:1856–1862

Anderson MC 1891 On Koch's treatment. Lancet i:651–652

Barnes PF, Lu S, Abrams JS, Wang E, Yamamura M, Modlin RL 1993 Cytokine production at the site of disease in tuberculosis. Infect Immun 61:3482–3489

Cohen IR 1996 Heat shock protein 60 and the regulation of autoimmunity. In: van Eden W, Young DB (eds) Stress proteins in medicine. Marcel Dekker, New York, p 93–102

Corlan E, Marica C, Macavei C, Stanford JL, Stanford CA 1997 Immunotherapy with *Mycobacterium vaccae* in the treatment of tuberculosis in Romania, 2. Chronic or relapsed disease. Respir Med 91:21–29

Donald PR, Beyers N, Rook GAW 1996 Adolescent tuberculosis. S Afr Med J 86:231–233

Escher G, Frey FJ, Frey BM 1994 11-β-Hydroxysteroid dehydrogenase accounts for low prednisolone/prednisone ratios in the kidney. Endocrinology 135:101–106

Escher G, Galli E, Vishwanath BS, Frey B, Frey FJ 1997 Tumour necrosis factor α and interleukin 1β enhance the cortisone/cortisol shuttle. J Exp Med 186:189–198
Fine PE 1995 Variation in protection by BCG: implications of and for heterologous immunity. Lancet 346:1339–1345
Fischer HP, Sharrock CE, Panayi GS 1992 High frequency of cord blood lymphocytes against mycobacterial 65 kDa heat shock protein. Eur J Immunol 22:1667–1669
Flynn JL, Chan J, Triebold KJ, Dalton DK, Stewart TA, Bloom BR 1993 An essential role for interferon gamma in resistance to *Mycobacterium tuberculosis* infection. J Exp Med 178:2249–2254
Flynn JL, Goldstein MM, Chan J et al 1995a Tumor necrosis factor-α is required in the protective immune response against *Mycobacterium tuberculosis* in mice. Immunity 2:561–572
Flynn JL, Goldstein MM, Triebold KJ, Sypek J, Wolf S, Bloom BR 1995b IL-12 increases resistance of BALB/c mice to *Mycobacterium tuberculosis* infection. J Immunol 155:2515–2524
Gibbs JH, Grange JM, Beck JS et al 1991 Early delayed hypersensitivity responses in tuberculin skin tests after heavy occupational exposure to tuberculosis. J Clin Pathol 44:919–923
Grange JM 1992 The mystery of the mycobacterial persister. Tubercle Lung Dis 73:249–251
Hernandez-Pando R, Rook GAW 1994 The role of TNFα in T cell-mediated inflammation depends on the Th1/Th2 cytokine balance. Immunology 82:591–595
Hernandez-Pando R, Pavon L, Arriaga K, Orozco H, Madrid-Marina V, Rook GAW 1997 Improvement and exacerbation of murine tuberculosis by previous exposure to low and high doses of an environmental mycobacterial saprophyte. Infect Immun 65:3317–3327
Hubbard WC, Bickel C, Schleimer RP 1994 Simultaneous quantitation of endogenous levels of cortisone and cortisol in human nasal and bronchoalveolar lavage fluids and plasma via gas chromatography-negative ion chemical ionization mass spectrometry. Anal Biochem 221:109–117
Kam JC, Szefler SJ, Surs W, Sher ER, Leung DY 1993 Combination IL-2 and IL-4 reduces glucocorticoid receptor-binding affinity and T cell response to glucocorticoids. J Immunol 151:3460–3466
Kardjito T, Beck JS, Grange JM, Stanford JL 1986 A comparison of the responsiveness to four new tuberculins among Indonesian patients with pulmonary tuberculosis and healthy subjects. Eur J Respir Dis 69:142–145
Koch R 1891 Fortsetzung über ein Heilmittel gegen Tuberculose. Deutsch Med Wochenschr 17:101–102
Lin YG, Zhang M, Hofman FM, Gong JH, Barnes PF 1996 Absence of a prominent Th2 cytokine response in human tuberculosis. Infect Immun 64:1351–1356
Lindblad EB, Elhay MJ, Silva R, Appelberg R, Andersen P 1997 Adjuvant modulation of immune responses to tuberculosis subunit vaccines. Infect Immun 65:623–629
Onyebujoh PC, Abdulmumini T, Robinson S, Rook GAW, Stanford JL 1995 Immunotherapy for tuberculosis in African conditions. Respir Med 89:199–207
Ramirez F, Fowell DJ, Puklavec M, Simmonds S, Mason D 1996 Glucocorticoids promote a Th2 cytokine response by $CD4^+$ T cells *in vitro*. J Immunol 156:2406–2412
Rook GAW, Hernandez-Pando R 1996 The pathogenesis of tuberculosis. Annu Rev Microbiol 50:259–284
Rook GAW, Honour J, Kon OM, Wilkinson RJ, Davidson R, Shaw RJ 1996 Urinary steroid metabolites in tuberculosis: a new clue to pathogenesis. Q J Med 89:333–341
Sanchez FO, Rodriguez JI, Agudelo G, Garcia LF 1994 Immune responsiveness and lymphokine production in patients with tuberculosis and healthy controls. Infect Immun 62:5673–5678
Schauf V, Rom WN, Smith KA et al 1993 Cytokine gene activation and modified responsiveness to interleukin-2 in the blood of tuberculosis patients. J Infect Dis 168:1056–1059

Surcel HM, Troye-Blomberg M, Paulie S et al 1994 Th1/Th2 profiles in tuberculosis based on proliferation and cytokine response of peripheral blood lymphocytes to mycobacterial antigens. Immunology 81:171–176

van Eden W, de Vries RR, Stanford JL, Rook GAW 1983 HLA-DR3 associated genetic control of response to multiple skin tests with new tuberculins. Clin Exp Immunol 52:287–292

Wang CC, Rook GAW 1998 Inhibition of an established allergic response to ovalbumin in Balb/c mice by killed *Mycobacterium vaccae*. Immunol 93:307–313

Wilson GS, Schwabacher H, Maier I 1940 The effect of the desensitisation of tuberculous guinea-pigs. J Pathol Bacteriol 50:89–109

Wynn TA, Cheever AW, Jankovic D et al 1995 An IL-12-based vaccination method for preventing fibrosis induced by schistosome infection. Nature 376:594–596

Yong AJ, Grange JM, Tee RD et al 1989 Total and anti-mycobacterial IgE levels in serum from patients with tuberculosis and leprosy. Tubercle 70:273–279

Zhang M, Lin Y, Iyer DV, Gong J, Abrams J, Barnes P 1995 T cell cytokine responses in human infection with *Mycobacterium tuberculosis*. Infect Immun 63:3231–3234

DISCUSSION

Russell: You administered a high dose of *Mycobacterium vaccae* to mice, and your interpretations were based on antigen-specific T cell responses. However, a large dose of *M. vaccae* will contain lipopolysaccharide, as well as bacterial cell wall lipids, and this is a pro-inflammatory environment which will likely contain a high level of interleukin (IL)-6. This environment will favour a T helper (Th) 2-type response, irrespective of the antigen given.

Rook: First, these preparations are suitable for human use and they do not contain endotoxin which, in any case, is not a component of mycobacteria. Second, the dose is not that large—it is 60 μg of material given two months before the infection. Therefore, your suggestion is unlikely.

Russell: You could administer an irrelevant antigen and look at the proportion of $CD4^+$ cells produced.

Rook: Yes, it would be interesting to administer ovalbumin in similar amounts and see the effect of the Th2 response.

Russell: Barry Bloom and colleagues have recently argued that mycobacteria escape into the cytosol (McDonough et al 1993), which is why a CD8 response is observed, and I would argue that there's no need to invoke that kind of route, because latex beads coupled with protein and inoculated into mice results in a response that includes $CD8^+$ T cells. Therefore, there is an intersection of the class I presentation pathway with the endocytic pathway, and this is not fully understood but it does generate a $CD8^+$ cytotoxic T cell response.

Rook: There are many examples where this occurs, but the point I was making was that killed Bacillus Calmette–Guérin (BCG) and killed *Mycobacterium tuberculosis* do not generate this type of response, whereas killed *M. vaccae* does (Skinner et al 1997, M. Skinner, S. Yuan, R. Prestidge, J. D. Watson & P. L. J.

Watson, unpublished paper, Vaccines beyond 2000, Queensland, Australia, 6–9 July 1997). There may be other environmental saprophytes which have not been studied that produce this response.

Russell: Leishmania also produces a CD8 response.

Orme: There is a general consensus that when *M. vaccae* is injected into mice interferon is produced, so how can the production of $CD8^+$ cytotoxic T cells be explained?

Rook: These pathways don't obey the current dogma, but the data are strong (Skinner et al 1997).

Blackwell: When you heat kill this bacteria are you also destroying all the DNA? Could you effectively be vaccinating them with DNA? Because this may lead to a CD8 response.

Rook: We're not destroying the DNA. The best way to extract DNA from this particular organism is to heat kill it. It is indeed possible to do PCR with the autoclaved material.

Ulmer: It would be highly unlikely that the DNA caused a CD8 response, unless some of this mycobacterial DNA was expressed by the eukaryotic cell. However, DNA itself may have some immunostimulatory effects. There are motifs in DNA of bacterial origin that do have direct effects on lymphocytes, although these tend to stimulate a Th1 response.

Kaplan: Can you clarify your hypothesis concerning the age-related effects of the immune response? Are you suggesting that young children (<5 years) are more susceptible to tuberculosis but children aged 5–10 years are less so because of age-related differences in the immune response?

Rook: There are many models in which an exact balance of glucocorticoids and anti-glucocorticoids seem to be essential. This is what is suggested in Table 1 of my chapter.

Kaplan: But whatever this balance is that determines whether cytokines are or are not made, which cytokines will give rise to differential susceptibility to tuberculosis in that population?

Rook: Mice that have high levels of corticosterone and are not given any dehydroepiandrosterone (DHEA) die because of a massive proliferation of *M. tuberculosis*. I suspect that the macrophages are further deactivated and the response is further pushed towards a Th2-type response (Hernandez-Pando et al 1996).

Kaplan: If cytokine production by macrophages is shut down why isn't cytokine production of T cells also shut down?

Rook: In an environment of mixed Th1 and Th2 cytokines, T cells become resistant to glucocorticoids because of a fall in the affinity of their glucocorticoid receptors (Kam et al 1993, Chrousos et al 1996). Moreover, glucocorticoids drive newly recruited T cells towards a Th2 response (Ramirez et al 1996).

I need to clarify our recent cytokine data from tuberculous mice subjected to endocrine manipulations, and the hypothetical relationship between these data and age-related changes in the pathogenesis and incidence of tuberculosis in children. This relationship is summarized in Table 1 of my chapter. Whereas glucocorticoid alone (corticosterone) accelerates death of tuberculous mice by causing bacterial overgrowth, a critical balance of corticosterone to 'anti-glucocorticoid' (DHEA or androstenediol, AED) increases survival and Th1 production (R. Hernandez-Pando & G. A. W. Rook, unpublished results 1997). However, when too much DHEA or AED is present, massive necrosis appears and the animals die rapidly. I am suggesting that these results correspond to three distinct age-related changes in childhood susceptibility to tuberculosis (Donald et al 1996). From birth until the onset of adrenarche at about 5 years, children are susceptible but die without the necrosis and cavitation seen in adult disease. During the onset of adrenarche DHEA levels start to rise, resulting in an intermediate ratio of glucocorticoid to DHEA. Such children, aged 5–10 years, are resistant to the infection. Finally, at puberty DHEA levels become high, and children become susceptible to the adult necrotic and cavitory forms of tuberculosis.

I suspect that this last effect is due to excessive production of tumour necrosis α (TNF-α) in an environment where there is insufficient glucocorticoid to down-regulate TNF-α production or to inhibit tissue damage. It is clear that TNF-α becomes particularly toxic in sites where there is a mixture of Th1 and Th2 cytokines, whether in models of tuberculosis (Hernandez-Pando et al 1997), schistosomiasis (Wynn et al 1995) or leishmaniasis.

Kaplan: What are the levels of cytokines in the final phase, where the incidence of tuberculosis is the highest?

Rook: There are high levels of Th1 cytokines and pro-inflammatory cytokines.

Kaplan: Are you suggesting, therefore, that in the middle phase there is a predominant Th1 response, rather than Th2, and that this is why there is less susceptibility to *M. tuberculosis* in this age group?

Rook: Yes, but the real question is: what is the actual Th1-dependent effector mechanism that is so successful at this point? And we don't yet know the answer to that.

Kaplan: How do you explain the susceptibility observed in the first phase?

Rook: In young children aged <5 years with low DHEA levels I suspect that the Th1 response is not able to function sufficiently. We know that there is a deficit of Th1 responsiveness in neonates (Shu et al 1994), but we know little about children between this stage and those aged less than 5 years.

Kaplan: So, in the first phase there's not enough of something, in the middle phase there's a balance between two things, and the good one overrides, and in the final phase there's too much of everything.

Rook: That is not how I would express the hypothesis. Glucocorticoid levels (cortisol in humans) are more or less constant throughout life, but DHEA levels change markedly. Therefore, our working hypothesis is that as far as the effector mechanisms of immunity to tuberculosis are concerned (whatever they may be), there is too little DHEA relative to cortisol in the first 5 years of human life, the optimal balance during the next five years, but too much DHEA relative to cortisol from puberty onwards. This high DHEA/cortisol ratio causes massive production of Th1 cytokines and of TNF and IL-1.

Kaplan: But these cytokines aren't necessarily regulated together.

Rook: They're not necessarily regulated together, but in this situation they are because 3β-17β-AED increases both Th1 cytokine production, and the pro-inflammatory cytokines TNF and IL-1 (R. Hernandez-Pando & G. A. W. Rook, unpublished results 1997).

Kaplan: The terminology is being used a little too loosely. When we talk about the Th1 response, I think we all agree we're talking about γ-interferon and IL-2. However, when you talk about the Th2 response, are you talking about inflammatory cytokines (i.e. IL-1 and TNF-α) or cytokines that drive antibody production (i.e. IL-4 and IL-5)?

Rook: We are looking at the presence of IL-4-producing cells, and we find them in the peripheral blood of tuberculosis patients. However, we are not suggesting that the IL-4 itself is the problem. We are saying that something which correlates with the presence of IL-4-secreting cells severely compromises Th1-mediated immunity to tuberculosis.

Orme: So why do IL-4 knockouts develop chronic disease just like normal animals? By your thinking one would imagine that mice without the Th2 component would be less susceptible.

Rook: All I can say is that cytokine knockouts often don't do what you expect them to do because the developing immune system adapts. You would need to inactivate the gene in the adult.

Kaplan: There is no evidence that in tuberculosis patients IL-4 is a major component of the *in vivo* response during infection.

Rook: You have demonstrated the presence of IL-4 mRNA in the peripheral blood T cells of tuberculosis patients (Schauf et al 1993). Tuberculosis patients also have activated Th2 cells, which can be made to make IL-4 *in vitro* by the addition of TPA (12-*O*-tetradecanoylphorbol 13-acetate) and calcium ionophore. This response is not observed in normal individuals. Tuberculosis patients also have IgE antibodies for *M. tuberculosis* (Yong et al 1989). It is a minor Th2 component, but in the mouse a minor Th2 component dramatically affects the ability of the mouse to cope with infection. We don't know the nature of the final effector pathway, but we do know that it is sensitive to the presence of even a minor Th2 component (Hernandez-Pando et al 1997, Lindblad et al 1997).

Kaplan: I still don't understand how the inflammatory cytokines fit into your model.

Rook: There are two different systems running in parallel: the Th1/Th2 shift and the regulation of the inflammatory cytokines, which are down-regulated by glucocorticoids. In a Th1 background TNF is good, whereas in the presence of a Th2 component TNF is toxic. I'm not saying that the Th2 response causes the release of the inflammatory cytokines. I'm saying that the presence of the Th2 response causes the change in the role of the inflammatory cytokines. The same situation occurs in the schistosomiasis model, where there's a mixture of Th1/Th2 cells and TNF. If the Th2 component is removed (Wynn et al 1995) then you no longer observe permanent tissue damage, suggesting that the role of TNF changes in the presence of Th2 cells, although we don't know the nature of that change.

Orme: Let's go back to the classical immunology experiments of 30 years or so ago. Lefford & McGregor (1974) injected a large dose of *M. tuberculosis* into mice, and waited a few months before administering isoniazid. Isoniazid results in bacterial death and bacterial presentation. We now know that this is accompanied by increased IL-4 levels and decreased γ-interferon levels. If the animal is then re-challenged with another dose of *M. tuberculosis*, it is highly immune. However, it seems to me that if the Th2 response was depressing the Th1 response, then the animals would not be resistant. How can you explain those classical experiments using your model?

Rook: It is not reasonable retrospectively to guess the Th1 or Th2 balance of those mice at the time of challenge.

Blackwell: In leishmanial diseases the condition under which mucocutaneous leishmaniasis is observed is a mixed Th1/Th2 response in the presence of high levels of TNF. We don't know the precise mechanism of that either, but the parallels are interesting.

Fine: I had the impression from your presentation that there is a male/female difference in plasma DHEA levels with age, such that after the 'golden age' there are higher levels in females.

Rook: There have been studies of this (de Peretti & Forest 1978), but I wouldn't take the male/female difference too seriously because DHEA levels are similar in males and females.

Fine: There are two interesting age-dependent sex differences in tuberculosis. One of these is that in most populations the prevalence of tuberculin sensitivity becomes higher among males from adolescence onwards, but paradoxically the risk of tuberculosis disease is higher among females than males at young adult ages. This has always struck me as a bizarre observation, and I wondered if you could tease this out.

Rook: Possibly, but we haven't yet tried.

Orme: Didn't Ray Daynes show that DHEA increases resistance to infection?

Rook: Ray Daynes has not used DHEA in a model of tuberculosis. We are the only group to have done so (see Rook et al 1997). It's a question of the balance with glucocorticoid. You could think of it as being like giving glucagon and insulin together, i.e. the blood sugar levels remain the same but you now have a completely new physiological state. When we give glucocorticoid and AED simultaneously to those animals, we observe a Th1-switching effect of the AED but not the pro-inflammatory effect, and we observe a anti-inflammatory effect of the glucocorticoid but not the Th2-switching effect.

Kaplan: What happens in old age, when individuals become even more susceptible to tuberculosis infection?

Rook: The cortisol levels remain constant and the DHEA levels decrease.

Orme: Do people with tuberculosis have small adrenals?

Rook: That is not well documented. Anecdotally, post mortem examination of patients with severe tuberculosis shows that some have small adrenals.

Bateman: What do the adrenals of mice that die of tuberculosis look like at post mortem?

Rook: Their adrenals increase in size during the early Th1-dominated phase of the disease, but then atrophy to about 50% of their normal weight during the late Th1+Th2 phase (Hernandez-Pando et al 1995).

Bateman: Do we take home the message that your hypothesis doesn't address this in the golden years?

Rook: No. It suggests that during the tuberculosis-resistant years ($\sim$5–10 years) there is exactly the right ratio of glucocorticoid to AED, which is probably the relevant derivative. The deaths either side of what you call the 'golden years' are for different reasons. In older children and adults it is immunopathology associated with the turning on of everything, and in infants prior to adrenarche it's an immunopathology associated with insufficient Th1 response. The situation in the middle is balanced.

Orme: The Th1/Th2 classification of $CD4^+$ T cell responses is nice in those terms, but they represent two ends of a wide spectrum, and all the rest of the T cell clones are mixtures of both of them.

Rook: I agree.

Kaplan: You are one of the few people in the field who feel that the Th2 hypothesis is still relevant in tuberculosis. Most immunologists in the field of tuberculosis believe that a Th2 response does not determine the outcome in *M. tuberculosis* infection.

Beyers: Phil Bardin (Department of Internal Medicine, University of Stellenbosch) has done some immunocytochemistry on lung tissue from tuberculosis patients and found not only IL-2, IL-12 and γ-interferon, but also IL-4 and IL-10 (personal communication 1997).

Rook: Similarly, Bergeron et al (1997) have shown Th0 cells in tuberculous but not sarcoid lymph nodes. One should take note that much of the work denying a role of a Th2 response in tuberculosis has been performed on either pleural effusion or lymph node tuberculosis, which are high resistance forms of the disease (Barnes et al 1993). Others use peripheral blood and look for cytokine production after adding antigen. In such experiments Th2 cytokine production is swamped by the Th1 response, which down-regulates much of the Th2 activity. It's not difficult to show Th2 cells in the peripheral blood of tuberculosis patients. If you block cytokine secretion completely with monensin or brefeldin-A and then stimulate the cells non-specifically, you find cells that you would not find in normal individuals which make Th2 cytokines.

Kaplan: But that's because the normal individual doesn't have an ongoing immune response. Infected individuals with active disease have ongoing immune responses—their leukocytes are activated and they express the cytokines and the surface markers that are associated with immune activation. We know antibody is made in tuberculosis patients, but does the Th2 response regulate outcome? And is there a Th1/Th2 balance that will determine whether the patient is going to do well or not? Many immunologists feel that there is no concrete evidence to date to prove that there is a Th2 determinant which regulates outcome. There was a period when the Th1/Th2 balance idea was popular, but those papers have not been confirmed. Indeed, a Th2 response does not appear to be the regulatory determinant for outcome in studies that have been done recently in animals and in humans (Lin et al 1996, Orme & Collins 1994, Kaplan & Freedman 1996). Therefore, is the Th1/Th2 balance the correct hypothesis on which to model our thinking in terms of what's happening during the immune response to tuberculosis?

Rook: I disagree. In our mouse model the Th2 response is crucial, and as I have already pointed out, if you look in the right way and in the right place, a Th2 component is clearly present in humans with tuberculosis.

Young: I would like to bring some genetics into this discussion. Would genetic polymorphisms of candidate genes associated with the endocrine axis be useful?

Bellamy: Possibly. What would be useful is the identification of polymorphisms that resulted in individuals developing a Th1 or a Th2 predominant response following infection. We have looked at polymorphisms in several cytokine genes believed to be important in the activation of a Th1 or a Th2 immune response, and we have not found any associations with tuberculosis. However, this does not exclude the possibility that these cytokine genes are important in determining which individuals will succumb to infection.

Rook: Hennebold & Daynes (1997) claim to have identified a polymorphism in the enzyme responsible for inactivating corticosterone to dehydrocorticosterone in the stromal cells of murine lymphoid tissue. They claim that the difference between BALB/c and C57BL/6 mice in their tendencies to drive a Th2 response is attributable to a major difference in the activity of this enzyme. The hypothalamopituitary-adrenal axis is rich in polymorphisms, and it could be a fruitful hunting ground.

Fourie: I would like to turn attention towards multidrug-resistant tuberculosis. Why would there be clinical efficacy in multidrug-resistant patients and not in ordinary tuberculosis patients?

Rook: Because we would not have to compete with an effective therapy. The World Health Organization and the Food and Drug Administration are going to have to look at how to test new therapies if they have to be superimposed upon a therapy which in hospitalized patients is already fully effective. Can one expect the immune system to contribute anything when it is superimposed upon daily therapy with fully effective drugs? It is not the same as trying to treat certain cancers or allergies, for example, where nothing else works. The problem with multidrug-resistant tuberculosis is that one has to tinker with the treatments for each patient, so it's going to be extremely difficult to design a study.

Fourie: Are you proposing immunotherapy with *Mycobacterium vaccae* as an alternative rather than an adjunct to chemotherapy?

Rook: No. We've been proposing immunotherapy ultimately as an adjunct, but in order to prove clinical efficacy one's going to have to turn to multidrug-resistant disease. In the field tuberculosis is rarely treated as effectively as it is under directly observed therapy short-course (DOTS) and I don't believe that DOTS is ever going to prove to be universally applicable.

Fourie: I expected that your basis for making that statement would be on study design rather than on other observations.

Rook: We do have anecdotal data on about 274 multidrug-resistant patients, although I am not the right person to assess those data. We need to do a proper study.

Kaplan: In principle, this is going to be the case for any so-called weak therapeutic intervention. By weak, I mean compared to the three most effective drugs that clear sputum in about two weeks and sterilize *M. tuberculosis* infection in about three to four weeks. Any weak intervention is going to be masked by something extremely powerful. The problem is how do you demonstrate that it is contributing to the response? You take a situation where the three most effective drugs don't work, i.e. multidrug-resistant tuberculosis patients, and you look at whether the interventions have an effect. It is not ethical to test any weak interventions in the absence of these powerful drugs, although this can be done in animal models.

Duncan: This is a critical issue, and I will be covering some of it in my presentation. When you start to think about any new therapy, i.e. one that differs from today's bactericidal agents, you have to face up to the ethical issues, otherwise these therapies will never go into clinical trials. We have to get over the hurdle of being able to design clinical studies in a way that will show efficacy. If you do these studies in multidrug-resistant patients, you have to face other issues; for instance, the patients may have severe side-effects from the second-line drugs they are taking, thus masking the benefit of a new agent.

Kaplan: In the general population there will always be individuals who for some reason do not respond, and those are the individuals who need the additional interventions. The problem is that this is a difficult situation to test because of inherent heterogeneity, i.e. it is not possible to test everyone with the same drugs under the same conditions because each patient has his/her own individual history of resistance and sensitivity to the drugs. This has to be accommodated in the study design.

Duncan: This is why these experiments should not be performed in patients with multidrug-resistant infections, rather against purely sensitive organisms.

Kaplan: But you can't ethically do this if you already have the best drugs available.

Duncan: Why not? It's possible as long as you have strong animal data which suggest that your therapy is going to work.

Kaplan: But your therapy will never be as good as the best drugs already available.

Duncan: So there's no point in even testing it.

Kaplan: Yes, there is, because there may come a time when the best available drugs can no longer be used; for example, if the patients become multidrug resistant.

Rook: I can't accept that one would not want to consider a weaker therapy under field conditions, even in patients taking effective chemotherapy. The actual cure rate of tuberculosis in South Africa at the moment is only about 50%, which is way below what we ought to be able to achieve. We need additional therapies that provide cover for patients who default or who have drug-resistant disease.

Colston: This isn't a problem that's unique to tuberculosis. For example, leprosy drugs are highly effective and the prevalence rates have decreased dramatically over the last 10 years, and yet a completely new regimen has just been introduced using none of those existing drugs. Therefore, it is possible to introduce a new regimen that does not use any of the first-line drugs. The way this is done, is that you first have to have strong experimental evidence of an effect, and then you have to design a short-term trial, which may not be therapeutically relevant but which will show an effect.

Kaplan: But in leprosy disease progression is so slow that it takes a month for the final diagnosis before drug therapy is initiated. This gives you a month during which new interventions can be tested. You can't wait that long in patients with tuberculosis; you have to use the best therapy available as soon as you suspect tuberculosis.

Fine: With reference to *M. vaccae*, the evidence from animal models is not, in my opinion, as strong as we would like. However, I just wanted to point out that a much larger formal randomized control trial of *M. vaccae* immunotherapy is now being undertaken by ourselves in northern Malawi, in collaboration with Alwyn Muringa at the University Teaching Hospital in Lusaka, Zambia. This trial involves more than 1200 patients, and is being conducted in areas where compliance is low.

Bateman: A study of *M. vaccae* immunotherapy has just been completed in Natal. To date, only sputum conversion at eight weeks has been analysed, and further analyses are awaited. If, however, this study were to show that in those who are non-compliant with chemotherapy, immunotherapy was associated with a better outcome, this would be another form of efficacy. The trouble with the multidrug-resistant group, which in the Western Cape amounts to over 4000 cases, is that the early conversion rate in the compliant patients is in excess of 80% (although sustained cure is much lower). Therefore, even in these cases it will be difficult to show the contribution, if any, of *M. vaccae*. In addition, one would have to demonstrate that *M. vaccae*, or any other immunotherapy, has an effect during the follow-up; that is, the first two years after the initial six-month treatment period. There is no substitute for doing classic intervention trials involving single and multiple doses for dose ranging, add-on trials and then substitution studies. A major problem when testing treatments for tuberculosis is that the endpoints are longer and must include sputum conversion, consolidation and relapse.

Duncan: Yes, and that makes them totally unrealistic to do. If each trial takes two years, then that's a total of six years, and no company will support this.

Anderson: Why are the ethics different from the HIV trials? These trials are running beyond two years, it's a chronic disease and the outcome is uncertain. There are three drugs that work well, and yet trials are going on at the moment to replace one of those drugs with new protease inhibitors as new products are developed.

Duncan: There are a number of issues. One is that in contrast to tuberculosis, AIDS is a disease for which there is a clear, high value market. The HIV trial you describe is a replacement trial, where new protease inhibitors are being tested that have better properties than those already available. I don't have a problem with a clinical trial that, for example, tests an isoniazid replacement, but the approach to testing something that may stimulate the immune system is completely novel and difficult because you can't compete with the best therapy.

Kaplan: There is another difference in that the HIV drugs do not cure HIV, whereas under optimal conditions the tuberculosis drugs cure tuberculosis. At the moment the three HIV drugs work together effectively, but sooner or later drug resistance is likely to develop, so we are going to need replacements. This gives us the ethical justification to design and test replacements.

Donald: I would like to give my support to the study of multidrug-resistant patients. They are a valuable resource for evaluating new therapies, and the outcomes can be more easily measured because they have a higher mortality and the conversion to sputum culture negativity is slower. What has never been accurately assessed is whether effective therapy will give earlier culture negativity and whether the speed with which culture negativity is achieved will correlate with ultimate sterilization. In addition, the early bactericidal activity of the drugs can be easily measured within a period of 48 to 72 h, so it will tell you quickly how fast your drug is killing the bacteria, although this doesn't help your problem with immunotherapy. Therefore, there is a fairly clear path towards evaluating new drugs.

Duncan: There's a clear path to evaluating new isoniazid replacement drugs, but anything else poses more of a problem. It comes back to the point of taking too long to conduct these trials.

Donald: Yes, I acknowledge the commercial imperative in these trials.

References

Barnes PF, Lu S, Abrams JS, Wang E, Yamamura M, Modlin RL 1993 Cytokine production at the site of disease in tuberculosis. Infect Immun 61:3482–3489

Bergeron A, Bonay M, Kambouchner M et al 1997 Cytokine patterns in tuberculous and sarcoid granulomas. J Immunol 159:3034–3043

Chrousos GP, Castro M, Leung DY et al 1996 Molecular mechanisms of glucocorticoid resistance/hypersensitivity. Potential clinical implications. Am J Respir Crit Care Med 154:2 Pt 2 S39–43; discussion S43–3

de Peretti E, Forest MG 1978 Pattern of plasma dehydroepiandrosterone sulfate levels in humans from birth to childhood: evidence for testicular production. J Clin Endocrinol Metab 47:572–577

Donald PR, Beyers N, Rook GAW 1996 Adolescent tuberculosis. S Afr Med J 86:231–233

Hennebold JD, Daynes RA 1997 Microenvironmental control of glucocorticoid functions in immune regulation. In: Rook GAW, Lightman SL (eds) Steroid hormones and the T cell cytokine profile. Springer-Verlag, London, p 101–133

Hernandez-Pando R, Orozco H, Honour JP, Silva J, Leyva R, Rook GAW 1995 Adrenal changes in murine pulmonary tuberculosis: a clue to pathogenesis? FEMS Immunol Med Microbiol 12:63–72

Hernandez-Pando R, Orozco H, Sampieri A et al 1996 Correlation between the kinetics of Th1/Th2 cells and pathology in a murine model of experimental pulmonary tuberculosis. Immunology 89:26–33

Hernandez-Pando R, Pavon L, Arriaga K, Orozco H, Madrid-Marina V, Rook GAW 1997 Improvement and exacerbation of murine tuberculosis by previous exposure to low and high doses of an environmental mycobacterial saprophyte. Infect Immun 65:3317–3327

Kam JC, Szefler SJ, Surs W, Sher ER, Leung DY 1993 Combination IL-2 and IL-4 reduces glucocorticoid receptor-binding affinity and T cell response to glucocorticoids. J Immunol 151:3460–3466

Kaplan G, Freedman VH 1996 The role of cytokines in the immune response to tuberculosis. Res Immunol 147:565–572

Lefford MJ, McGregor DD 1974 Immunological memory in tuberculosis. I. Influence of persisting viable organisms. Cell Immunol 14:417–428

Lin YG, Zhang M, Hofman FM, Gong JH, Barnes PF 1996 Absence of a prominent Th2 cytokine response in human tuberculosis. Infect Immun 64:1351–1356

Lindblad EB, Elhay MJ, Silva R, Appelberg R, Andersen P 1997 Adjuvant modulation of immune responses to tuberculosis subunit vaccines. Infect Immun 65:623–629

McDonough K, Kress Y, Bloom BR 1993 Pathogenesis of tuberculosis: interactions of *Mycobacterium tuberculosis* with macrophages. Infect Immun 61:2763–2774

Orme IM, Collins FM 1994 Mouse models of tuberculosis. In: Bloom BR (ed) Tuberculosis: protection, pathogenesis and control. ASM Press, Washington DC, p 113–134

Ramirez F, Fowell DJ, Puklavec M, Simmonds S, Mason D 1996 Glucocorticoids promote a Th2 cytokine response by $CD4^+$ T cells *in vitro*. J Immunol 156:2406–2412

Rook GAW Hernandez-Pando R, Baker R et al 1997 Human and murine tuberculosis as models for immuno-endocrine interactions. In: Rook GAW, Lightman SL (eds) Steroid hormones and the cell cytokine profile. Springer-Verlag, London, p 193–220

Schauf V, Rom WN, Smith KA et al 1993 Cytokine gene activation and modified responsiveness to interleukin-2 in the blood of tuberculosis patients. J Infect Dis 168:1056–1059

Shu U, Demeure CE, Byun D-G, Podlaski F, Stern A, Delespesse G 1994 Interleukin 12 exerts a differential effect on the maturation of neonatal and adult human $CD45RO^-$ $CD4^+$ T cells. J Clin Invest 94:1352–1358

Skinner MA, Yuan S, Prestidge R, Chuk D, Watson JD, Tan PLJ 1997 Immunization with heat-killed *Mycobacterium vaccae* stimulates $CD8^+$ cytotoxic T cells specific for macrophages infected with *Mycobacterium tuberculosis*. Infect Immun 65:4525–4530

Wynn TA, Cheever AW, Jankovic D et al 1995 An IL-12-based vaccination method for preventing fibrosis induced by schistosome infection. Nature 376:594–596

Yong AJ, Grange JM, Tee RD et al 1989 Total and anti-mycobacterial IgE levels in serum from patients with tuberculosis and leprosy. Tubercle 70:273–279

Recombinant interleukin 2 adjunctive therapy in multidrug-resistant tuberculosis

Barbara Johnson, Linda-Gail Bekker*, Stan Ress* and Gilla Kaplan[1]

*Laboratory of Cellular Physiology and Immunology, The Rockefeller University, 1230 York Avenue, New York, NY 10021–6399, USA and *The University of Cape Town, Department of Medicine, Observatory, Cape Town 7925, South Africa*

Abstract. Multidrug-resistant tuberculosis patients respond poorly to antituberculosis therapy and therefore require new modalities of treatment to overcome the infection. Administration of low dose recombinant human interleukin 2 (rhuIL-2) in combination with chemotherapy to multidrug-resistant tuberculosis patients resulted in reduced or cleared sputum acid-fast bacilli in about 60% of the patients in association with enhanced activation of the immune system. Daily rhuIL-2 administration for 30 days induced increases in $CD25^+$ and $CD56^+$ cells in the blood. rhuIL-2 therapy also resulted in increased expression of γ-interferon and IL-2 mRNA at the site of a delayed-type hypersensitivity (DTH) response to purified protein derivative of tuberculin. Differential display reverse transcriptase PCR revealed several genes expressed at the DTH skin test site that were up- or down-regulated during rhuIL-2 treatment. The differentially regulated genes included components of endocytic vacuoles, enzymes of the respiratory pathway and other regulators of cellular function. The physiological importance of the differential expression of these genes is under investigation to determine their roles in leukocyte activation and in the development of an antimycobacterial response.

1998 Genetics and tuberculosis. Wiley, Chichester (Novartis Foundation Symposium 217) p 99–111

The host protective response to *Mycobacterium tuberculosis* infection

The cell-mediated immune response to tuberculosis infection involves a complex network of leukocytes and a cascade of soluble immune mediators. Following exposure to *Mycobacterium tuberculosis* the majority of individuals mount an

[1]This chapter was presented at the symposium by Gilla Kaplan.

immune response that presumably prevents active disease manifestations (protective immune response; Fig. 1). However, some individuals (about 10%) develop active disease. Moreover, when individuals do develop active disease, there is a wide range in symptoms, rate of clinical deterioration and duration of survival. In individuals with active tuberculosis, the protective immune response may still be operative, although obviously unable to prevent development of disease. Alternatively, active disease may be due to the absence of any protective response in some infected individuals.

During therapy, even if the infecting organism is sensitive to antimycobacterial drugs, the cell-mediated immune response may be ineffective and consequently the infection may not resolve, which will lead to chronic active tuberculosis. When individuals are infected with multidrug-resistant strains of *M. tuberculosis*, the antibiotics cannot sterilize the infection even in the presence of an adequate cell-mediated immune response, which will result in progressive and often fatal disease.

Our studies have concentrated on the characterization of the cytokines that regulate the protective response to *M. tuberculosis*. Here, we describe an adjunctive immune therapeutic intervention that may enhance the immune response to infection, thereby improving clinical outcome in patients who are not responding adequately to conventional antituberculosis therapy.

The cells and cytokines of the host response to tuberculosis

When the host is infected with *M. tuberculosis*, the development of a protective immune response depends on the activation of T helper (Th)1 cells and cytotoxic

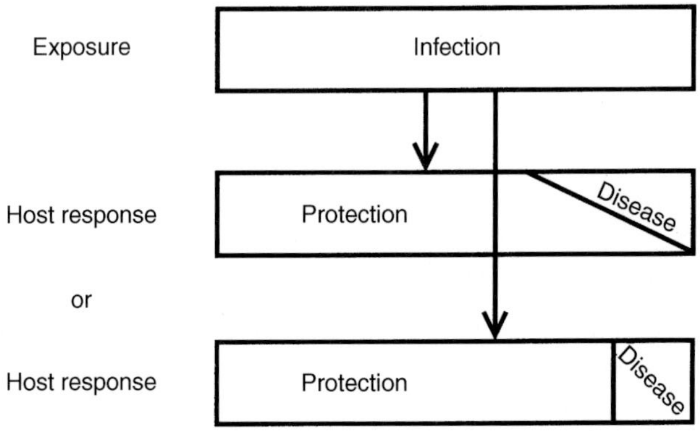

FIG. 1. Infection and development of active disease following exposure to *Mycobacterium tuberculosis*.

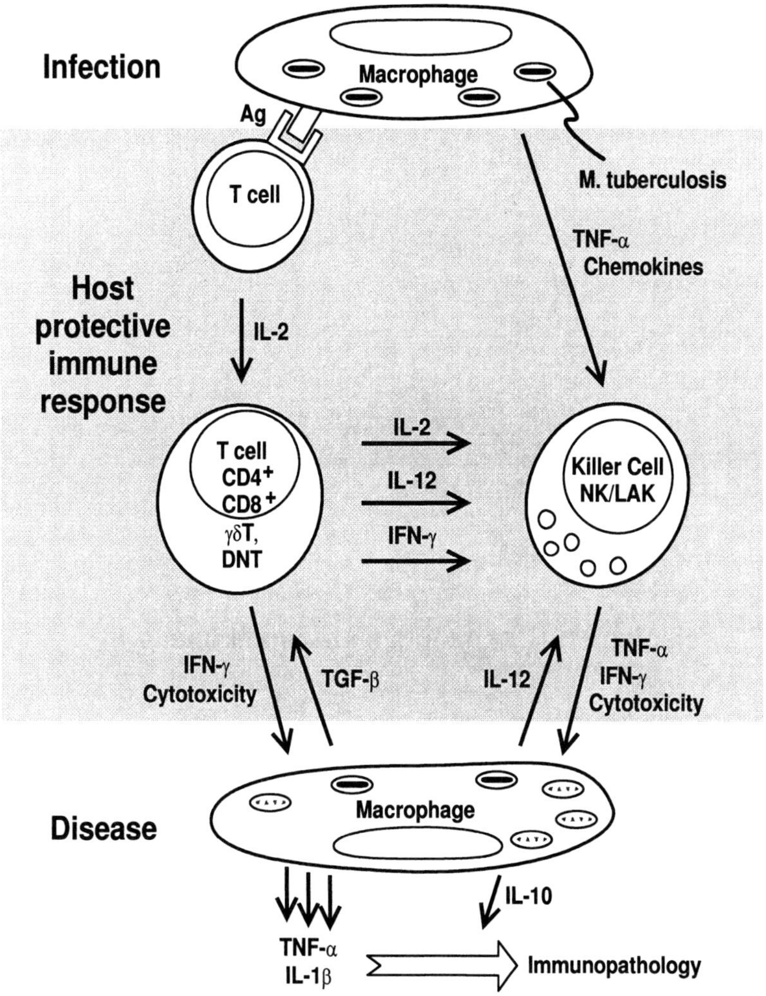

FIG. 2. The cells and cytokines involved in the host immune response to *Mycobacterium tuberculosis* infection. Ag, antigen; DNT, double-negative T cells; IFN-γ, γ-interferon; IL, interleukin; LAK cell, lymphokine-activated killer cell; NK cell, natural killer cell; TGF-β, transforming growth factor β; TNF α, tumour necrosis factor α.

effector cells (Fig. 2). Following exposure of these cells to mycobacteria and their antigens, tumour necrosis factor α (TNF-α) and interleukin (IL)-12 are produced predominantly by macrophages, whereas IL-2 is produced by T cells (Orme & Collins 1994, Boom et al 1992, Barnes et al 1993, Balaji et al 1995, Orme et al 1993).

IL-2 is a central regulator of the Th1 response, and it stimulates leukocyte proliferation and differentiation (Kaplan et al 1992; Fig. 2). This cytokine induces γ-interferon (IFN-γ) production by lymphoid cells, which in turn activates macrophages to better control the growth of intracellular mycobacteria. IL-2 can also be shown to act directly by inducing cytotoxic T cells specific for *M. tuberculosis* antigens (Hancock et al 1989). Thus, the presence of IL-2 at the site of a mycobacterial infection stimulates the appropriate effector cells and thus limits the course of the infection. The Th2 type cytokines, including IL-4 and IL-5, which are associated with an enhanced humoral immune response, do not appear to contribute significantly to the development of protective immunity against tuberculosis infection (Lin et al 1996).

Recombinant human interleukin 2 adjunctive therapy

It has long been noted that cell-mediated immunity may be impaired in patients with tuberculosis. This abnormality is associated with a deficiency in IL-2-induced T cell proliferation, reduced numbers of IL-2-responsive cells in the circulation and decreased IL-2 receptor expression on blood leukocytes (Toossi et al 1986, Schauf et al 1993, Wallis & Ellner 1994). Our previous studies using recombinant human (rhu) IL-2 adjunctive therapy in patients with leprosy and with *Leishmania* infections suggest that IL-2 may be effective in enhancing cell-mediated immunity (Hancock et al 1989, 1991, Kaplan et al 1989, 1991, Converse et al 1991, Akuffo et al 1990). When low dose rhuIL-2 was administered to leprosy patients, there was an increased dermal cell-mediated immune response, as well as an increased production of IFN-γ, resulting in a more rapid clearance of *Mycobacterium leprae* from the skin as compared to multidrug therapy alone (Kaplan et al 1989, 1991). Since IFN-γ is an important inducer of the Th1 protective immune response, the observation that rhuIL-2 treatment of patients induces the production of this cytokine suggests that IL-2 may co-ordinate the Th1 type protective response in patients.

Recombinant human interleukin 2 therapy in tuberculosis patients

We conducted a patient-based investigation using rhuIL-2 as an adjunct to drug therapy in tuberculosis to evaluate the safety of this approach and to determine whether IL-2 can affect the response and outcome in these patients. Our results showed that rhuIL-2 administration in combination with multidrug therapy is safe (Johnson et al 1995). Daily rhuIL-2 administered at a low dose (12.5 μg twice daily) for 30 days to patients was associated with immune activation, as manifested by increased numbers of $CD25^+$ and $CD56^+$ leukocytes in the peripheral blood. The proliferative response of lymphocytes to purified protein derivative of tuberculin

(PPD) and the percentage of IL-2-responsive cells in the circulation of the treated patients was estimated by assaying the [^{3}H] thymidine incorporation into peripheral blood mononuclear cells (PBMC) *in vitro* following exposure to exogenous PPD and/or IL-2. At the pre-study time point, the per cent of responsive cells was the same as that seen in PBMC from PPD$^+$ normal controls. However, following rhuIL-2 therapy for 30 days, the PBMC of treated patients showed a statistically significant increase in the frequency of cells capable of proliferation in response to PPD ($p < 0.006$) and IL-2 ($p < 0.005$) (Johnson et al 1995).

Plasma soluble interleukin 2 receptor levels

Elevated soluble IL-2 receptor (sIL-2R) levels have been associated with immune activation (Deehan et al 1995) and may correlate with the patient clinical response to rhuIL-2 immunotherapy (Deehan et al 1994, Gooding et al 1995, Mangge et al 1995). We therefore assayed sIL-2R levels in multidrug-resistant tuberculosis patient plasma at pre-, mid- and post-study time points. There was a significant increase in plasma sIL-2R of treated patients at the mid-study time point, which decreased to baseline after the last injection of rhuIL-2 (Johnson et al 1998a).

Effect of recombinant human interleukin 2 therapy on cytokine mRNA levels in patient leukocytes

To investigate the effect of exogenous rhuIL-2 administration on the expression of cytokine mRNA in patient leukocytes, we used semi-quantitative reverse transcriptase (RT)-PCR to assay cytokine mRNA levels. IFN-γ mRNA was observed in PBMC obtained from all tuberculosis patients tested, while IL-2 mRNA levels were lower (Fig. 3). Following rhuIL-2 therapy, IFN-γ mRNA levels in patient PBMC showed decreases at both the mid - and post -study time points when compared to the pre-ILγ-2 treatment time point (Johnson et al 1995, 1998b). When the ratio of IFN-γ mRNA to CD3δ mRNA was calculated, a decrease in the level of IFN-γ mRNA expression per T cell during and immediately following rhuIL-2 therapy was observed (Johnson et al 1998b). No change was observed in control untreated patients. Thus, rhuIL-2 therapy was associated with a reduction of IFN-γ levels in blood leukocytes. Other cytokine mRNAs, including TNF-α, IL-4, IL-10 and IL-12, did not show obvious changes in PBMC in response to rhuIL-2 therapy.

mRNA expression of IFN-γ and IL-2 was higher in the PPD skin test site than in cells of the peripheral blood of the same patients (Fig. 3). IFN-γ mRNA expression was also higher than IL-2 mRNA expression in the cells at this site. The effect of rhuIL-2 administration on cytokine gene expression at the PPD site was studied. In

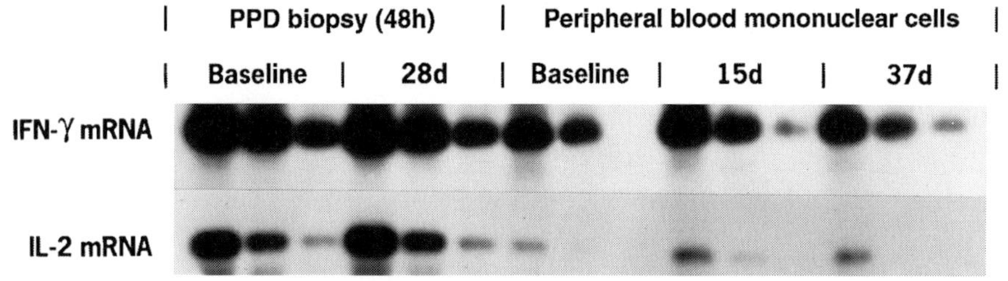

FIG. 3. A representative Southern blot of serially diluted RNA isolated from peripheral blood mononuclear cells and purified protein derivative of tuberculin (PPD) biopsy sites of multidrug-resistant tuberculosis patients. Total RNA from the samples was amplified by reverse transcriptase-PCR and hybridized to radiolabelled probes specific for γ-interferon (IFN-γ) and interleukin (IL)-2 mRNA.

contrast to the results obtained with PBMC we observed that the mean mRNA expression levels for the T cell cytokines IL-2 and IFN-γ were increased during rhuIL-2 therapy relative to the pre-study time point.

Differential regulation of other genes in response to recombinant human interleukin 2 treatment of multidrug-resistant tuberculosis patients

We have recently begun to utilize differential display RT-PCR to facilitate the identification of changes in expression of unselected genes as a result of rhuIL-2 treatment. PCR amplifications of RNA from biopsies of PPD skin tests placed before and during rhuIL-2 administration to multidrug-resistant tuberculosis patients (or the corresponding time points in control patients not treated with rhuIL-2) were performed. Comparison of cDNA displays generated from the RNA isolated from rhuIL-2-treated patient biopsies and from control patient biopsies were carried out (Liang et al 1993, 1995). A number of bands were found to be consistently differentially expressed (Johnson et al 1998b). These genes included components of endocytic vacuoles, enzymes of the respiratory pathway and other regulators of cellular function. The physiological importance of the differential expression of these genes is under investigation to determine their roles in leukocyte activation and in the development of an anti-mycobacterial response.

Effect of recombinant human interleukin 2 on disease manifestations

Multidrug-resistant tuberculosis presents serious problems in that patients remain infectious despite treatment. Thus, any therapy that would enhance the rate of clearance of sputum bacterial load and lead to clinical improvement would have great potential in patient management. We therefore evaluated sputum bacterial load and changes in chest radiographs during the 30 days of daily adjunctive rhuIL-2 treatment. Although this study consisted of a small number of patients, a trend toward sputum bacterial clearance was seen (Johnson et al 1998a). We observed that five out of the eight multidrug-resistant tuberculosis patients who entered the study sputum acid-fast bacilli positive and received 30 days of daily rhuIL-2 injections showed reduction or clearance of bacterial burden by the end of the study. In contrast, only three of nine placebo control patients showed decrease in bacterial load. A similar trend toward improvement was also noted in the chest radiographs. Seven of 12 patients treated with rhuIL-2 adjunctive therapy demonstrated definite improvement in the chest radiographs; four patients showed highly significant improvement.

Conclusions

Our pilot studies with rhuIL-2 adjunctive therapy in multidrug-resistant tuberculosis patients suggest that the cytokine may be a useful therapeutic agent. Our evaluation of patient responses indicated some decrease in the bacillary load and improvement in chest radiographs following daily rhuIL-2 therapy, justifying further and more extensive investigations of the effects of rhuIL-2 adjunctive therapy in multidrug-resistant tuberculosis.

Acknowledgements

These studies were supported by grants from the US Public Health Service (AI 40314 and AI 42056) and by Chiron Corporation, Emeryville, CA. We thank Victoria H. Freedman for help in preparation of the manuscript.

References

Akuffo H, Kaplan G, Kiessling R et al 1990 Administration of recombinant interleukin-2 reduces the local parasite load of patients with disseminated cutaneous leishmaniasis. J Infect Dis 161:775–780

Balaji KN, Schwander SK, Rich EA, Boom WH 1995 Alveolar macrophages as accessory cells for human $\gamma\delta$ T cells activated by *Mycobacterium tuberculosis*. J Immunol 154:5959–5968

Barnes PF, Shuzhuang L, Abrams JS, Wang E, Yamamura M, Modlin RL 1993 Cytokine production at the site of disease in human tuberculosis. Infect Immun 61:3482–3489

Boom WH, Chervenak KA, Mincek MA, Ellner JJ 1992 Role of the mononuclear phagocyte as an antigen-presenting cell for human $\gamma\delta$ T cells activated by live *M. tuberculosis*. Infect Immun 60:3480–3488

Converse P, Ottenhoff THM, Teklemarian SW et al 1991 Intradermal recombinant interleukin 2 enhances peripheral blood T-cell responses to mitogen and antigens in patients with lepromatous leprosy. Scand J Immunol 32:83–91

Deehan DJ, Heys SD, Simpson WG, Broom J, Franks C, Eremin O 1994 *In vivo* cytokine production and recombinant interleukin 2 immunotherapy: an insight into the possible mechanisms underlying clinical responses. Br J Cancer 69:1130–1135

Deehan DJ, Heys SD, Simpson W, Broom J, McMillan DN, Eremin O 1995 Modulation of the cytokine and acute-phase response to major surgery by recombinant interleukin-2. Br J Surg 82:86–90

Gooding R, Riches P, Dadian G, Moore J, Gore M 1995 Increased soluble interleukin-2 receptor concentration in plasma predicts a decreased cellular response to IL-2. Br J Cancer 72:452–455

Hancock GE, Cohn ZA, Kaplan G 1989 The generation of antigen-specific major histocompatibility complex-restricted cytotoxic T lymphocytes of the $CD4^+$ phenotype. Enhancement by the cutaneous administration of interleukin 2. J Exp Med 169:909–919

Hancock GE, Molloy A, Kale AB et al 1991 *In vivo* administration of low-dose human interleukin-2 induces lymphokine-activated killer cells for enhanced cytolysis in vitro. Cell Immunol 132:277–284

Johnson BJ, Ress SR, Willcox P et al 1995 Clinical and immune responses of tuberculosis patients treated with low dose IL-2 and multidrug therapy. Cytokines Mol Ther 1:185–196

Johnson BJ, Bekker L-G, Rickman R et al 1998a rhuIL-2 Adjunctive therapy in multidrug resistant tuberculosis: a comparison of two treatment regimens and placebo. Tubercle Lung Dis, in press

Johnson BJ, Estrada I, Shen Z et al 1998b Differential gene expression in response to adjunctive rhuIL-2 immunotherapy in multidrug resistant tuberculosis patients. Infect Immun, in press

Kaplan G, Kiessling R, Teklemariam S et al 1989 The reconstitution of cell-mediated immunity in the cutaneous lesions of lepromatous leprosy by recombinant interleukin 2. J Exp Med 169:893–907

Kaplan G, Britton WJ, Hancock GE et al 1991 The systemic influence of recombinant interleukin 2 on the manifestations of lepromatous leprosy. J Exp Med 173:993–1006

Kaplan G, Cohn ZA, Smith KA 1992 Rational immunotherapy with interleukin 2. Biotechnology 10:157–162

Liang P, Averboukh L, Pardee AB 1993 Distribution and cloning of eukaryotic mRNAs by means of differential display: refinements and optimization. Nucleic Acids Res 21:3269–3275

Liang P, Bauer D, Averboukh L et al 1995 Analysis of altered gene expression by differential display. Methods Enzymol 254:304–321

Lin YG, Zhang M, Hofman FM, Gong JH, Barnes PF 1996 Absence of a prominent TH2 cytokine response in human tuberculosis. Infect Immun 64:1351–1356

Mangge H, Kenzian H, Gallistl S et al 1995 Serum cytokines in juvenile rheumatoid arthritis: correlation with conventional inflammation parameters and clinical subtypes. Arthritis Rheum 38:211–220

Orme IM, Anderson P, Boom WH 1993 T cell response to *Mycobacterium tuberculosis*. J Infect Dis 167:1481–1497

Orme IM, Collins FM 1994 Mouse model of tuberculosis. In: Bloom BR (ed) Tuberculosis: pathogenesis, protection and control. ASM Press, Washington DC, p 113–134

Schauf V, Rom WN, Smith KA et al 1993 Cytokine gene activation and modified responsiveness to interleukin-2 in the blood of tuberculosis patients. J Infect Dis 168:1056–1059

Toossi Z, Kleinhenz ME, Ellner JJ 1986 Defective interleukin 2 production and responsiveness in human pulmonary tuberculosis. J Exp Med 163:1162–1172

Wallis RS, Ellner JJ 1994 Cytokines and tuberculosis. J Leuk Biol 55:676–681

DISCUSSION

Colston: You showed a Southern blot of RNA taken from a purified protein derivative of tuberculin (PPD) site of a tuberculosis patient that had been probed with γ-interferon (IFN-γ) and interleukin (IL)-2. If you did the same experiment with someone who has been exposed to *Mycobacterium tuberculosis* for a long time but has not developed disease would you see a difference, i.e. is IL-2 a quantitative indicator?

Kaplan: IL-2 and IFN-γ should not be expressed in the blood of someone who doesn't have active disease. However, at the PPD site IL-2 and IFN-γ should be expressed at similar levels.

Colston: So it's not a simple situation in which you are adding IL-2 to someone who is IL-2 insufficient.

Kaplan: If I gave IL-2 to someone who didn't have an ongoing immune response, the T cells would not become activated, because antigen would be required. It is not a simple situation because we are looking at patients who have a certain level of immune activation, and we are assuming that a stronger immune activation will be better and that IL-2 will achieve this.

Colston: Am I correct in saying that you have differentiated between an immunopathological pathway and an immunoprotective pathway?

Kaplan: Operationally, we've separated them, but they occur simultaneously at the same site. The extent of pathology or the extent of inflammatory cytokines versus the usefulness of the protective immune response is going to determine the outcome. They are interconnected, and if you change one you may also change the other. The question is, can you manipulate the system to boost immunity without damaging the lungs? You can't treat the patient with an agent that induces increased levels of tumour necrosis factor α (TNF-α), because TNF-α will give you more pathology. You have to activate the macrophages gently without inducing a sudden explosion in TNF-α production. We've been looking at immune modulation in different ways, and we've been using thalidomide to inhibit TNF-α levels. To our amazement, thalidomide and some of the thalidomide analogues activate T cells, induce IL-2 and IFN-γ production, expand lymphoid cells and give a T helper (Th) 1 response whilst inhibiting TNF-α production.

Colston: These types of studies started in leprosy, where immunopathology is more invisible than in tuberculosis. Can you show that you were not inducing an immunopathological response in your work on leprosy?

Kaplan: Yes. We were looking for induction of reaction in leprosy, which is what we expected to see if the cellular immune response was enhanced. The patients did not develop any reaction. Leprosy is easier to study because the lepromatous leprosy patients, for whom we have used IL-2 therapy, do not have an antigen-specific T cell response. Therefore, an IL-2-induced, non-antigen-specific T cell response is contributing to clearing the lesions. If they had a vigorous T cell response we might have expected to see too much T cell activation and a resultant reversal reaction, but we didn't see any toxicity. We were using low doses of IL-2, which saturate the high affinity receptors but not the low affinity

IL-2 receptors. We have also carried out IL-2 therapy on HIV patients, but the doses were at least 100-fold higher.

Hopewell: What new role of IL-2 are you trying to demonstrate?

Kaplan: The rationale is that in order to cure tuberculosis we need to ensure sufficient immune response together with effective antibiotic therapy. If the antibiotics are effective then it doesn't matter how strong the immune response is: patients will be cured because the antibiotics will kill the bacilli. But if the antibiotics are not effective, either because the patient is multidrug resistant or the disease is too extensive, then the immune response has to be boosted. IL-2 is one intervention that will enable us to determine whether it is possible to modulate the immune response in a patient who has active disease.

Hopewell: Multidrug-resistant disease is fairly straightforward, but because of the heterogeneity in treatment regimens it is somewhat difficult to dissect out the effect. My question is, what are you going to do with the result of the Ugandan trial in which it is likely that there will be no evident effect?

Kaplan: These patients will be newly diagnosed drug-sensitive tuberculosis patients who will respond to therapy so efficiently that the addition of IL-2 might not have a reasonable effect. This study needs to be done, even though it will be difficult to show efficacy. If we find that a higher percentage of individuals clear their sputum by three months with adjunctive IL-2 treatment, compared to without treatment, we'll be delighted because we will have shown that a strong immune response helps. In principle, under ideal circumstances, I would like to be able to shorten the length of time necessary for elimination of the infection by adding IL-2 to standard therapy.

Hopewell: Do you envision IL-2 therapy as a means of shortening the duration of therapy in patients with drug-susceptible organisms?

Kaplan: In most of the patients this will not be necessary because the drugs are so good. However, immune modulation could shorten the duration of therapy. I have been told that in Uganda 75–80% of patients clear their sputum within two months. This is not necessarily what always happens in the field, but if we can add another 15% on top of that, we would be happy.

Hopewell: But that means little in terms of the outcome, unless it translates into the ability to shorten the overall duration of therapy.

Kaplan: The preliminary studies indicate that IL-2 accelerates clearance and/or reduces the number of patients who reactivate, so immune modulation combined with therapy may be better than therapy without immune modulation.

Ryffel: One issue that you didn't touch on is the adverse effects of IL-2 therapy. I would like to know which type of formulation you are using, and whether you have explored a subcutaneous approach. Because it's well known that IL-2 induces a vascular leak syndrome and toxicity, although perhaps not at the low concentrations you are using. Ultimately, you want to be able to induce IFN-γ, so would an alternative strategy be to use IL-12 to induce IFN-γ?

Kaplan: First, the dose of IL-2 is so low that we do not observe toxicities. The only side-effect we've seen is the local induration, which we use to monitor whether IL-2 is biologically active. We haven't used subcutaneous administration, we specifically use intradermal administration because we can then make sure the drug is active.

Second, we could use other immunomodulatory molecules, but we selected IL-2 because it is the least toxic in our hands at that dose. IL-12 is toxic. I'm not sure that IL-12 would necessarily be better or easier to deliver, but maybe someone will do a study so that we can compare IL-2 and IL-12.

Ryffel: I do know that IL-12 is more effective at inducing IFN-γ than IL-2.

Wadee: You suggested that lymphokine-activated killer cell (LAK) activity increases early following administration of IL-2. What is the time frame before LAK activity is evident and for how long is it sustained?

Kaplan: We haven't yet looked at how long the increase in activity is sustained. LAK activity increases 14–28 days following initiation of IL-2 treatment. This does depend on which day the blood is collected and it does vary from individual to individual. We have looked again 1–2 weeks post-therapy and found that the levels are still increased, but later they probably decrease.

Wadee: What read-out are you using for LAK activity?

Kaplan: Chromium release from the labelled target cells.

Rook: I have three points. First, you said that if the bacteria are drug sensitive, it doesn't matter how wimpy the immune response is because the patient will still get better. In extreme cases we know that's not true, i.e. children without IFN-γ receptors cannot be treated even when they're infected with drug-sensitive Bacillus Calmette–Guérin (BCG; Newport et al 1996). Therefore, the drugs do require something from the immune system.

Second, you said you were going to increase the numbers of the drug-resistant cases that you treat with IL-2. We have been down that route, largely for compassionate reasons. John Stanford has now treated about 275 drug-resistant cases with *Mycobacterium vaccae* (J. L. Stanford, personal communication 1997), but we still don't know for certain whether the treatment works because some drug-resistant patients do lose sputum positivity, and historical controls are of little use because treatments keep changing. Therefore, you must do a placebo-controlled study.

Third, although it makes sense to me to try IL-2 in the belief that they are IL-2 deficient, I believe that there's more to the immunopathology of tuberculosis than that. Something else is getting turned on, which is causing the cytokines to start causing immunopathology. In the mouse model, whatever it is that is getting turned on correlates with IL-4 production, and in humans it correlates with the appearance of cells in the peripheral blood that will make IL-4. Therefore, it makes sense to not only turn on a Th1 response, but also turn off whatever it is that's getting turned on.

Kaplan: We have monitored the production of all the cytokines, and found that TNF-α, IL-1β, IL-4, IL-12 and IL-10 mRNAs are induced at low levels or not at all. Only IL-2 and IFN-γ are further induced in response to IL-2 treatment. We have no indication that IL-4 is expressed in the blood cells or in the cells of the PPD delayed-type hypersensitive site, or that the protein is measurable in plasma. Thus, the levels of IL-4 are extremely low. This is not to say that we couldn't take these cells, manipulate them *ex vivo* and induce them to release IL-4.

Rook: It may be useful to monitor whether the low level expression of IL-4 you do detect can be switched off. It is not unusual for obvious effects of IL-4 to be present *in vivo* even when concentrations of IL-4 are unmeasurably low.

Kaplan: We have no evidence for this, given the limitations of these assays.

Russell: Have you looked at IL-6? Does it have a suppressive effect in an *in vivo* infection rather than just *in vitro*? And is the recovery of T cell responses after treatment with thalidomide mediated via suppression of IL-6?

Kaplan: IL-6 can be detected in the plasma, but there are no changes in IL-6 expression in response to IL-2 treatment. We stopped looking at it because we believed we could not use it as a monitor for immune activation. We could probably push the system and show that IL-6 plays a role, but I'm not sure in which direction. IL-6 is not directly affected by thalidomide, a drug that inhibits TNF-α production. After short-term treatment with thalidomide, i.e. for 2–3 weeks, the plasma levels of IL-6 decrease, but this is true for all the pro-inflammatory cytokines. I am not sure what the role of IL-6 is in infection. I believe that this cytokine is telling us something about what's happening to immune activation and the inflammatory response.

Anderson: As an immunologist, do you ever see a stage emerging where one can put quantitative detail on effector mechanisms so that you know if you perturb a system in a particular manner you will illicit a predictable quantitative response in a defined compartment of the immune system? At the moment it seems extraordinary that immunologists, who are designing therapies for seriously ill patients, perturb systems with no quantitative understanding of what the outcomes of the perturbation will be.

Kaplan: I agree. In essence, we know that we're not going to be able to give yes/no answers for particular molecules, although we are looking at quantitative changes. A further complication is that these factors are interregulated. We don't understand how increased IFN-γ results in the killing of organisms, so what we need is a molecule that can kill organisms and that can be measured.

Anderson: Another aspect of quantification is the often-made assumption that if the concentration of cytokine x or cell type y is low, it's unlikely to play an important role as a mediator or an effector mechanism. This argument is often false because the density of a cytokine can be exceedingly low but it can be extremely important. It is all related to rates of turn-over.

Kaplan: My assumption is that a key regulatory cytokine will be modified as the immune response is changed. For example, IL-4 is not detected, nor is it induced, in response to IL-2, so it's not playing an essential role in whatever potential clinical readout we're looking at, given the limitations of the assays we're using.

Anderson: It's not the quantity you want to measure, it's the rate of turn-over.

Kaplan: It's not the quantity I want to measure, it's the change induced by intervention that mediates the next change and so forth, ultimately leading to the cure.

Reference

Newport M, Huxley CM, Huston S et al 1996 A mutation in the interferon-γ receptor gene and susceptibility to mycobacterial infection. New Engl J Med 335:1941–1949

Cellular and genetic mechanisms underlying susceptibility of animal models to tuberculosis infection

Ian Orme

Mycobacteria Research Laboratories, Department of Microbiology, Colorado State University, Fort Collins, CO 80523, USA

Abstract. I shall propose a new working hypothesis regarding the immunopathogenesis of tuberculosis infections in the mouse and guinea-pig models of pulmonary disease which differs from the current dogma that the development of caseous and potentially liquefied lesions in the susceptible guinea-pig lung is due to excessive expression of delayed-type hypersensitivity. I intend to show that the mouse is resistant because T cells entering the granulomatous lesions create highly organized wedges into the epithelioid macrophage fields, thus potentially saturating the tissues with cytokines. Where local necrosis does occur, it is contained by an efficient fibrotic response. In the guinea-pig, however, T cells remain in a peripheral mantle and do not invade the lesion. As a result, central macrophages are not activated, allowing bacterial growth and destructive caseation. I will describe data from the mouse system which show that certain mouse strains (C57BL/6, C5BL/10) are highly resistant to aerosol challenge, whereas other strains (DBA/2, AKR, CBA) are less so. These latter strains initially control the infection, but are subject to reactivation disease 50–200 days later. The histological data so far tend to support the hypothesis that these strains only induce acquired immunity during the reactivation phase, and upon initial exposure rely mainly on innate immunity to control the infection. This may be an important factor underlying the onset of reactivation disease in humans.

1998 Genetics and tuberculosis. Wiley, Chichester (Novartis Foundation Symposium 217) p 112–119

Why are guinea-pigs much more susceptible to pulmonary tuberculosis than mice?

Mice are in general able to control and contain pulmonary infection with *Mycobacterium tuberculosis* by the formation of granulomas, whereas a similar mechanism of protection in guinea-pigs and rabbits is less successful, with these structures gradually developing central areas of caseous necrosis and (in the rabbit) even liquefaction of the lesion and cavity formation, similar to that seen in humans with advanced tuberculosis disease. For some time now, the accepted hypothesis to explain these events has been based on an acquired, biphasic T cell response, with protective T cell-mediated immunity activating macrophages to destroy bacilli, and cytolytic T cells mediating delayed-type hypersensitivity leading to the development of caseous necrosis and tissue damage (Dannenberg 1991).

In comparing the mouse and guinea-pig granulomatous response, however, it is clear that lymphocytes accumulate differently in the two species. In the guinea-pig mixtures of lymphocytes and macrophages accumulate in the mantle of the granuloma, and do not associate to any extent with epithelioid macrophages occupying the centre of this structure. Then, when this centre begins to degenerate and become necrotic, the zone of lymphocytes is further distanced from the centre by a layer of foamy macrophages (indicating they have ingested cell debris) and another layer composed of neutrophils (Fig. 1).

In contrast, in the mouse incoming lymphocytes create organized wedges or rafts of cells that are distributed more centrally within the fields of epithelioid macrophages. As these coalesce they may create circlets of lymphocytes, further surrounding any potentially infected macrophages.

It is reasonable to assume that at least a proportion of the incoming lymphocytes are $CD4^+$ T cells secreting γ-interferon (IFN-γ), a key cytokine in tuberculosis immunity (Cooper et al 1993, Flynn et al 1993). In the mouse lesion the juxtaposition of macrophages and lymphocytes suggests that the great majority of epithelioid macrophages in the granuloma will be exposed to reasonable concentrations of this cytokine. (A bacteriostatic rather than bactericidal concentration, given the 'chronic' disease state at this time?)

In the guinea-pig, however, the T cells remain in the mantle. If the granuloma is relatively large, I would propose that it is possible that insufficient cytokine reaches the centre, and as a result the infection eventually reactivates in this region, triggering gradual degeneration and eventual necrosis of the structure.

If this hypothesis is correct, it still does not identify the underlying basis. It could simply involve chemokines; because if cells in the centre are not exposed to IFN-γ they fail to produce tumour necrosis factor, which in turn fails to induce local chemokine production (Rhoades et al 1995) and hence fails to attract mononuclear cells into the lesion centre. Other possibilities could reflect the poor invasive properties of guinea-pig T cells in terms of their movement through the extracellular matrix, which may be due to the poor expression of appropriate integrin molecules.

Reactivation in tuberculosis: a genetic predisposition?

For well over a decade it has been believed that innate resistance to mycobacterial infection is controlled by a single gene locus, initially identified in leishmaniasis

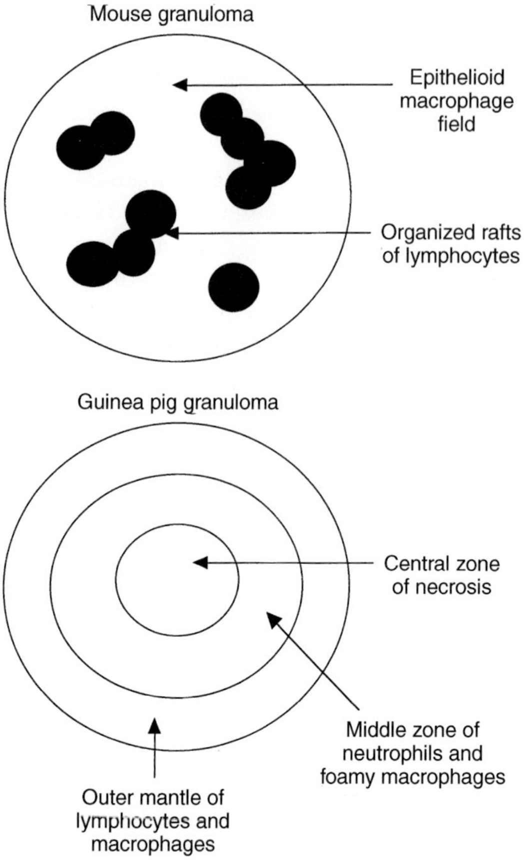

FIG. 1. A cartoon representing lesions in the mouse and guinea-pig. In the guinea-pig lymphocytes preferentially occupy the outer mantle, whereas in the mouse these cells invade the epithelioid macrophage fields creating organized rafts of cells.

(*Lsh*) and salmonellosis (*Ity*) but now more generally known as *Bcg*, due to the growth of the Montreal strain of Bacillus Calmette–Guérin (BCG) in different mouse strains (Skamene & Forget 1988). More recently, a candidate gene from within this locus, designated *Nramp* (for 'natural resistance-associated macrophage protein', despite the fact that evidence is slim that it does such a thing) has been identified (Vidal et al 1993).

Despite the fact that *Nramp* effects are only evident with certain mycobacterial strains (BCG Montreal and *Mycobacterium avium*), the inference has gradually developed that all mycobacterial strains are similarly affected, including

M. tuberculosis. That this is incorrect has been clearly shown by North & Medina (1996), who have found that this gene does not influence the growth of *M. tuberculosis*.

Even worse, ironically, mouse strains identified as being *Bcg*^{r} are in fact more susceptible to this infection. We have looked at several such strains in the low dose aerosol infection model, and find that all mouse strains seem to control the infection over the first 40–50 days or so. The infection then appears to become 'chronic' but thereafter the infectious load in the lungs surges (reactivation disease) and the animal dies.

There appears to be a clear hierarchy within strains in terms of this susceptibility. AKR mice undergo reactivation about 100 days post-aerosol, DBA/2 a little later, and CBA between 150 and 200 days. In contrast, *Bcg*^{s} mice such as C57BL/6 and C57BL/10 do not undergo this event. A comparison of the course of infection in C57BL/6 and DBA/2 mice is shown in Fig. 2.

Histological examination of these two mouse strains has revealed a significant difference between them. In both there is a good macrophage response in the infected lungs, but in the DBA/2 mice there are virtually no lymphocytes present over the initial 40 days. Some lymphocyte infiltration can be seen when the infection then reactivates later, but this appears to be 'too little, too late'.

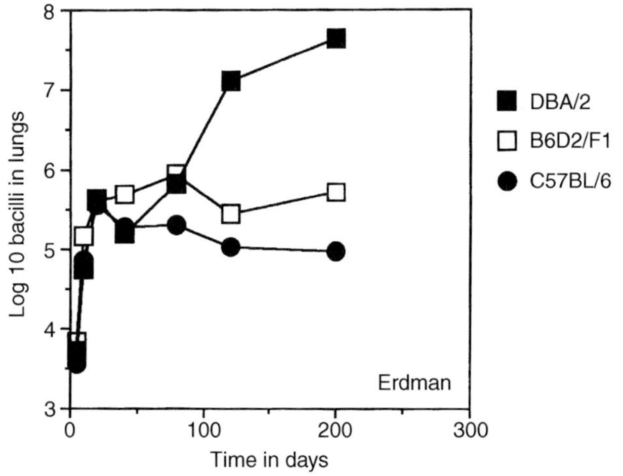

FIG. 2. The course of *Mycobacterium tuberculosis* Erdman infection in C57BL/6 and DBA/2 mouse strains.

What these models may predict for human disease

The above observations appear to support the working hypothesis that certain strains of mice react to low dose aerosol infection by relying more heavily on innate mechanisms of immunity than other strains in which a prominent lymphocyte influx ensues. In both strain types we have observed an initial spike of mRNA for IFN-γ in the lungs at this time (which rapidly decays in reactivating strains), suggesting that something (possibly $\gamma\delta$ or natural killer cells) is doing this, and that this is enough initially to control the infection. However, such a response scenario would not generate antigen-specific T cells, and (perhaps more importantly) memory T cells. In the absence of these effector cells the infection at some point reactivates, killing the mouse.

This may suggest that reactivation disease in humans has a genetic basis (excluding secondary infections, pulmonary vascular diseases, immunosenescence, etc. as extraneous causes of reactivation). If correct, the above hypothesis implies that early decisions made in the lungs in terms of the triggering of innate versus acquired immunity subsequently influence/drive events leading to whether or not the individual has a memory T cell immunosurveillance mechanism in place (Orme & McMurray 1996) and can hence prevent reactivation disease from occurring. Thus, the more the individual depends on innate mechanisms early during the infection, the more likely reactivation may occur at a later date.

I would also propose that this predisposition is not because *Bcg*^{r} macrophages slow the infection down so less antigen is available; after all, the infection grows at an identical rate over this initial period. On the other hand, antigen presentation may be the key; macrophages from reactivation-prone individuals may be more catabolic or they may express co-stimulatory molecules differently (either failing to sensitize T cells or even inducing apoptosis in them). The mechanism does not seem to apply at higher bacterial loads when reactivation has begun; under these conditions some lymphocyte accumulation can be seen in the *Bcg*^{r} mouse strains.

Acknowledgements

This work was supported by National Institutes of Health grants AI-40488 and AG-06946. I thank Tony Frank, Beth Rhoades, Joanne Turner and Bernie Saunders for helpful discussions.

References

Cooper AM, Dalton DK, Stewart TA, Griffin JP, Russell DG, Orme IM 1993 Disseminated tuberculosis in interferon gamma gene-disrupted mice. J Exp Med 178:2243–2247

Dannenberg AM 1991 Delayed-type hypersensitivity and cell-mediated immunity in the pathogenesis of tuberculosis. Immunol Today 12: 228–233

Flynn JL, Chan J, Triebold KJ, Dalton DK, Stewart TA, Bloom BR 1993 An essential role for interferon gamma in resistance to *Mycobacterium tuberculosis* infection. J Exp Med 178:2249–2254
North RJ, Medina E 1996 Significance of the antimicrobial resistance gene, *Nramp1*, in resistance to virulent *Mycobacterium tuberculosis* infection. Res Immunol 147:493–499
Orme IM, McMurray DN 1996 The immune response to tuberculosis in animals. In: Rom WN, Garay S (eds) Tuberculosis infections. Little, Brown & Co., New York, p 269–280
Rhoades ER, Cooper AM, Orme IM 1995 Chemokine response in mice infected with *Mycobacterium tuberculosis*. Infect Immun 63:3871–3877
Skamene E, Forget A 1988 Genetic basis of host resistance and susceptibility to intracellular pathogens. Adv Exp Med Biol 239:23–37
Vidal SM, Malo D, Vogan K, Skamene E, Gros P 1993 Natural resistance to infection with intracellular parasites: isolation of a candidate for *Bcg*. Cell 73:469–485

DISCUSSION

Cole: I was interested by the Bacillus Calmette–Guérin (BCG) protection studies in guinea-pigs and the studies of the subunit vaccine. Would you care to tell us what was in the subunit vaccine?

Orme: We believe there is a multiple T cell response to multiple antigens. Rather than extracting the proteins from the filtrate we injected the whole lot. The adjuvant is monophospholipid, which is made by Ribi ImmunoChem Research Inc. (Hamilton, MT) and is a mild adjuvant. None of the animals become skin test positive, and they don't react to purified protein derivative of tuberculin (PPD) when we give them the filtrate-based vaccine. We also include Chiron (Emeryville, CA) Peg-IL-2 in the vaccine. We mix the filtrate, adjuvant and Peg-interleukin (IL)-2 together, inject the animals, wait a month and inject them again. After a further six weeks we challenge, then follow the weights of the guinea-pigs for 40 weeks or so. We found that the histopathologies were almost identical to those for BCG. Some animals died for reasons we don't quite understand, whereas in the BCG control none of the animals died. We also used IL-12, and observed that this caused a reduction in bacterial counts within one month, but 10 weeks after that they all died, whereas the IL-2 group did not show any protection at 30 days but lived merrily for another year or so.

Brennan: The weak link is the adjuvant, monophospholipid, which is a segment of the lipid A portion of lipopolysaccharide and is not acceptable to the Food and Drug Administration.

Orme: It has been used in clinical trials, and the initial results are good.

Ryffel: I had the naïve understanding that the expression of intercellular adhesion molecule (ICAM)-1 is important for granuloma formation and for the containment of tuberculosis infection. The data from the ICAM-1 knockout show that the bacterial elimination is not affected despite the absence of granulomas in the knockout mice. These data would suggest that granulomas play no role in the control of infection. What is the long-term fate of the ICAM-1-deficient mice?

Orme: The data suggest that protective immunity is an early event involving γ-interferon (IFN-γ), and that the reason you have this granuloma formation or delayed-type hypersensitive (DTH) response is to wall off the infection and prevent dissemination. We looked at the ICAMs for about 80 days and they looked OK, but we then did a second experiment, which we haven't finished yet, and found that 150 days following infection the bacteria are crawling down the interstitial spaces and starting to disseminate, and the bacterial node is definitely increasing. Therefore, the DTH component is essential for long-term survival, although it's not critical for initial expression and protection because if you don't have it the bugs still survive and disseminate.

Bellamy: I was interested in the differences in susceptibility to tuberculosis reactivation in different mouse strains which you presented. However, I would have to dispute your suggestion that *Bcg*^{r} strains are more susceptible than the *Bcg*^{s} strains. If you wanted to test the hypothesis that *Nramp1* is involved you need to compare *Nramp1* knockout mice with control mice of the same inbred strain. Which inbred strains of mice are the most susceptible to tuberculosis is probably determined by a large number of interacting genes.

Blackwell: You wouldn't expect to see any differences between *Nramp1* congenic mouse strains on a B6 background in the regulation of this later effect. But if you looked at a series of BXD recombinant inbred strains you would probably be able to map the DBA/2 versus B6 late effect easily. There are also a series of strains that are recombinant between AKR and B6 that would allow you to determine whether the same gene is causing this effect in this strain combination, whether they reactivate or not, and whether multiple genes are affecting the time at which different mouse strains reactivate. It would be relatively easy to sort out the genetics of that late response.

Orme: One reason why I'm not too enthusiastic about this is because we have injected a small number of different *Mycobacterium tuberculosis* strains (about 50 bacteria; because *Bcg* is supposed to be more effective against low doses) and over the first 10–15 days we see no differences in growth rates in any of these strains of mice. The differences become apparent much later. The concept of some animals having more innate resistance than others is consistent with this, because it looks to me as though these mice rely purely on attracting macrophages and inducing natural killer cells to make IFN-γ. This is all that's required to kill 100 or 1000 bacteria, but if those bacteria are not killed and suddenly they jump up in numbers to 10 000 or 100 000, and a memory T cell response has not been generated by then, then this becomes a serious situation in which reactivation disease is likely.

Colston: Can you detect lymphocytes in the DBA mice that generate a protective response early?

Orme: No. Protection occurs at about 20–30 days, but at that stage we do not observe an accumulation of lymphocytes.

Colston: Can you transfer protection in the DBA mice at that time?

Orme: We haven't yet done that experiment.

Susceptibility to tuberculosis as a complex genetic trait: analysis using recombinant congenic strains of mice

Igor Kramnik, Peter Demant* and Barry B. Bloom

*Department of Microbiology and Immunology, Howard Hughes Medical Institute, Albert Einstein College of Medicine of Yeshiva University, 1300 Morris Park Avenue, Bronx, NY 10461, USA and *The Netherlands Cancer Institute, Amsterdam, Netherlands*

Abstract. Previous advances in the genetics of infectious diseases derived principally from identification of single genes and their isolated effects on the progression of infection. Modern genetic analysis represents a powerful means of understanding the interplay among different pathways activated in the course of infection, their hierarchy and interactions in terms of the development of an optimal protective strategy. By utilizing both whole-genome scanning of (C3HxC57BL/6)F2 and a set of the recombinant congenic strains, produced by backcrossing B10 onto a C3H background, we demonstrated that susceptibility to tuberculosis is a multigenic trait. We have identified two distinct groups of susceptible mice: one that dies within four to six weeks of infection (supersusceptible) and another that dies within seven to 10 weeks (comparable to the susceptible parental strain). Our preliminary genetic analysis suggests that the susceptibility of those groups is controlled by different genetic factors. Supersusceptible mice exhibit dramatic lung pathology, not observed in either parental strain, and their survival after infection with virulent *Mycobacterium tuberculosis* is comparable to that of mice rendered immunodeficient by disruption of essential immune genes. Further genetic and functional analyses of these strains offer possibilities for understanding the control of transmission, preferential growth of the pathogen in the lung, and mechanisms of local and systemic protective immune responses.

1998 Genetics and tuberculosis. Wiley, Chichester (Novartis Foundation Symposium 217) p 120–137

Understanding the mechanisms of virulence of *Mycobacterium tuberculosis* is central to the development of approaches that optimize host protective responses and thus prevent or contain the disease caused by this pathogen. Extensive tissue remodelling and damage caused by the persisting *M. tuberculosis* infection, and a lack of direct toxicity of its constituents on host cells, have been interpreted as an indication that the pathogenesis of tuberculosis is dependent on the ability of mycobacteria to survive the attack of host protective mechanisms and manipulate the immune responses in order to create a local environment favourable for its multiplication and spread to another host (Rook & Bloom 1994). From this point of view, the virulence factors of *M. tuberculosis* could only be revealed through their interactions with the immune system of the host. The complexity of this situation is increased when the heterogeneity within the host population is taken into account. Indeed, the success rate of the strategy adopted by *M. tuberculosis* is limited, with fewer than 10% of infected individuals in modern human populations developing the transmissible disease (Stead 1989, Raviglione et al 1995). The use of genetically standardized experimental animals allows one to minimize the impact of the environmental factors and reduce the complexity of the system to controlled genetic variation. The work of Lurie is the most compelling evidence that comparative functional studies of experimental animals selected on the basis of their resistance or susceptibility to tuberculosis are able to provide important insights into the complex pathogenesis of the disease (Lurie et al 1952, Allison et al 1962). The precise knowledge of multiple genetic factors underlying the differences in the course of infection between resistant and susceptible animals, as well as the possibility of studying their effect in isolation, would contribute significantly to our understanding of the disease. Studies of mycobacterial infections in mice rendered immunodeficient by targeted mutagenesis is a recent example of the successful implementation of this strategy (Flynn et al 1992,1993,1995, Cooper et al 1993, MacMicking et al 1997). Nevertheless, this approach has an important limitation in that the essential components of the normal immune system are completely eliminated from the mutant animal, which affects the function of their immune system irrespective of the interactions with the specific pathogen. As a result, these mice are usually severely compromised in their ability to control infection with a number of unrelated pathogens as well as an infection with the otherwise avirulent *Mycobacterium bovis* Bacillus Calmette–Guérin (BCG). Alternatively, the ideal system that would allow the analysis of the virulence factors of *M. tuberculosis* would highlight those aspects of host–parasite interactions specific for that pathogen, and distinguish them from those specific for avirulent mycobacteria. Moreover, the genetic variations underlying the differences in susceptibility to infection with virulent *M. tuberculosis* should be detectable and amenable to study using modern genetic analysis, allowing the identification of host resistance genes and analysis of their functional roles both in isolation and through their interactions. Here we present preliminary findings of our attempts to develop a mouse model that complies with the above criteria and may prove useful for dissecting complex host interactions with *M. tuberculosis* and understanding the mechanisms of its virulence.

Analysis of parental strains

Previous studies by other groups have already reported the differences in resistance to *M. tuberculosis* among the inbred strains of mice (Sever & Youmans 1957, Gray 1958, Lynch et al 1965, Nikonenko et al 1985, Musa et al 1987). The data presented by those groups are somewhat inconsistent in respect to the strain distribution pattern as well as the criteria used to determine susceptibility. Attempts to identify genes in those studies have led to the view that a multigenic inheritance pattern is involved in tuberculosis resistance (Buschman et al 1988). No linkages to mouse chromosomes have yet been reported for susceptibility to *M. tuberculosis*, although *Bcg* (also known as *Nramp1*) on chromosome 1 controls early growth of BCG strains. We have compared survival times of several inbred strains of immunocompetent mice after infection with the laboratory strain of *M. tuberculosis* (Erdman strain). Mice were infected with 1 million colony-forming units (CFU) intravenously. C57BL/6J, C57BL/10J, C57BL/10SnDem as well as 129/SvJ survived the infection for more than 20 weeks, whereas median survival times (MSTs) of BALB/cJ, DBA/2J and several substrains of C3H did not exceed 10 weeks. Among the susceptible strains C3H and DBA/2 carry the resistant allele of *Nramp1*, which confers the ability to control early replication of avirulent *M. bovis* BCG (Skamene et al 1982, Vidal et al 1993), and thus comply with criteria discussed above. C3H and B6 strains were chosen for subsequent analysis as a susceptible and resistant parental strain, respectively.

To characterize the infection in C57BL/6J and C3HeB/FeJ parental strains, we determined the growth of *M. tuberculosis* in lungs, livers and spleens after intravenous infection at weekly intervals (Fig. 1). Initially, the bacteria grew progressively in the spleens and livers of both strains and the CFU did not differ significantly between the two at one week post-infection. Few bacteria were detected in the lungs of either strain at that time point. Two weeks after infection the growth pattern in the spleens and livers differed between C3H and B6, both quantitatively and qualitatively. Whereas the growth of *M. tuberculosis* in the spleens and livers of the susceptible (C3H) mice continued at about the same doubling time up to four weeks after infection, the numbers of the bacteria in the spleens and livers of the resistant mice decreased as compared to the Day 7 time point. In the lungs, the bacterial loads at two weeks after infection were similar in both strains. While they remained at the same low level in the lungs of resistant mice within the first four weeks of infection and slowly increased during next six months of experiment, the bacterial growth in the lungs of susceptible mice was more rapid than in the other organs. The bacterial load in the lungs exceeded that in

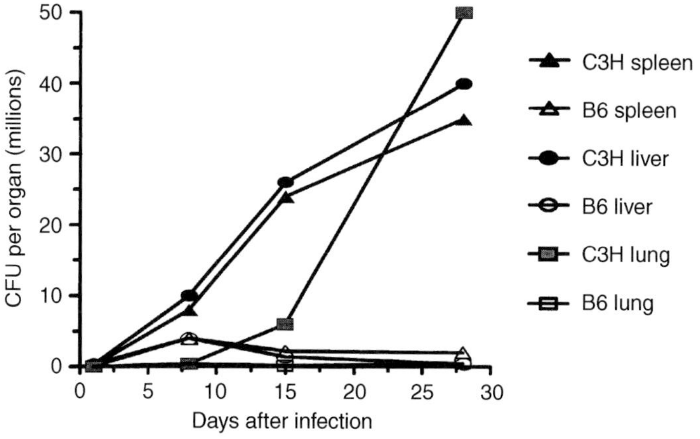

FIG. 1. Kinetics of the growth of *Mycobacterium tuberculosis* Erdman in the organs of C57BL/6J (B6) and C3HeB/FeJ (C3H) mice after infection with 10^6 colony-forming units (CFU) intravenously.

the spleens and livers by the fifth week after infection. Mice of the C3HeB/FeJ substrain died within five to seven weeks after infection. Histopathological analysis revealed macroscopic lesions in the lungs, kidneys and hearts of those mice. No visible pathology in the same organs of the B6 mice sacrificed at the same time was observed.

Under microscopical examination acid-fast bacilli (AFB) were detected in the spleens, livers, lungs, kidneys and hearts of both resistant and susceptible mice at various times after infection in the lungs two weeks after infection (Fig. 2). AFB were localized within separate foci associated with mononuclear infiltrates and the numbers of AFB per focus were somewhat greater in the susceptible mice. By the fifth week of infection, the lungs of C3H mice displayed prominent areas of necrosis containing the AFB. The bacteria were also spread throughout the interstitial lung tissue. The lungs of the B6 mice showed interstitial infiltration with macrophages and lymphocytes as well as isolated foci of AFB associated mostly with large epithelioid-like cells. At six months after infection, the alveolar space in the lungs of the resistant mice was greatly reduced due to the massive interstitial mononuclear infiltrates, although no necrotic lesions were observed. The infected cells contained few bacteria and were surrounded by numerous mononuclear cells.

Overall the data in our model are consistent with the following four inferences. (i) Early growth of virulent *M. tuberculosis* is equal in both strains, and hence there is no discernible difference in the efficiency of natural resistance against the virulent *M. tuberculosis* between the two strains. (ii) Differences in CFU emerge

(A)

(B)

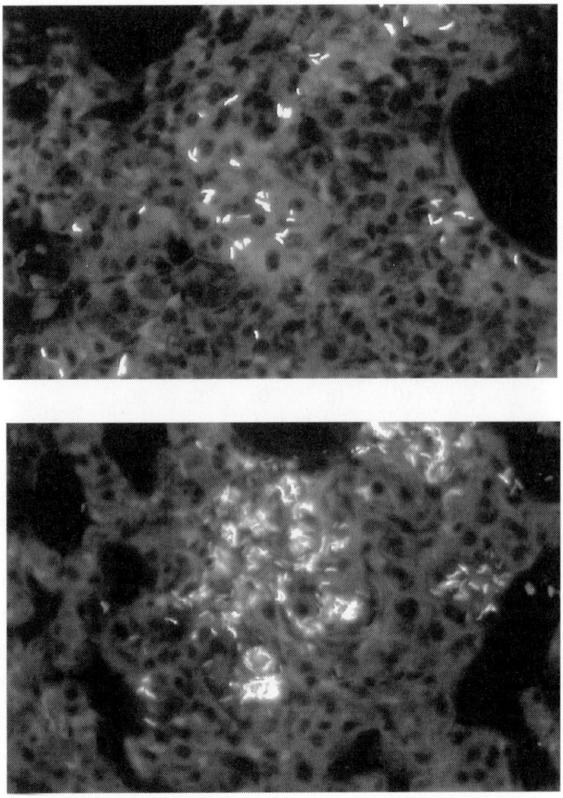

FIG. 2. Acid-fast fluorescent staining of the organs of mice infected with *Mycobacterium tuberculosis* Erdman. (A) and (B) are lungs of B6 and C3H, respectively, two weeks after infection. Magnification = × 400.

at the time that immune responses develop. The resistant B6 mice control the growth of *M. tuberculosis* in lungs, liver and spleen, at times when *M. tuberculosis* growth in C3H is progressive in all organs, but is most rapid in lungs. (iii) The inability to control bacterial growth is reflected in disorganized granuloma formation in the lungs, liver, kidney and heart of C3H mice. (iv) The resistant strain does develop a chronic disease with the slow progressive development of the lung pathology, while the susceptible C3HeB/FeJ strain succumbs to fulminating disseminated tuberculosis. Therefore, our model deals with the mechanisms involved in the control of the disease progression and is not likely to address the natural resistance mechanisms preventing the disease at the onset of infection.

Analysis of crosses between resistant and susceptible strains

To assess the genetic factors underlying the observed differences in the course of infection between C57BL/6J and C3HeB/FeJ strains, we infected F1 and F2 hybrids of those strains intravenously with the standard dose (10^6 CFU) of *M. tuberculosis*. The bacterial growth in the lungs, livers and spleens of the F1 hybrids, as well as their survival, followed the pattern of the resistant parent, suggesting the dominance of resistance. We have analysed almost 400 F2 generation mice for survival in four independent experiments. Twenty-three per cent of the progeny succumbed to tuberculosis in less than eight weeks, a percentage that does not differ significantly from the expectation of 25% for a character caused by a single recessive gene. In order to begin to identify this gene, we carried out a whole-genome scan with 130 microsatellite markers (minimum six per chromosome at 10–20 cM intervals) on the group of mice that died within 10 weeks of infection (Fig. 3). That analysis has revealed a segment of chromosome 1 in which homozygosity for the C3H allele associated significantly with early death (four to six weeks) after infection. In a second step all F2 mice were mapped with microsatellite markers at this region spanning a 25–99 cM interval of chromosome 1. The best association (χ^2 with two degrees of freedom = 184.4505; $P < 1 \times 10^{-4}$) was found with the marker D1mit49 located at approximately 45 cM (Table 1). Remarkably, the cohort of the susceptible mice could be divided into

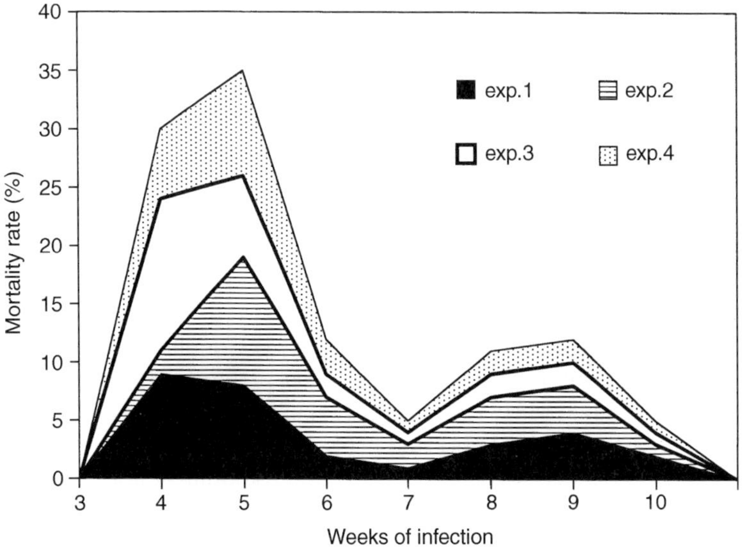

FIG. 3. Kinetics of death of (C3HXB6)F2 mice after intravenous infection with 10^6 colony-forming units of *Mycobacterium tuberculosis* in four independent experiments.

TABLE 1 Linkage of susceptibility to tuberculosis with the marker on chromosome 1 (D1mit49)

Survival	*hh*	*hb*	*bb*	*Total*
<8 weeks	73	9	4	86
>8 weeks	19	149	71	239
Total	92	158	75	325

two groups based on their mortality and homozygosity for the C3H allele on chromosome 1. One group of mice died within four to six weeks and their susceptibility could be explained by homozygosity for the C3H allele on chromosome 1 (the B6 resistant allele is dominant). The mice that died within seven to 10 weeks after infection represent a smaller group and their susceptibility could not be associated with the same locus. Moreover, not every mouse that is homozygous for the C3H allele at chromosome 1 died early after infection; the mice that died within four to six weeks represented 70–80 % of all homozygotes. The incomplete penetrance of the chromosome 1 locus suggests that other genes may be influencing the phenotype. Consistent with this is the finding that the majority of the F2 mice which survived the first 10 weeks of infection gradually died within the next three months with fewer than 10% of the animals surviving for as long as the F1 hybrids (the expectation based on the Mendelian distribution for a single autosomal gene would be 75%).

No significant association of early susceptibility with either parental H-2 alleles was detected in our screen. These data do not exclude a role for MHC genes in tuberculosis infection, but would indicate that their phenotypic expression is affected by epistatic interactions with other genetic factors. For example, a role for H-2 in the control of multiplication of *M. tuberculosis* in chronically infected mice has been reported (Brett et al 1992), but such an effect may be revealed only by determination of the viable bacteria in the lungs of long-term survivors. Finally, our data showed no evidence for a protective effect of the resistant allele of *Nramp1*, which is another important gene implicated in the control of the multiplication of a number of intracellular pathogens (Skamene et al 1982, Vidal et al 1995, Blackwell 1996). Indeed, C3H mice carry the resistant allele of *Nramp1* on chromosome 1, but the locus on chromosome 1 identified in our experiments is of C3H origin and confers susceptibility, not resistance. This result suggests that experiments using *Bcg* congenic strains to study *Nramp1* activity in *M. tuberculosis* infection (Medina & North 1996) may be confounded by effects of closely linked genes. The effect of *Nramp1* might possibly be revealed if its resistant allele, which is present on the chromosome 1 of the C3H mice (39.2 cM), could be dissociated from the tuberculosis susceptibility locus on the same chromosome (30–70 cM), which was detected in our study. Another intriguing possibility is that the same functional activity of *Nramp1* that suppresses the multiplication of avirulent BCG could be harmful to the host during the course of infection with the virulent *M. tuberculosis*. Further genetic studies involving marker-directed backcrosses will be required to distinguish between the two possibilities.

Analysis of a set of recombinant congenic strains

Our preliminary data thus far suggest that susceptibility to tuberculosis in our strain combination is a multigenic trait. Major advances in the analysis of the multigenic traits were made possible recently with the development of the high density genome maps of humans and mice using microsatellite markers (reviewed by Lander & Schork 1994). A different strategy is based on the premise that when the trait of interest is controlled by a limited number of genes they could be separated by random breeding, and thus can be studied either individually or in a setting in which the complexity of interactions is significantly reduced. The concept and production of the unique set of inbred recombinant congenic strains (RCS) is described in detail by Demant and colleagues (Demant & Hart 1986, Groot et al 1992). Briefly, an initial cross was made between two distinct inbred strains C3H/HeSnDem (which is susceptible to *M. tuberculosis*) and C57Bl/B10. The next two generations were produced by backcrossing, without selection, to one of the parental strains (recipient, C3H). This was followed by strict brother–sister mating for more than 14 generations in order to produce a set of homozygous inbred strains (HcB/Dem series), which have mosaic genomes. On average each strain carries a random 12.5% fraction of the donor genome (B10 in our case) and the rest of the genome is of recipient origin (C3H). The RCS analysis offers two unique advantages over classical genetic analysis: (i) epistatic effects of other genes that can confound analysis of multiple genetic pathways that affect the resistant or susceptible phenotype can be eliminated by separation of the parts of the donor genome (Frankel & Schork 1996, Van Wezel et al 1996, Fijneman et al 1996); and (ii) the RCS are homozygous inbred strains, which can be studied for functional activities long before the classical approach would permit meaningful immunological and physiological analysis.

The HcB/Dem series consists of 42 strains, of which 28 strains were obtained by inbreeding after two backcrosses to C3H background and thus theoretically contain about 12.5% of the B10 genome, on average. Another 14 strains are the progeny of N4 (three backcrosses to C3H) and thus contain half the amount of B10 genome (6.25% on average). All 42 strains were tested for their susceptibility to infection with 10^6 CFU of *M. tuberculosis* at least twice in three independent experiments

TABLE 2 Distribution of survival phenotypes of recombinant congenic strains of HcB/Dem series

Phenotype	*Mean survival time (days)*	*No. of recombinant congenic strains*
As resistant as B10	<150	0
More resistant than C3H	150–100	4
Intermediate susceptibility	100–50	19
Supersusceptible	<50	5

(Table 2). The 21 strains that were the most resistant or susceptible were genotyped using microsatellite markers in order to obtain genetic maps with the marker intervals of 5–10 cM. According to computational analysis, more than 90% of donor chromosomal segments should be detected at that density of the markers. From the maps it is clear that the strains differ in the proportion of donor genome: the genotyped strains have between four and 10 chromosomes containing donor segments of varying sizes. It was clear that no B10 chromosomal segment is common to all resistant or all susceptible mice, and that no identical or overlapping patterns for RCS, either resistant or susceptible to tuberculosis, were observed.

The results of these experiments confirm our previous finding in (C3HXB6)F2 of a multigenic control of resistance/susceptibility to tuberculosis in our strain combination. Were there single-gene control, all strains would be segregated into two groups with polar phenotypes representing the phenotypes of the parental strains. In reality, survival time forms a continuous distribution pattern that is the hallmark of multigenic control. Also, it was clear that a number of strains are significantly more susceptible than our susceptible parental strain: strains 10,15,18, 22 and 31 died within 31–47 days after infection in two independent experiments, whereas the MST for the parental C3H/HeSnDem substrain in our experiments was 65–75 days. Importantly, the survival times of these supersusceptible strains in our experiments are not significantly different from the survival times of mice rendered immunodeficient by γ-interferon or inducible nitric oxide synthase (iNOS) gene knockouts (Flynn et al 1993, MacMicking et al 1997). This observation suggests that some segments of B10 chromosomes introduced into those particular strains contain gene(s) which dramatically exacerbate susceptibility to *M. tuberculosis* upon interaction with the C3H background. Therefore, the F2 hybrids between the supersusceptible strains HcB-22 and C3H are being used for the mapping of that gene(s).

In order to correlate the survival of RCS with other parameters of the disease progression, we compared the bacterial growth in the lungs and spleens of two resistant and two supersusceptible RCS. The virulent *M. tuberculosis* grew progressively in the spleens and livers of all strains of mice for the initial seven days of infection. Few bacteria were detected in the lungs on the seventh day of infection. This early seeding of the lungs was detected in both resistant and susceptible mice and was confirmed by acid-fast fluorescent staining of the lung tissue. Once the bacteria seeded the lungs, they grew progressively in all strains of mice including the F1 hybrids (most resistant), although the growth rates, being the best correlate of the survival times, differed significantly among resistant and susceptible mice. Two strains of supersusceptible mice showed indistinguishable kinetics of the CFU in their lungs, albeit the growth in the spleens was different. The bacterial loads in the spleens and livers of HcB-18 were higher than the number of CFU per lung. This strain is characterized by the

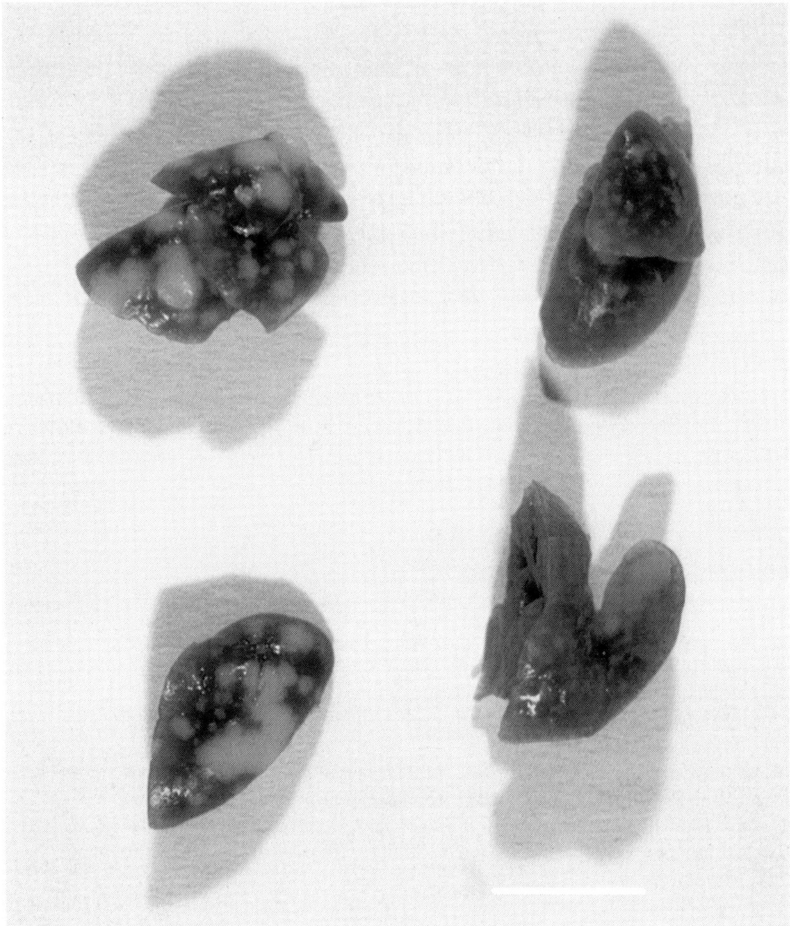

FIG. 4. Lungs of two mice of supersusceptible recombinant congenic strains at a time of death. Left, HcB-22; right, HcB-18. Bar = 2 cm.

generalized progressive infection. On the contrary, in HcB-22 the number of viable bacteria per organ did not increase in the spleens and livers between Day 7 and 25 of infection, whereas *M. tuberculosis* grew progressively only in the lungs of mice that developed dramatic macroscopical lesions (Fig. 4). Microscopic analysis revealed large areas of necrosis in the lungs of that strain, with tissue debris loaded with *M. tuberculosis* in the lumen of the airways, which is unusual for murine tuberculosis. Therefore, we assume that this strain (HcB-22) could be particularly useful for studying the mechanisms and underlying genetic control of the higher vulnerability of the lung tissue to *M. tuberculosis*.

Conclusion

We believe that our preliminary results have confirmed the utility of RCS as a genetic tool that may not only alleviate the task of identifying the genes involved in the control of tuberculosis, but will also allow us to dissect complex aspects of the infectious process using different RCS as well as their hybrids. This may lead to a better understanding of the hierarchy of host–resistance mechanisms, the role of their mutual control in balancing the protective and destructive tendencies of immune response, and thus the virulence strategy of *M. tuberculosis*.

Acknowledgements

This work was supported by National Institutes of Health grants AI07118 and AI23545, grant CT930262 from the European Commission (P.D.) and by the Howard Hughes Medical Institute (B.R.B.).

References

Allison MJ, Zappasodi P, Lurie MB 1962 Host–parasite relationships in natively resistant and susceptible rabbits on quantitative inhalation of tubercle bacilli: their significance for the nature of genetic resistance. Am Rev Respir Dis 85:553–569

Blackwell JM 1996 Genetic susceptibility to leishmanial infections: studies in mice and man. Parasitology 112 (suppl):S67–S74

Brett S, Orrell JM, Beck JS, Ivanyi J 1992 Influence of H-2 genes on the growth of *Mycobacterium tuberculosis* in the lungs of chronically infected mice. Immunology 76:129–132

Buschman E, Apt AS, Nickonenko BV, Moroz AM, Averbakh MM, Skamene E 1988 Genetic aspects of innate resistance and acquired immunity to mycobacteria in inbred mice. Springer Semin Immunopathol 10:319–336

Cooper AM, Dalton DK, Stewart TA, Griffin JP, Russell DG, Orme IM 1993 Disseminated tuberculosis in interferon gamma gene-disrupted mice. J Exp Med 178:2243–2247

Demant P, Hart AAM 1986 Recombinant congenic strains: a new tool for analyzing genetic traits determined by more than one gene. Immunogenetics 24:416–422

Fijneman RJ, de Vries SS, Jansen RC, Demant P 1996 Complex interactions of new quantitative trait loci, Sluc1, Sluc2, Sluc3 and Sluc4, that influence the susceptibility to lung cancer in the mouse. Nat Genet 14:465–467

Frankel W, Schork NJ 1996 Who's afraid of epistasis? Nat Genet 14:371–373

Flynn JL, Goldstein M, Triebold KJ, Koller B, Bloom BR 1992 Major histocompatibility complex class I-restricted T cells are required for resistance to *Mycobacterium tuberculosis* infection. Proc Natl Acad Sci USA 89:12013–12017

Flynn JL, Chan J, Triebold KJ, Dalton DK, Stewart TA, Bloom BR 1993 An essential role for interferon gamma in resistance to *Mycobacterium tuberculosis* infection. J Exp Med 178: 2249–2254

Flynn JL, Goldstein MM, Chan J et al 1995 Tumor necrosis factor-α is required in the protective immune response against *Mycobacterium tuberculosis* in mice. Immunity 2:561–572

Gray DF 1958 Immunity, natural anergy, and artificial desensitization in experimental tuberculosis. Am Rev Tuberc Pulm Dis 78:235–250

Groot PC, Moen CJ, Dietrich W, Stoye JP, Lander ES, Demant P 1992 The recombinant congenic strains for analysis of multigenic traits: genetic composition. FASEB J 6:2826–2835

Lander ES, Schork NJ 1994 Genetic dissection of complex traits. Science 265:2037–2048

Lurie MB, Zappasodi P, Dannenberg AM, Weiss GH 1952 On the mechanism of genetic resistance to tuberculosis and its mode of inheritance. Am J Hum Genet 4:302–314

Lynch CJ, Pierce-Chase CH, Dubos R 1965 A genetic study of susceptibility to experimental tuberculosis in mice infected with mammalian tubercle bacilli. J Exp Med 121:1051–1070

MacMicking JD, North RJ, LaCourse R, Mudgett JS, Shah SK, Nathan CF 1997 Identification of nitric oxide synthase as a protective locus against tuberculosis. Proc Natl Acad Sci USA 94:5243–5248

Medina E, North RJ 1996 Evidence inconsistent with a role for the *Bcg* gene (*Nramp1*) in resistance of mice to infection with virulent *Mycobacterium tuberculosis*. J Exp Med 183:1045–1051

Musa SA, Kim Y, Hashim R, Wang G, Dimmer C, Smith DW 1987 Response of inbred mice to aerosol challenge with *Mycobacterium tuberculosis*. Infect Immun 55:1862–1866

Nikonenko BV, Apt AS, Moroz AM, Averbakh MM 1985 Genetic analysis of susceptibility of mice to H37Rv tuberculosis infection: sensitivity versus relative resistance. In: Skamene E (ed) Genetic control of host resistance to infection and malignancy. Alan R Liss, Toronto, p 291–298

Raviglione MC, Snider DE, Kochi A 1995 Global epidemiology of tuberculosis. Morbidity and mortality of a worldwide epidemic. JAMA 27:3220–3226

Rook GAW, Bloom BR 1994 Mechanisms of pathogenesis in tuberculosis. In: Bloom BR (ed) Tuberculosis: pathogenesis, protection and control. ASM Press, Washington, DC, p 485–501

Sever JL, Youmans GP 1957 Enumeration of viable tubercle bacilli from the organs of nonimmunized and immunized mice. Am Rev Tuberc Pulm Dis 76:616–635

Skamene E, Gros P, Forget A, Kongshavn PAL, St Charles C, Taylor BA 1982 Genetic regulation of resistance to intracellular pathogens. Nature 297:506–510

Stead WW 1989 Pathogenesis of tuberculosis: clinical and epidemiological perspective. Rev Infect Dis 11:S366–S368

Van Wezel T, Stassen A, Moen C, Hart AAM, van der Valk, Demant P 1996 Gene interaction and single gene effects in colon tumour susceptibility in mice. Nat Genet 14:468–470

Vidal SM, Malo D, Vogan K, Skamene E, Gros P 1993 Natural resistance to infection with intracellular parasites: isolation of a candidate for *Bcg*. Cell 73:469–485

Vidal SM, Tremblay ML, Govoni G et al 1995 The *Ity/Lsh/Bcg* locus: natural resistance to infection with intracellular parasites is abrogated by disruption of the *Nramp1* gene. J Exp Med 182:655–666

DISCUSSION

Quesniaux: What is the response of the two different mice strains to lipopolysaccharide?

Kramnik: C3H/HeJ mice carry the lipopolysaccharide mutation, but our parental strain, C3HeB/FeJ, doesn't have this mutation, so both of our parental strains are not deficient in lipopolysaccharide responsiveness.

Orme: You injected these mice with 10^6 colony-forming units (CFU) of *Mycobacterium tuberculosis*, which is close to a lethal dose for any strain of mouse. Low doses are generally sufficient to show these genetic traits, whereas for large doses the genetic background of the mouse is irrelevant because the acquired immune response will involve thousands of genes.

I would like to play devil's advocate and say that I didn't interpret your slides the same way as you did. If you displayed your data in terms of log curves there would only be a log difference in the lungs, and there would be no differences in the spleen and liver. The histology of what you call the supersensitive strain looked to me like an acquired response in the lungs. A granuloma was present with organized lymphocytes, so I worry that your animals were dying not specifically for that reason.

Kramnik: The pathologist who looked at the supersusceptible strain said that the morphology of granulomas and the character of intracellular replication of *M. tuberculosis* is reminiscent of AIDS patients infected with atypical mycobacteria, whereas in the resistant strain the epithelioid cell formation was observed.

Orme: They all have epithelioid cells, but it's how lymphocytes are then reorganized in that lesion that is important. I interpreted the results as exactly the reverse.

Kramnik: In the supersusceptible strain we see good granuloma formation but the macrophages lack the ability to control intracellular replication, suggesting that the effect of the gene is at the level of macrophage activation. We will investigate this with markers of macrophage activation in the future.

Colston: You are looking at tuberculosis clinical disease, which can be reactivation disease or primary disease, and also at mouse survival, which can be affected by uncontrolled bacterial growth and by the immunopathology resulting from control of bacterial growth. It seems to me, therefore, that these phenotypes are too complex for this type of analysis, and you should be focusing on a more precise phenotype.

Kramnik: But to focus on a more precise phenotype you either have to choose which phenotype you think is the most important phenotype and will have the most predictive value, or you have to find the best correlate for survival in terms of function. We opted for the latter.

Bellamy: There are two different aims in relation to your choice of mouse phenotype. One aim is to use a phenotype that is likely to optimize the power of your linkage study, and the other is to define a phenotype that is likely to be due to a single major gene effect. You selected mice with the most extreme phenotypes, in terms of *M. tuberculosis* susceptibility, and this is why you were successful in finding the chromosome 1 linkage. In order to have sufficient power in your linkage study, you needed to include individuals who are not typical of the average population. These individuals represent the extreme ends of a normal distribution and they have genes that are not common in the general population. This means that if two related individuals possess the same rare disease-associated gene mutation they will almost certainly have inherited it from a common ancestor, thereby providing useful linkage information. However, if you look at common phenotypic variables then the gene variants associated with the phenotype will probably also be common. Relatives who share these common gene variants are more likely to have inherited them from different ancestors and will not provide any useful linkage information. Your linkage study was designed to map a disease susceptibility gene variant that might be relatively uncommon in the general mouse population, and it was a successful strategy.

The other phenotypic strategy that can be employed is to assume that if you can focus on a highly specific phenotype it will be determined by a single major genetic effect. This strategy was also successful for the identification of the *Nramp* gene (Vidal et al 1993).

Colston: But if there are multiple phenotypes then it will be less likely that you will find them.

Bellamy: Yes, that's true unless there's a common pathology to the multiple phenotypes.

Kaplan: You are doing a genetic analysis of the host response, i.e. the mouse response to a massive bacteraemia that is induced by high doses of *M. tuberculosis* given intravenously rather than a genetic analysis of *M. tuberculosis* infection.

Kramnik: I agree. And the genetics of intratracheal or aerosol infection would be different because there are more levels of control.

Kaplan: They would be different because they are a different disease.

Kramnik: It may not be a different disease, rather a different aspect of the disease progression. We have focused specifically on the mechanisms controlling disease dissemination and the formation of secondary lesions in the lungs, so we are avoiding another level of complexity. We can then compare our results with those from aerosol challenges and hopefully identify novel genetic factors that regulate this pathogenesis.

Kaplan: I can see your logic, but what's happening in your system is that those which succumb early, succumb before the cellular immune response kicks in.

Orme: Is protective immunity expressed in the supersusceptible mice that die within a few weeks?

Kramnik: We don't have any data on this. We are now analysing the RNA from the lungs and spleens of resistant and susceptible mice.

Orme: But your cytology results suggest to me that protective immunity is expressed. Also, in your spleen data the bacterial load increases then decreases, which is again suggestive of protective immunity.

Kramnik: This is true for strain 22, but not for strain 18, in which the growth of *M. tuberculosis* is uncontrolled in all organs, suggesting we are dealing with different mechanisms.

Kaplan: You are suggesting that you can look at different components of the host response to infection by setting up different models of infection to show that, for example, gene A on chromosome B controls dissemination and gene C on chromosome D controls granuloma formation. Then you can piece together the results of looking at different models of infection under different conditions to obtain a bigger picture of how different genes contribute to different aspects of the disease.

Kramnik: Yes, but so far we have not varied the experimental conditions, only the genetic background.

Bellamy: It is important to realize that geneticists must carry out studies that have maximal power to map disease susceptibility genes. These may not always involve the disease phenotypes that mouse immunologists or human population epidemiologists are the most interested in. In most diseases there is a normal distribution throughout the population in terms of disease susceptibility. A few individuals are highly susceptible and will develop the disease regardless of other factors, a few individuals are highly resistant, and the majority of the population are of intermediate susceptibility and will only develop the disease given specific environmental conditions. It is the individuals who are at the extreme ends of the distribution who are of the most interest to population geneticists because they will presumably possess gene variants that contribute to their unusual phenotype and are relatively rare in the population. For example, if you wanted to find genes that determined height, you would need to do a study on tall and short people. People who are of average height would contribute relatively little information. This is analogous to Igor Kramnik's study where he has first identified mice that show extreme susceptibility and extreme resistance to *M. tuberculosis* and then used these to carry out a successful linkage study. The problem is that these mice may not be representative of murine susceptibility to tuberculosis following a different challenge experiment. However, if he had designed a different challenge experiment it is probable that he would not have been able to define sufficiently resistant and susceptible mice to have adequate power to carry out a worthwhile linkage study. The same problems occur in human linkage studies. The breast cancer susceptibility genes *BRCA1* and *BRCA2* were identified by studies on families where multiple female family members developed breast cancer at an unusually early age. However, the mutations that cause breast cancer in these highly susceptible families are rare in the general population, and *BRCA1* and *BRCA2* gene mutations do not explain the majority of cases of this disease. It has so far proved difficult to identify the genes that determine why people develop common multifactorial diseases, but this is a rapidly advancing field.

Kramnik: Our primary goal is not to identify murine genes that represent diagnostically interesting genes in humans, because mice are not subject to the same degree of natural selection as humans. Also, different loci could have different prognostic values in different human populations depending on their exposure to *M. tuberculosis*. It is important, therefore, to generate better models that allow us to untangle major pathogenesis steps.

Donald: Early childhood mortality and morbidity seems entirely related to disseminated disease. We've heard a lot about local reactions that concentrate upon cellular immunity, but another interesting point about epidemiology in young children is that there is a predominance in males. Is there any merit, therefore, in looking at humoral immunity, as opposed to cellular immunity, at this age? Because humoral immunity has a sex linked aspect, which would fit with the increased susceptibility to dissemination in young males. Amongst all the various models that we have looked at, is there any evidence that the humoral immunity might play a part in preventing dissemination, as opposed to the local lesion that one sees in the lungs? Because this might be important for vaccine design.

Kaplan: There is one physiological aspect that we have not discussed. Mycobacteria are not free-swimming organisms, rather they are within macrophages when they disseminate. There is no evidence that aside from the highly necrotic centre of granulomas mycobacteria are anything but intracellular within monocytes and macrophages.

Donald: During the process of dissemination are the mycobacteria transported by macrophages from the lungs to other tissues?

Kaplan: Yes, and macrophages can even cross the blood–brain barrier. Humoral immunity *per se* and antibody opsonization of mycobacteria do not necessarily have much to do with the crossing of the blood–brain barrier. The question is, why do infected macrophages cross the blood–brain barrier more or less easily in some children?

Donald: Disseminated disease is almost entirely limited to early childhood, when humoral immunity is depressed.

Kaplan: One of the things that boosts humoral immunity is Bacillus Calmette–Guérin (BCG) vaccination. One sometimes observes systemic BCG disseminated in children who have been vaccinated with BCG, which indicates that the cellular immune response is not yet functional. Dissemination occurs before the cellular immune response is generated. I'm not sure that the levels of humoral immune response are going to give any answers. The question is, what influences the ability of infected macrophages to leave the original site of phagocytosis and disseminate? Factors that might be important in this include the level of T cell activation, the generation of cytokines and how efficiently the rest of the immune system responds to infection.

Anderson: What is the evidence that one sees dissemination of infected macrophages, rather than invasion of new macrophages and further replication cycles?

Kaplan: There is a further bacillary replication cycle, but it's a question of time.

Anderson: Then the humoral response could be important in restricting the invasion of new macrophages.

Orme: That's a good point. What people call foamy macrophages may actually represent debris from other macrophages that is picked up by macrophages that are still alive. In caseous necrosis in the lungs of γ-interferon gene knockout mice bacteria are clearly extracellular.

Ryffel: There have also been gene knockout studies reviewed by Kumararatne (1997), who showed that B cell-deficient mice when infected with *M. tuberculosis* are more susceptible than controls, suggesting that the antibodies play a role. I would like to ask Ian Orme to comment on this.

Orme: We have done the same experiment, but we obtained different results. They showed an increased bacterial load in the spleen following a large dose, and they therefore suggested that the B cells were doing something. We infected mice with *M. tuberculosis* in the lungs so as to generate a local mucosal antibody response, and then we waited a few months before we challenged. When we did the same experiment in a B cell knockout mouse, we didn't see any differences in protection. There are immune complexes in people with tuberculosis that seem to be more specifically raised against lipoglycans, and these immune complexes can stick in the kidneys.

Rook: We also know nothing about whether bacteria from the lungs of young children disseminate inside or outside cells. There are no data, so it is possible that antibodies play a role.

Ehlers: Forcing *M. tuberculosis* into monocytes and macrophages via the Fc receptor is unfavourable for the pathogen, but may be decisive for protective host immunity, especially when numbers of mycobacteria are low and it becomes a question of control at metastatic sites.

Russell: There are also the old studies of Armstrong & Hart (1975), who looked at the fusion of mycobacterial phagosomes with lysosomes, and the effect of opsonizing the antibody. When they opsonized *M. tuberculosis* with antibody, they found that the number of fusion events increased. However, when they plated out those infected macrophages and looked at CFU, they found that there was no difference. We have extended those studies, and we argue that the mycobacterial vacuole is highly dynamic, such that there may be fusion at early time points, but then the bacteria are able to exercise their influence over macrophages. The vacuoles then change back into the non-acidic vacuole phenotype typical of intracellular bacteria.

Young: I would now like to take advantage of the general discussion period by inviting David Russell to give a short resume of this work.

References

Armstrong JA, Hart PD 1975 Phagosome–lysosome interactions in cultured macrophages infected with virulent tubercule bacilli. J Exp Med 142:1–16

Kumararatne DS 1997 Tuberculosis and immunodeficiency — of mice and men. Clin Exp Immunol 107:11–14

Vidal SM, Malo D, Vogan K, Skamene E, Gros P 1993 Natural resistance to infection with intracellular parasites: isolation of a candidate gene for *Bcg*. Cell 73:469–485

General discussion I

Endocytic trafficking and the mycobacterial vacuole

Russell: The data I am going to present are predominantly for *Mycobacterium avium*. We do have the same data for *Mycobacterium tuberculosis*, but they are not as extensive. In a repeat of the classic experiments of Hart et al (1972), we incubated mycobacteria-infected macrophages with biotin dextran as a fluid-phase marker and chased it to the lysosomes. We then fixed the cells in order to do immunoelectromicroscopy. We found that the bacilli and the lysosomes loaded with biotin dextran did not co-localize, suggesting that material directed to the lysosome does not traffic through the mycobacterial vacuole and that fusion between mycobacterial vacuoles and lysosomes is limited (Xu et al 1994). The idea that the mycobacterial vacuole is a sequestered compartment that doesn't interact with other intracellular vesicles and is relatively inert became generally fixed in the literature. However, in my opinion it is incorrect.

The pH of the mycobacterial vacuole is between 6.2 and 6.3 (Sturgill-Koszycki et al 1994), and this high pH is abnormal for an intracellular compartment within the endocytic lysosomal network. A so-called terminal lysosome has a pH in the order of 4.5. pH 6.2 to 6.3 is actually the pH of the re-cycling endosomal network, which is involved in the process whereby transferrin enters the cell, iron is released and then transferrin, still attached to its receptor, is recycled to the cell surface. We have shown, both using immunoelectromicroscopy and cell fractionation techniques, that isolated mycobacterial vacuoles are rich in transferrin. If we give a transferrin pulse to the cell, we can show that the transferrin enters and exits the vacuole (Sturgill-Koszycki et al 1996). This is interesting from the point of view of drug trafficking. A hydrophilic drug endocytosed by an infected macrophage would traffic through a cell with restricted access to the mycobacterial vacuole if the mycobacterial vacuole sits in the recycling arm of the pathway.

We also have data relevant to a study that was published by Ralph Steinman and colleagues in *Science* a few years ago (Pancholi et al 1993). They indicated that the mycobacterial vacuoles are privileged compartments that do not interact with the antigen presentation pathway. There is a problem with that interpretation, however, because it does not address where material goes after it exits the mycobacterial vacuole, and we find that bacterial lipids and proteins traffic out of this vacuole efficiently (Xu et al 1994).

I mentioned above that the mycobacterial vacuole doesn't interact with material going to the lysosome. However, it is a highly dynamic compartment, and it gets material from the synthetic pathway of the *trans* Golgi network and from the plasmalemma. We looked at a different marker of trafficking, i.e. glycosphingolipids, by binding the cholera toxin B subunit to the cell surface on ice, and then warming up the cells so that the cholera toxin could redistribute. In these preparations every *M. tuberculosis*-containing vacuole was positive for the GM1 ganglioside complexed with cholera toxin (Russell et al 1996). We looked at the kinetics of cholera toxin entering the mycobacterial vacuole, and found that it reaches a steady state with the surface of the cell within 5–10 minutes of the cell being warmed up. If the experiment is repeated using an IgG bead containing phagolysosome, steady state isn't even reached after 60 minutes, suggesting that the mycobacterial compartment is readily accessible to certain elements of the host cell plasmalemma.

We have also looked at trafficking out of the vacuole. Initially, we were interested in interleukin (IL)-6 induction, because mycobacterial vacuoles produce large amounts of IL-6, as both Wadee and ourselves have described (Sussman & Wadee 1992, VanHeyningen et al 1997), and this correlates with the release and transfer of bacterial cell wall components. We labelled mycobacteria via terminal oxidizable carbohydrate residues, i.e. periodate treatment followed with Texas Red hydrazide, and then we labelled acceptor macrophages with fluorescein diacetate, which is cleaved by serine esterases in the cytosol. Therefore, acceptor cells (green) were added to the donor cells which had been infected previously with Texas Red-labelled mycobacteria. The intensely labelled bacteria are in the central perinuclear region, and the tubular endosomes that are trafficking material out of the mycobacterial vacuoles are visible. We believe that lipidoglycan moieties on the cell wall are labelled predominantly by this technique, although I'm trying to persuade Patrick Brennan to do some 2D thin layer chromatography on the labelled moieties so we can determine what is actually being labelled. If we label the proteins instead, using n-hydroxysuccinimide fluorescein, then we don't see this picture. Mycobacteria do not release large amounts of surface proteins inside the macrophage. Also, because we only see this copious release of cell wall lipids with live bugs we believe it to be a metabolically active process. Therefore, cell wall constituents are released from infected macrophages and are acquired by bystander macrophages. Any inducer effects, in terms of IL-6 production or immunomodulation, are not only restricted to the infected macrophages but are also expressed by bystander macrophages.

This indicates that the mycobacterial vacuole should not be thought of as a little bag of membrane that sits inside the cell and does nothing. It's an extremely dynamic structure. The transferrin trafficking in allows the bacteria access to iron, and the trafficking out allows the infected macrophages to redistribute bacterial cell wall lipids that are involved in the modulation of the environment, not only in the infected cells but also in the bystander cells within the granuloma structure. We're now isolating these vesicles and trying to find out exactly what they contain.

Rook: Did you do all these experiments in *M. avium*?

Russell: No. We did the trafficking experiments with *M. tuberculosis* because *M. avium* is poor at releasing cell wall lipids.

Rook: Is the type of vacuole the same for the two species?

Russell: Yes. The phenotype of the *M. tuberculosis* vacuole is much more robust than it is with *M. avium*, so *M. tuberculosis* appears much better at maintaining its compartment as a non-acidic compartment.

Rook: When the macrophages are successfully activated are the organisms transferred into phagosomes?

Russell: We've done a lot of work on this. It's a chicken and egg question. The most accepted view in the literature is that when macrophages are activated, mycobacteria are killed and then transferred into a degradative compartment or lysosome and digested. We feel that this idea is wrong. When macrophages are activated the vacuole is able to acidify, and the mycobacteria are transported into a functionally different compartment (Russell et al 1997, Schaible et al 1998), which has acquired proton ATPases, contains activated hydrolases and is no longer accessible to transferrin. However, this alteration precedes bacterial death. This is important because experiments on nitric oxide showed that it was central in killing mycobacterium, so people stopped looking at activated macrophages in terms of their killing mechanisms. Nitric oxide is an essential component, but when macrophages are activated a cascade of other events take place. The translocation of mycobacteria into an acidic, hydrolytically competent compartment either potentiates or enables nitric oxide to kill, because nitric oxide in an acidic environment will be converted into nitrous acid. If superoxide is present peroxynitrite is produced. And if iron is present hypervalent iron is produced. This is in addition to the hydrolases, which will attack the surface and thus expose more targets. It's a complex series of events, and nitric oxide is required but it is not sufficient.

Rook: It may be required as a signalling molecule rather than as a killing molecule. Are you absolutely sure it's involved in killing?

Russell: It is more likely to be involved in killing because the translocation of the vacuole is independent of nitric oxide. Translocation of the vacuole occurs in activated macrophages from inducible nitric oxide synthase knockout mice, yet there is no change in the viability of the organism within the time period of those assays.

Blackwell: Does the vacuole fuse with the late endosomal compartment?

Russell: Yes, mycobacterial vacuoles in activated macrophages fuse with preformed endosomes and lysosomes.

Kaplan: Have you looked at human monocytes?

Russell: No.

Wadee: We have. We find that some *M. tuberculosis* organisms inside macrophages are degraded somewhat, but that the lipid component of these degraded organisms inhibits/partially inhibits phagosome–lysosome fusion within cells. This may allow other viable organisms to survive within the host.

Kaplan: Since the lipid is acting as a scavenger, and since nitric oxide is essential for the killing in the murine cell and there's less nitric oxide in the human monocyte, the human monocyte is going to be less efficient at killing mycobacteria.

Orme: One of the things we can't figure out is why only 1% of epithelial macrophages in the mouse contain acid-fast bacilli. If lipoarabinomannan was transported out into other surrounding cells all sorts of cytokines would be produced, which could be a trigger for granuloma formation.

Russell: From the IL-6 induction point of view, certainly lipoarabinomannan in solution is an inducer, but it is not nearly as potent as live mycobacteria. It's likely that there is something else present which is a stronger inducer of IL-6 induction and macrophage modulation.

Ehlers: What is the evidence that the materials that are shed comprise lipids?

Russell: We have affinity isolated the vesicles involved in trafficking using biotin hydroxide and a streptavidin affinity isolation procedure. We then did some immunoelectromicroscopy on these vesicles with gold particles to show that they contain lipoarabinomannan, and we also showed that they contain host-derived lipid by labelling them with antibodies against lysobisphosphatidic acid, which is the fatty acid produced by membrane fusion events inside cells. It accumulates in multivesicular, autophagosomal compartments, such as the antigen presentation compartment, and cells subsequently exocytose these lipids because they can't degrade them.

Brennan: But this antibody also reacts with the extracellular polysaccharide, so you have to keep an open mind as to whether it's polysaccharide or whether it's phospholipid.

Russell: The structure is so typical of a lipid vesicle that it is difficult to imagine it being anything else.

Ulmer: Could you speculate on how mycobacterial protein antigens are presented by MHC class I molecules, given that they're released poorly from unactivated macrophages? Does presentation occur after the macrophages are activated, when proteins are easily released by the hydrolytic compartment?

Russell: This is too large and complex a question to answer briefly, because there are many levels at which antigen presentation in infected macrophages breaks down. These include: the down-regulation of class I and class II molecules; the down-regulation of co-stimulatory molecules; the suppression of T cell proliferation through IL-6 production; and the changes in the intercellular trafficking machinery.

Ulmer: You could circumvent this by transferring the antigen to another cell that's not infected.

Russell: Yes, but it becomes problematic if you also transfer the lipid involved in modulation of the cell.

Ress: I would like to raise a point about David Russell's re-interpretation of Ralph Steinman's data (Pancholi et al 1993). They showed that Bacillus Calmette–Guérin-infected Day 6 macrophages contain mycobacteria that are sequestered from the immune response. One interpretation is that these macrophages don't secrete the kind of protein antigens that activate the indicator T cells they were using, i.e. immune $CD4^+$ T cell clones. This would suggest that there's no protein secretion. I assume that any lipid secreted in their model system would not have activated the T cells because there wouldn't have been any $\gamma\delta^+$ or double-negative, lipid-restricted T cells present.

Kaplan: But David Russell is suggesting that the reason T cell proliferation does not occur is not because there is no antigen presentation, but because IL-6 down-regulates T cell activity. There is no information as to whether the antigen is present or not because the responding cells are prevented from proliferating (the readout for T cell stimulation by antigen is T cell proliferation).

Ress: We don't know if IL-6 is playing a role in Ralph Steinman's model.

Kaplan: Ralph Steinman looked at proliferation, and David Russell is arguing that proliferation will be inhibited under those conditions because IL-6 will be induced. We don't know if protein is released, and we don't know how much is required for antigen presentation. If you block the antigen response you are not going to be able to monitor antigen presentation.

Russell: If we add an exogenous antigen to the infected macrophages we don't observe T cell proliferation, unless we add a neutralizing antibody to IL-6 and then we get T cell proliferation. This suggests IL-6 is suppressing T cell proliferation in this assay, and we need to get rid of this before we can start asking the question.

Ress: It would be fascinating to know, for example, if cytotoxic T lymphocytes (CTLs) would recognize an antigen in that model (i.e. chronically infected macrophages).

Kaplan: Is CTL activity independent of IL-6?

Wadee: No. CTL activity is inhibited by IL-6. Introduction of purified IL-6 into *in vitro* cultures inhibits CTL activity, and this can be abrogated by the addition of anti-IL-6 antibodies (Wadee et al 1994).

Russell: Another point I would like to make is that Rincon et al (1997) took naïve $CD4^+$ T cells and expanded them with an antigen in the presence of IL-6. They found that IL-6 preferentially drove the system towards T helper (Th) 2 cell production. All our IL-6 suppression data were generated using an IL-2-

dependent response, so we are now wondering whether the IL-6 suppression data are specific for IL-2-dependent Th1-type cells, and whether we would see the same suppression with a Th2-type cell response. Ahmed Wadee, have you looked at the differences between IL-2 and IL-4 responses to IL-6?

Wadee: IL-6 turns on IL-4 production, thereby turning on Th2 responses, but IL-2 production is markedly suppressed by the introduction of IL-6. These effects can be abrogated by the introduction of anti-IL-6 antibodies (Wadee et al 1994).

Orme: We had a curious result in our IL-6 knockout experiments with low dose aerosols. We found that the bacterial counts surged for about two weeks, and then a compensatory mechanism is activated and the levels decrease to normal control values. This suggests that IL-6 is important in the first couple of weeks.

Kaplan: An important emerging concept is that not only are the infected macrophages involved in the regulation of the host immune response, but other cells including uninfected macrophages in the granuloma are also contributing. All the cells are physically sealing off the infected site and producing cytokines and reactive metabolites. The more information we generate on the exact regulation of bystander effector mechanisms at the infection site, the clearer the picture will be of the physiological role of the granuloma. We need to know how easily molecules can penetrate through the granuloma to the infected cells.

Kramnik: Another important aspect is, for how long do mycobacteria stay within a single macrophage? Once the bacteria start to proliferate inside macrophages, the macrophages become disabled in terms of their ability to produce effector reactions. At this point it may be important to transfer the bacteria to a different macrophage that has already been exposed to γ-interferon.

Kaplan: The assumption is that in a dynamic granuloma newly recruited cells are more competent in terms of their ability to be activated than the old resident cells, which have been sitting there for a while and have lost some of their ability to be activated. If the organisms just sit in the macrophages for weeks and months, and do not replicate rapidly, they are within a cellular environment with only a low level of activation. In contrast, in a dynamic environment in which macrophages are dying and new macrophages are re-phagocytosing the old macrophages and the organisms, the new macrophages are immune activated and may be more efficient at handling the intracellular pathogen. In air spaces of the lung there would be little if antibody or complement. But as soon as bacilli gain access to plasma and serum they are exposed to opsinins including antibodies and complement.

References

Hart PD, Armstrong JA, Brown CA, Draper P 1972 Ultrastructural study of the behaviour of macrophages toward parasitic mycobacteria. Infect Immun 5:803–807

Pancholi P, Mirza A, Bhardwaj N, Steinman RM 1993 Sequestration from immune $CD4^+$ T cells of mycobacteria growing in human macrophages. Science 260:984–986

Rincon M, Anguita J, Nakamura T, Fikrig E, Flavell RA 1997 Interleukin (IL)-6 directs the differentiation of IL-4-producing $CD4^+$ T cells. J Exp Med 185:461–469

Russell DG, Dant J, Sturgill-Koszycki S 1996 *Mycobacterium avium*- and *Mycobacterium tuberculosis*-containing vacuoles are dynamic, fusion-competent vesicles that are accessible to glycosphingolipids from the host cell plasmalemma. J Immunol 156:4764–4773

Russell DG, Sturgill-Koszycki S, VanHeyningen T, Collins H, Schaible U 1997 Why intracellular parasitism need not be a degrading experience for *Mycobacterium*. Philos Trans R Soc Lond B Biol Sci 352:1303–1310

Schaible UE, Sturgill-Koszycki S, Sclesinger P, Russell DG 1998 Cytokine activation leads to acidification and increased maturation of *Mycobacterium avium*-containing phagosomes in murine macrophages. J Immunol 160:1290–1296

Sturgill-Koszycki S, Schlesinger PH, Chakraborty P et al 1994 Lack of acidification in *Mycobacterium* phagosomes produced by exclusion of the vesicular proton-ATPase. Science 263:678–681

Sturgill-Koszycki S, Schaible UE, Russell DG 1996 *Mycobacterium*-containing phagosomes are accessible to early endosomes and reflect a transitional state in normal phagosome biogenesis. EMBO J 15:6960–6968

Sussman G, Wadee AA 1992 Supernatants derived from $CD8^+$ lymphocytes activated by mycobacterial fractions inhibit cytokine production. The role of IL-6. Biotherapy 4:87–95

VanHeyningen T, Collins HL, Russell DG 1997 IL-6 produced by macrophages infected with *Mycobacterium* species suppresses T cell responses. J Immunol 158:330–337

Wadee AA, Sussman G, Kuschke RH, Reddy SG 1994 Suppression of cytokine production by supernatants from $CD8^+$ lymphocytes activated by mycobacterial fractions: the role of interleukins 4 and 6. Biotherapy 7:125–136

Xu S, Cooper A, Sturgill-Koszycki S 1994 Intracellular trafficking in *Mycobacterium tuberculosis*- and *Mycobacterium avium*-infected macrophages. J Immunol 153:2568–2578

Signals that regulate the host response to *Mycobacterium tuberculosis*

Albert D. Beyers, Annelies van Rie*, Joanita Adams, Gael Fenhalls, Robert Gie* and Nulda Beyers*

*MRC Centre for Molecular and Cellular Biology, and Departments of Medical Biochemistry, *Paediatrics and Child Health, Faculty of Medicine, University of Stellenbosch, PO Box 19063, Tygerberg 7505, South Africa*

Abstract. An appropriate T helper (Th) 1 immune response is required for the elimination of *Mycobacterium tuberculosis*. The factors regulating the polarization of mouse or human T cells to produce Th1 or Th2 cytokines are briefly reviewed. These factors include host genetics, cytokines present at the site of T cell activation, the type, dose and localization of antigen, the type of antigen-presenting cells, the engagement of different co-stimulatory molecules, steroid hormones and age. T cells of children produce low levels of γ-interferon and we hypothesize that this may partly explain the differences in the clinical manifestations of tuberculosis in children and adults. Given that Th2 cytokines inhibit Th1 responses, the question arises of whether individuals mounting prominent Th2 responses, manifested by high serum IgE levels, are more susceptible to *M. tuberculosis*. In a community with a high incidence of tuberculosis, serum IgE levels, a marker of prominent Th2 responses, correlate with disease incidence and with socioeconomic deprivation. We propose that Th2 immune dominance, probably induced by intestinal parasites, enhances susceptibility to tuberculosis. Furthermore, our finding that serum IgE declines in patients following active tuberculosis argues that tuberculosis down-regulates Th2 responses.

1998 Genetics and tuberculosis. Wiley, Chichester (Novartis Foundation Symposium 217) p 145–159

Mycobacterium tuberculosis is a facultative intracellular organism, and the major mechanism for elimination of the organism is by cell-mediated immunity. Macrophages infected with *M. tuberculosis* secrete interleukin (IL)-12, which induces the development of T helper (Th) 1 lymphocytes secreting IL-2 and γ-interferon (IFN-γ; reviewed by Seder & Paul 1994, Abbas et al 1996). IFN-γ, in turn, activates macrophages and enhances their microbicidal activity. An appropriate Th1 response is therefore crucially important for an effective cell-mediated immune response to *M. tuberculosis*.

A Th2 response, on the other hand, is manifested by the secretion of IL-4, -5, -6 and -13, and subsequently by the production of non-opsonic antibodies, including

IgE, and by the activation of mast cells and eosinophils. This response is required for efficient humoral immunity and for the elimination of some but not all parasites (Seder & Paul 1994, Abbas et al 1996, Finkelman et al 1997). IL-10 was originally described as a Th2 cytokine, but it is also secreted by human Th1 cells and by activated macrophages (Abbas et al 1996). Type 1 and type 2 cells negatively cross-regulate each other. IFN-γ inhibits Th2 cells, whereas IL-4 and IL-10 have potent anti-inflammatory effects and inhibit the development and response of Th1 cells (Seder & Paul 1994, Abbas et al 1996). Abbas et al (1996) proposed that a key function of Th2 cells, which often accumulate during chronic diseases, is to inhibit chronic inflammation. The mutual cross-regulation of Th1 and Th2 responses are readily demonstrated *in vitro* and can also be seen in experimental mouse models, for example of *Leishmania* infection (Seder & Paul 1994). In some circumstances Th2 responses develop with slower kinetics than Th1 responses and serve to antagonize the former, keeping tissue damage in check (Abbas et al 1996). In murine tuberculosis models a pure Th1 response is found in the first three weeks following infection and a mixed response is observed after 50 days. In this model, the Th2 component of the immune response appears to sensitize tissue to tumour necrosis factor (TNF) and may contribute to the Koch phenomenon (Hernandez-Pando & Rook 1994, Rook & Hernandez-Pando 1998, this volume). In animals with anti-tuberculin IgE, a reflection of a Th2 component in the specific immune response, Doppler studies showed a reduction in blood flow to the centre of tuberculin skin lesions (Gibbs et al 1991).

The concept of Th1 and Th2 responses is an oversimplification of the complex interplay of various immune and non-immune cells producing a large variety of cytokines. Polarization of cytokine responses is not only found in Th cells, but also in cytotoxic cells and in $\gamma\delta$ lymphocytes, and these responses should perhaps rather be referred to as type 1 and type 2 rather than Th1 and Th2. Furthermore, there is increasing evidence that T cells do not differentiate neatly into only two clear subsets, each producing a particular package of cytokines. Individual T cells, sometimes designated Th0 cells, produce a heterogeneous mixture of cytokines *in vivo*, and a population of T cells produces a wide array of cytokines, of which the global distribution may be skewed to the Th1 or Th2 side (Kelso 1995, Abbas et al 1996). Furthermore, most pathogens are eliminated by multiple overlapping immune mechanisms (Allen & Maizels 1997). In some cases, pathogens manipulate the cytokine response to their advantage, e.g. to allow long-term survival of the organisms within the host (reviewed by Kelso 1995, Allen & Maizels 1997). In spite of all these caveats, the Th1/Th2 paradigm still provides a useful conceptual framework for considering the regulation of cytokine responses. In this chapter, genetic and environmental factors regulating cytokine responses will be reviewed, with emphasis on the immune response to *M. tuberculosis*. We mention our studies on cytokine responses in childhood versus adulthood and on IgE responses in a community with a high incidence of tuberculosis, and we discuss possible implications.

The choice between Th1 and Th2 responses: animal models

Little is known of genetic influences on cytokine production in animals. BD10.D2 or C57BL/6 mice, which are resistant to *Leishmania*, have predominantly Th1 responses; whereas BALB/c mice, which are susceptible to *Leishmania*, produce predominantly Th2 cytokines. The relevant locus maps to a region on chromosome 11, syntenic to human chromosome 5q31.1, a region known to control IgE levels (see below; Gorham et al 1996).

Cytokines exert a powerful influence on the polarization of naïve T cells. Th cells become committed to the Th1 or Th2 phenotype within a few days following infection or antigen administration, and the polarizing cytokines must therefore be present at the site of interaction of the T cell with the antigen-presenting cell (APC) within hours following T cell activation. IL-12, secreted by macrophages infected with intracellular organisms such as *M. tuberculosis*, induces the secretion of IFN-γ by natural killer (NK) cells and activates Th1 lymphocytes secreting IL-2 and IFN-γ (reviewed by Abbas et al 1996, Constant & Bottomly 1997). IL-4, on the other hand, induces Th2 responses and may be derived from basophils, mast cells or 'natural' T cells expressing the NK1.1 cell surface marker (Constant & Bottomly 1997). IL-12 knockout mice cannot mount efficient Th1 responses and are highly susceptible to viral infection (Abbas et al 1996). IL-4 knockout mice, on the other hand, have abrogated Th2 responses (Abbas et al 1996) and may be less susceptible to certain viral infections (murine AIDS; Kanagawa et al 1993).

The type, dose and route of antigen can influence subsequent T cell responses. Mycobacterial antigens such as purified protein derivative of tuberculin (PPD) induce a Th1 response, whereas parasite antigens such as Toxocara excretory substance induce a Th2 response. Low doses of parasites (e.g. *Leishmania major* or *Trichiuris muris*) tend to induce Th1 responses, whereas high doses induce Th2 responses (Abbas et al 1996, Constant & Bottomly 1997). Similarly, immunization of mice with 10^7 *Mycobacterium vaccae* induces a Th1 response, whereas 10^9 organisms induce a mixed response (Hernandez-Pando & Rook 1994). Efficient uptake and presentation by dendritic cells and macrophages, which both produce IL-12, probably contribute to the induction of Th1 responses. Low or high doses of soluble protein, on the other hand, tend to induce Th2 responses (reviewed by Constant & Bottomly 1997). In contrast to the parenteral administration of antigen, oral administration suppresses immune responses in several autoimmunity and transplantation models. The suppression appears to be mediated by T cells, activated in the Peyer patches, that secrete transforming growth factor β (TGF-β; Weiner et al 1994). The regulatory cells inhibit bystander cells, specific for another antigen, in their immediate vicinity. There are isolated examples of helminth-induced Th2 responses modulating cytokine responses against different antigens. If mice are infected with helminths prior to immunization with PPD, they do not only produce IFN-γ but also IL-4 and IL-5 in response to the PPD (Pearlman et al 1993). Helminth infection may also decrease the IFN-γ response to viruses (Actor et al 1993). These studies raise the question of whether pre-existing Th2 responses, elicited for example by chronic parasitic infestation, may impair cell-mediated immunity and resistance to *M. tuberculosis*.

The choice of cytokine responses can also be influenced by the type of APC. Antigen presentation by macrophages tends to induce type 1 responses, whereas B cells tend to induce type 2 responses (Abbas et al 1996) or tolerance (Matzinger 1994). The number and type of APCs also determine whether newborn mice will become tolerant or mount an immune response. In a model in which male spleen cells, expressing the HY antigen, have been transferred to female neonatal mice, immunization with large numbers of spleen cells induced tolerance, whereas small numbers of spleen cells or dendritic cells induced cytotoxic T lymphocyte responses (Ridge et al 1996). APCs can regulate T cell responses by engaging different co-stimulatory molecules (in certain studies the CD28 ligand B7–1 induces type 1 responses, whereas B7–2 induces type 2 responses; reviewed by Constant & Bottomley 1997), by secreting different cytokines (IL-12 secreted by macrophages and dendritic cells is a powerful Th1 inducer) or by secreting other mediators, such as PGE_2, that inhibit secretion of IFN-γ (Edwards et al 1986). Th1 responses are impaired in CD40 and CD154 knockout mice, and these mice are highly susceptible to *Leishmania major* (Grewal et al 1997). Taken together, there is increasing evidence that subtle differences in the density of ligands for the T cell receptor (TCR) and for co-stimulatory molecules play an important role in the regulation of immune responses.

Cytokine responses are also regulated by steroid hormones. Glucocorticoids and calcitriol (1α,25-dihydroxycholecalciferol) induce Th2 responses, whereas dehydroepiandrosterone (DHEA) induces Th1 responses (Daynes et al 1991, Rook & Hernandez-Pando 1998, this volume). The main factors that regulate the choice between Th1 and Th2 responses are shown in Fig. 1.

The choice between Th1 and Th2 responses in humans: nature versus nurture

Racial differences, twin and adoption studies show that host genetics contribute to susceptibility to infectious diseases such as tuberculosis. HLA-A10, -B8 and DR2 are associated with tuberculosis, whereas recently identified polymorphisms in the vitamin D receptor and mannose binding protein genes confer resistance to

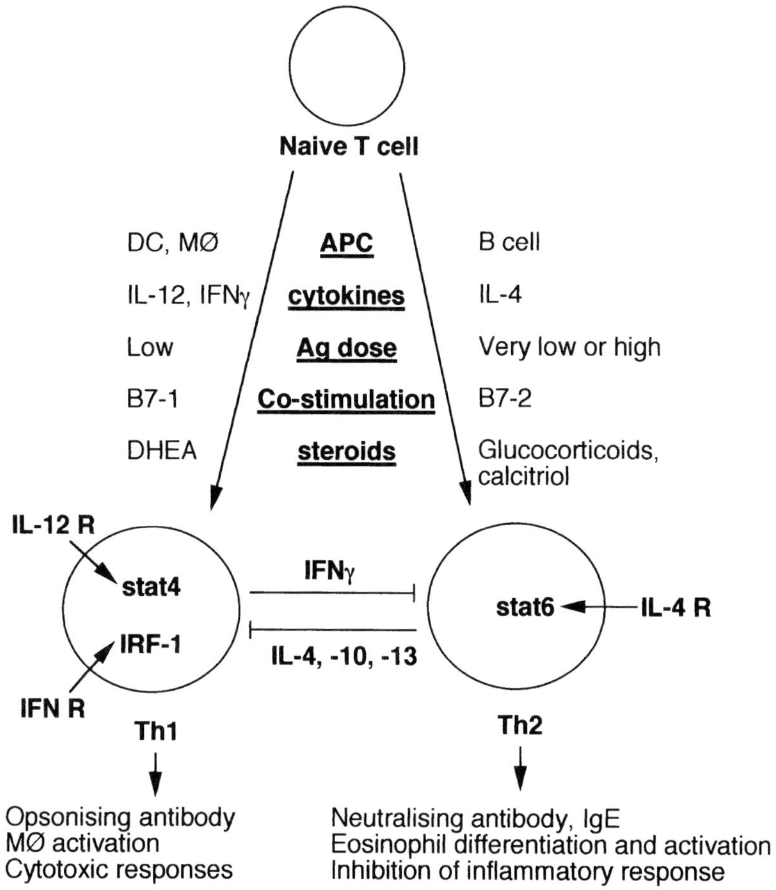

FIG. 1. Factors that regulate the development and effector functions of T helper (Th) 1 and Th2 cells. The most important factor for the induction of Th1 cells is the presence of interleukin (IL)-12 (produced by macrophages and dendritic cells), whereas IL-4 (produced by $CD4^+$ $NK1.1^+$ T cells, mast cells and basophils) is the most potent stimulus for the development of Th2 cells. The transcription factors Stat4 and IRF-1 are required for the development of Th1 cells, whereas Stat6 is needed for the development of Th2 cells. γ-Interferon (IFN-γ) inhibits the growth and effector functions of Th2 cells; IL-4, IL-10 and IL-13 inhibit Th1 cells. APC, antigen-presenting cell; DC, dendritic cell; DHEA, dehydroepiandrosterone; MØ, macrophage; R, receptor.

tuberculosis (Bellamy & Hill 1998, this volume). Newport et al (1996) identified a mutation in the IFN-γ receptor gene in a Maltese family with extreme susceptibility to infection with mycobacteria and other intracellular organisms. A host of other genetic defects lead to primary immune deficiencies. These rare genetic defects are beyond the scope of this review.

Asthma and atopy have been linked to genes on chromosome 11q13 (FcεRI), 5q31 (cytokine gene cluster) and 14q (TCR-α genes; Sandford et al 1996). Total serum IgE links to markers in chromosome 5q31.1, where the genes of IL-3, -4, -5, -9, -13, interferon regulatory factor 1 (IRF-1) and the β chain of the IL-12 receptor are clustered (Marsh et al 1994). The occurrence of atopic disease is also regulated by environmental factors. The prevalence of asthma has doubled in the western world over the last two decades. Increased exposure to environmental allergens such as house dust mite may contribute to the asthma epidemic. Several studies indicate that childhood infections may regulate cytokine responses and confer a degree of protection to atopy, while modern vaccination strategies induce predominantly Th2 responses (Shirakawa et al 1997). In Third World populations chronic intestinal parasite infestation has been shown to induce Th2 immune deviation. Bentwich et al (1995) proposed that chronic parasite-induced Th2 activation, manifested by increased secretion of Th2 cytokines, high serum immunoglobulin (especially IgE) levels and an eosinophilia, contributes to a greater susceptibility for HIV and to greater viral loads following infection. Parasite infestation also exacerbates asthma and in a recent interventional study, regular anti-helmintic treatment reduced IgE levels and led to clinical improvement of asthma (Lynch et al 1997).

Neuroendocrine factors have a powerful effect on immune responses. Populations stressed by war or natural disasters have an increased incidence of infections such as tuberculosis and typhus, but under these circumstances it is difficult to estimate the relative contributions of defective public health and of increased host susceptibility to infection. Bernton et al (1995) studied the immunological and endocrine changes in military recruits under conditions of ‘mental and physical stress approaching that found in combat’, and found raised cortisol levels and reduced DHEA/cortisol ratios in these recruits. Testosterone levels, delayed-type hypersensitive responses and T cell mitogenic responses decreased, while IgE levels increased. The authors suggested that stress induced a Th1 to Th2 shift.

Another steroid that modulates immune responses is the active metabolite of vitamin D, calcitriol, of which a deficiency enhances susceptibility to tuberculosis (Davies 1985). Calcitriol activates monocytes and stimulates cell-mediated immunity, and vitamin D metabolites enhance the killing of intracellular *M. tuberculosis* by monocytes (Rook et al 1986).

Signalling pathways

Signals from the TCR induce an array of different responses, depending on the stage of differentiation of T cells. In thymocytes weak signals via the TCR activate the MAP kinase cascade and lead to positive selection; whereas stronger signals, due to prolonged occupancy of individual receptors (the kinetic model) or possibly conformational changes (the allosteric model), activate additional pathways leading to negative selection (Alberola-Ila et al 1997). In mature T cells, TCR triggering induces proliferation, unresponsiveness (anergy) to subsequent stimulation or cell death (Alberola-Ila et al 1997). The mechanisms by which a particular TCR can induce different responses have been studied using altered peptide ligands (APLs), in which residues contacting the TCR have been altered. APLs can function as agonists, partial agonists or antagonists. In these models, APLs inducing anergy lead to incomplete phosphorylation of immunoreceptor tyrosine activation motifs (ITAMs) of the ζ chain and/or reduced signalling via ZAP-70 (Sloan-Lancaster & Allen 1996, Madrenas et al 1995).

The strongest factor regulating the polarization of naïve T cells to produce type 1 or type 2 cytokines is the presence or absence of IL-12 or IL-4. The IL-12 and IL-4 receptors associate with members of the JAK kinase family. Ligand binding activates the kinases, which phosphorylate the cytoplasmic domains of the cytokine receptors on tyrosine residues. Cytoplasmic Stat (signal transducer and activator of transcription) molecules bind to the phosphorylated receptors, become phosphorylated on tyrosine residues by JAK kinases and subsequently dimerize via their SH2 domains. IL-12 induces the activation of Stat1, Stat2 and Stat4, whereas IL-4 signals via Stat6 and phosphorylation of IRS-2. The Stat dimers translocate to the nucleus where they bind specific response elements and contribute to the expression of new genes. Knockout mice lacking IL-12 or Stat4 cannot mount efficient Th1 responses, whereas those lacking IL-4 or Stat6 cannot mount Th2 responses (Abbas et al 1996). Th2 cytokine expression is also regulated by the transcription factor c-Maf and GATA-3 (reviewed by Szabo et al 1997). The transcription factor IRF-1, which is induced by IFN-α, -β and -γ, plays a crucial role in the development of Th1 responses. IRF-1 is required for the production of IL-12 and inducible nitric oxide synthase (iNOS) by macrophages, as well as for the development of NK cells (Taki et al 1997, Lohoff et al 1997). IRF-1 therefore induces Th1 responses by regulating the function of cells of the innate immune system.

Responsiveness to cytokines can be regulated at the level of expression of cytokine receptors. Differentiated Th2 cells lose expression of IL-12 receptors and therefore become resistant to IL-12 (Abbas et al 1996). It is tempting to speculate that polymorphisms in the genes of cytokines (such as IL-4 and IL-12), of their receptors or of Stat proteins may modify Th1/Th2 responses.

Tuberculosis during childhood and adulthood

Children with tuberculosis usually present with lymphadenopathy or the complications thereof, or with systemic spread of the organism. In contrast, adults usually have pronounced systemic effects (such as weight loss and high fever) and immunopathology (such as cavitation and fibrosis). Adulthood disease ('secondary' tuberculosis) is usually ascribed to a secondary immune response following re-infection or re-activation (the Koch phenomenon). Other factors may however contribute to the differences in the immune response of children versus adults. We have seen a number of children presenting with culture proven childhood ('primary') tuberculosis, who were fully treated but developed culture proven tuberculosis again several years later, before the onset of puberty. These children were HIV negative and did not have any evidence for immune deficiency. On second presentation, they again had the childhood form of disease. Differences in the immune response between adults and children are also seen in viral diseases. Young children are more susceptible to viral infections than adults, but adults develop immunopathology, such as viral pneumonitis, more frequently than children during infection with measles or chicken pox. Factors contributing to the decreased immune responses in children include decreased chemotaxis and microbicidal activity of macrophages, decreased antigen presentation by dendritic cells, a larger proportion of naïve cells (requiring more efficient co-stimulation), reduced expression of CD40 ligand by activated neonatal T cells and a reduced secretion of cytokines other than IL-2 (Smith et al 1997).

Several studies showed that peripheral blood mononuclear cells (PBMC) from cord blood produce markedly reduced levels of IFN-γ following stimulation with mitogen *invitro* (Holt 1995, Smith et al 1997). Adult levels of IFN-γ production are only reached by the age of three to four years. Given that adrenal and sex steroids can modulate cytokine responses, we decided to compare the cytokine responses of children to those of adults following phytohaemagglutinin stimulation in serum-free medium. PBMC of healthy children (two to five years of age) produced barely detectable levels of IFN-γ, whereas PBMC from most adults produced large amounts. No significant differences were detected in IL-2, IL-4, IL-10 and TNF secretion. The decreased production of IFN-γ by PBMC from children is not only due to a higher ratio of naïve versus experienced T cells in infants, but also to other mechanisms including intrinsic defects of the T cells themselves and reduced APC function (Holt 1995, Smith et al 1997).

Immune responses in a high incidence community

The epidemiology and molecular epidemiology of tuberculosis have been studied in great detail in two adjacent suburbs of Cape Town: Ravensmead and Uitsig (Beyers et al 1996). These suburbs have a population of 34 000 with a high incidence of tuberculosis (1192 per 100 000). More than 90% of the population have been vaccinated with Bacillus Calmette–Guérin (BCG) and less than 2% are HIV positive. There was a remarkable variation in the incidence of tuberculosis (78 to 3150 per 100 000; Beyers et al 1996), socioeconomic status, literacy and crowding in enumerator subdistricts (A. van Rie, unpublished observations 1997). The incidence of tuberculosis correlated with socioeconomic status ($r = -0.57$, $p < 0.001$). Infestation with intestinal parasites, especially *Ascaris lumbricoides* and *Trichuris trichuria*, is rampant in deprived communities in the Western Cape, where up to 90% of children are infested with *A. lumbricoides*. Parasite infestation induces high polyclonal IgE responses and the mean serum IgE levels of 33 healthy children (<5 years) and 227 healthy adults from this community was 421 and 481 IU/ml, respectively. (Normal IgE levels in Caucasians are <100 IU/ml, and <200 IU/ml in the Coloured population of the Western Cape, the subjects of this study.) Median log IgE levels correlated with the incidence of tuberculosis in subdistricts ($r = 0.57$, $p < 0.001$; Fig.2) and with socioeconomic status ($r = -0.45$, $p < 0.001$). Although the association between IgE levels and tuberculosis incidence does not necessarily imply a causal relationship, we hypothesize that intestinal parasites and stress in this community, particularly in the most disadvantaged subdistricts, contribute to chronic Th2 stimulation, manifested by high IgE levels and an enhanced susceptibility to tuberculosis. In this community, we have previously shown that 34% of children exposed to an index case of

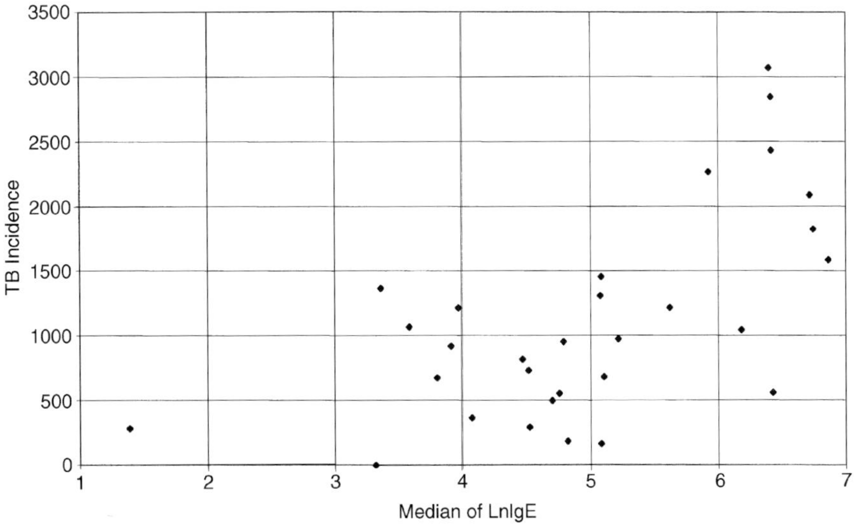

FIG. 2. Correlation of the incidence of tuberculosis (TB) in adults (>15 years of age) per enumerator subdistrict with median log IgE level per enumerator subdistrict of the Ravensmead and Uitsig suburbs of Cape Town ($r = 0.57$, $p < 0.001$, Spearman rank correlation).

tuberculosis in the household develop tuberculosis, compared to 5–10% in other studies (Beyers et al 1997).

In mouse models, the initial response to *M. tuberculosis* is type 1 and the response changes to a mixture of type 1 and type 2 during the course of disease. We were interested to measure how IgE levels, a surrogate marker for Th2 activation, changed during the course of tuberculosis. In each of 29 patients with tuberculosis, IgE levels at presentation were higher (mean = 463±86 IU/ml) than at two months following completion of treatment (mean = 166±30 IU/ml, $p < 0.0001$). Parasite loads have not been determined in these individuals, but standard tuberculosis therapy is not known to affect intestinal parasites. We hypothesize that chronic Th1 immune stimulation during the course of infection reduced the Th2 response driving IgE production in these individuals. If this is the case, a Th2 to Th1 immune modulation may explain why tuberculosis and asthma (both common diseases in the Western Cape) are rarely seen in the same patient (A. D. Beyers, R. Gie, N. Beyers, J. R. Joubert

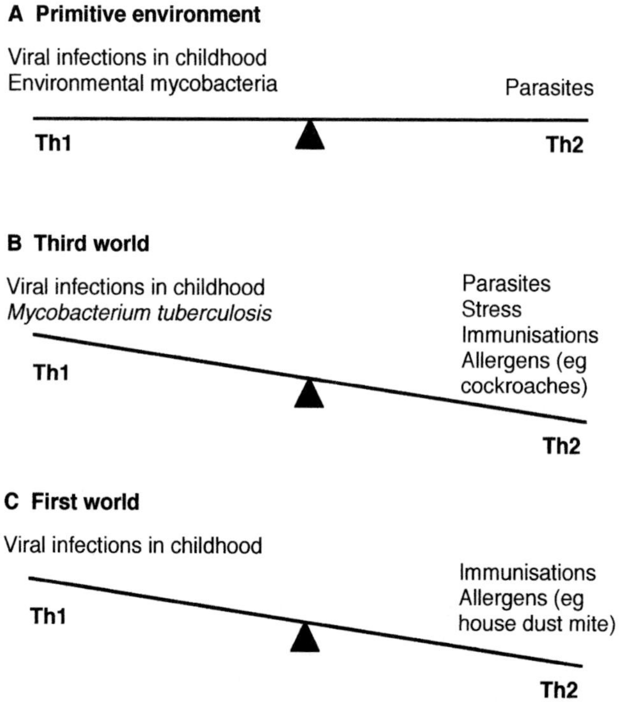

FIG. 3. A hypothetical view of factors regulating the balance between T helper (Th) 1 and Th2 immune responses in various communities.

& P. G. Bardin, unpublished observations 1997). Perhaps tuberculosis is good treatment for asthma!

In conclusion, PBMC from healthy children produce markedly less IFN-γ than those of adults, but no consistent differences are found in the production of other cytokines. It is tempting to propose that the immunopathology found in adults is not only due to a vigorous memory response (the Koch phenomenon), but also to age-related changes in cytokine secretion. Secondly, in a Western Cape community with rampant tuberculosis, disease incidence was linked not only to socioeconomic deprivation, but also to serum IgE levels. It remains to be determined if this association simply reflects higher parasite loads in the more disadvantaged subdistricts or whether it implies a causal relationship, i.e. whether chronic Th2 stimulation predisposes a community to tuberculosis (Fig. 3). Thirdly, the decrease in IgE levels following tuberculosis may be due to chronic Th1 stimulation during disease. This finding supports the argument that effective anti-tuberculosis vaccination, inducing a long-lasting Th1 response, might have beneficial effects such as a reduction in atopy.

Acknowledgements

A.D.B. is a Wellcome Trust Senior Research Fellow in South Africa. The work was supported by the Glaxo Wellcome Action TB initiative.

References

Abbas AK, Murphy KM, Sher A 1996 Functional diversity of helper T lymphocytes. Nature 383:787–793

Actor JK, Shirai M, Kullberg MC, Buller RM, Sher A, Berzofsky JA 1993 Helminth infection results in decreased virus-specific $CD8^+$ cytotoxic T-cell and Th1 cytokine responses as well as delayed virus clearance. Proc Natl Acad Sci USA 90:948–952

Alberola-Ila J, Takaki S, Kerner JD, Perlmutter RM 1997 Differential signaling by lymphocyte antigen receptors. Annu Rev Immunol 15:125–154

Allen JE, Maizels RM 1997 Th1–Th2: reliable paradigm or dangerous dogma? Immunol Today 18:387–392

Bellamy RJ, Hill AVS 1998 Genetics of susceptibility and resistance to tuberculosis. In: Genetics and tuberculosis. Wiley, Chichester (Novartis Found Symp 217) p 3–23

Bentwich Z, Kalinkovich A, Weisman Z 1995 Immune activation is a dominant factor in the pathogenesis of African AIDS. Immunol Today 16:187–191

Bernton E, Hoover D, Galloway R, Popp K 1995 Adaptation to chronic stress in military trainees. Adrenal androgens, testosterone, glucocorticoids, IGF-1, and immune function. Ann N Y Acad Sci 774:217–231

Beyers N, Gie RP, Zietsman HL et al 1996 The use of a geographical information system (GIS) to evaluate the distribution of tuberculosis in a high-incidence community. S Afr Med J 86:40–44

Beyers N, Gie RP, Schaaf HS et al 1997 A prospective evaluation of children under the age of five years living in the same household as adults with recently diagnosed pulmonary tuberculosis. Int J Tuberc Lung Dis 1:38–43
Constant SL, Bottomly K 1997 Induction of Th1 and Th2 $CD4^+$ T cell responses: the alternative approaches. Annu Rev Immunol 15:297–322
Davies PD 1985 A possible link between vitamin D deficiency and impaired host defence to *Mycobacterium tuberculosis*. Tubercle 66:301–306
Daynes RA, Meikle AW, Araneo BA 1991 Locally active steroid hormones may facilitate compartmentalization of immunity by regulating the types of lymphokines produced by helper T cells. Res Immunol 142:40–45
Edwards CK, Hedegaard HB, Zlotnik A, Gangadharam PR, Johnston RB, Pabst MJ 1986 Chronic infection due to *Mycobacterium intracellulare* in mice: association with macrophage release of prostaglandin E_2 and reversal by injection of indomethacin, muramyl dipeptide, or interferon-gamma. J Immunol 136:1820–1827
Finkelman FD, Shea-Donohue T, Goldhill J et al 1997 Cytokine regulation of host defence against parasitic gastrointestinal nematodes: lessons from studies with rodent models. Annu Rev Immunol 15:505–533
Gibbs JH, Grange JM, Beck JS et al 1991 Early delayed hypersensitivity responses in tuberculin skin tests after heavy occupational exposure to tuberculosis. J Clin Pathol 44:919–923
Gorham JD, Guler ML, Steen RG et al 1996 Genetic mapping of a murine locus controlling development of T helper 1/T helper 2 type responses. Proc Natl Acad Sci USA 93:12467–12472
Grewal IS, Borrow P, Pamer EG, Oldstone MBA, Flavell RA 1997 The CD40–CD154 system in anti-infective host defense. Curr Opin Immunol 9:491–497
Hernandez-Pando R, Rook GAW 1994 The role of TNFα in T-cell mediated inflammation depends on the Th1/Th2 cytokine balance. Immunology 82:591–595
Holt PG 1995 Postnatal maturation of immune competence during infancy and childhood. Pediatr Allergy Immunol 6:59–70
Kanagawa O, Vaupel BA, Gayama S, Koehler G, Kopf M 1993 Resistance of mice deficient in IL-4 to retrovirus-induced immunodeficiency syndrome (MAIDS). Science 262:240–242
Kelso A 1995 Th1 and Th2 subsets: paradigms lost? Immunol Today 16:374–379
Lohoff M, Ferrick D, Mittrucker HW et al 1997 Interferon regulatory factor-1 is required for a T helper 1 immune response *in vivo*. Immunity 6:681–689
Lynch NR, Palenque M, Hagel I, Diprisco MC 1997 Clinical improvement of asthma after antihelminthic treatment in a tropical situation. Am J Respir Crit Care Med 156:50–54
Madrenas J, Wange RL, Wang JL, Isakov N, Samelson LE, Germain RN 1995 Zeta phosphorylation without ZAP-70 activation induced by TCR antagonists or partial agonists. Science 267:515–518
Marsh DG, Neely JD, Breazeale DR et al 1994 Linkage analysis of IL4 and other chromosome 5q31.1 markers and total serum immunoglobulin E concentrations. Science 264:1152–1156
Matzinger P 1994 Tolerance, danger, and the extended family. Annu Rev Immunol 12:991–1045
Newport MJ, Huxley CM, Huston S et al 1996 A mutation in the interferon-gamma-receptor gene and susceptibility to mycobacterial infection. N Engl J Med 335:1941–1949
Pearlman E, Kazura JW, Hazlett FE, Boom WH 1993 Modulation of murine cytokine responses to mycobacterial antigens by helminth-induced T helper 2 cell responses. J Immunol 151:4857–4864
Ridge JP, Fuchs EJ, Matzinger P 1996 Neonatal tolerance revisited: turning on newborn T cells with dendritic cells. Science 271:1723–1726
Rook GAW, Hernandez-Pando R 1998 Immunological and endocrinological characteristics of tuberculosis that provide opportunities for immunotherapeutic intervention. In: Genetics and tuberculosis. Wiley, Chichester (Novartis Found Symp 217) p 73–98

Rook GA, Steele J, Fraher L et al 1986 Vitamin D_3, gamma interferon, and control of proliferation of *Mycobacterium tuberculosis* by human monocytes. Immunology 57:159–163
Sandford A, Weir T, Pare P 1996 The genetics of asthma. Am J Respir Crit Care Med 153:1749–1765
Seder RA, Paul WE 1994 Acquisition of lymphokine-producing phenotype by $CD4^+$ T cells. Annu Rev Immunol 12:635–673
Shirakawa T, Enomoto T, Shimazu S, Hopkin JM 1997 The inverse association between tuberculin responses and atopic disorder. Science 275:77–79
Sloan-Lancaster J, Allen PM 1996 Altered peptide ligand-induced partial T cell activation: molecular mechanisms and role in T cell biology. Annu Rev Immunol 14:1–27
Smith S, Jacobs RF, Wilson CB 1997 Immunobiology of childhood tuberculosis: a window on the ontogeny of cellular immunity. J Pediatr 131:16–26
Szabo SJ, Glimcher LH, Ho I-C 1997 Genes that regulate interleukin 4 expression in T cells. Curr Opin Immunol 9:776–781
Taki S, Sato T, Ogasawara K et al 1997 Multistage regulation of Th1-type immune responses by the transcription factor IRF-1. Immunity 6:673–679
Weiner HL, Friedman A, Miller A et al 1994 Oral tolerance: immunologic mechanisms and treatment of animal and human organ-specific autoimmune diseases by oral administration of autoantigens. Annu Rev Immunol 12:809–837

DISCUSSION

Rook: I would like to mention that there is at least one paper in which specific IgE against *Mycobacterium tuberculosis* itself has been measured in tuberculosis (Yong et al 1989). They showed that patients with tuberculosis do have specific IgE antibodies against *M. tuberculosis*, which is a T helper (Th) 2 cytokine-dependent phenomenon.

Beyers: Yong et al (1989) found that in total serum IgE levels in patients with tuberculosis were 968 U/ml compared to the controls, which were 122 U/ml and 237 U/ml. The implications are not clear, but perhaps we shouldn't look at tuberculosis in isolation from other immunological factors that may also influence Third World communities.

Anderson: Did you divide the patients who were being treated for tuberculosis, and in whom you measured IgE level, into two groups, and treat one with anti-helmintics?

Beyers: No, and a limitation in the interpretation of the present data is that we did not examine stool samples for parasites on presentation and after treatment.

Anderson: There is a lot of information in the literature about how IgE levels change with age, according to the intensity of intestinal helminth infection. It is clear that following anti-helmintic treatment there are major perturbations in IgE levels. Unfortunately, if you were going to do this sort of study then you would need a large sample size because the parasitic burden per host is highly variable.

Beyers: I agree that a large sample size is needed.

Ryffel: Did you look at specific IgG isotypes in addition to IgE? Because this would give further support of a preferential Th1 versus Th2 response.

Beyers: No, we have not done that.

Ress: We have a tuberculosis epidemic and high levels of asthma, but we don't see at the epidemiological level the inverse relationship that was described in the Japanese study (Shirakawa et al 1997). We don't know whether these asthmatics, many of whom are receiving corticosteroid treatments, are more susceptible to tuberculosis. Can anyone comment on this?

Bateman: I don't know of a study that has looked at the coexistence of asthma and tuberculosis. The Japanese study was performed in a low prevalence area, and they only measured purified protein derivative of tuberculin responsiveness. We must ensure that we are examining the same phenomenon, because to extrapolate directly is dangerous. If tuberculosis protects against asthma there is no evidence for it in the Western Cape. One of the best epidemiological assessments of asthma prevalence suggests that asthma in 9–13-year-old children of the Cape Flats is as high as 13%, which is probably one of the highest figures for any region. It is possible that the social risks for both of these diseases are high and operate in parallel. An inverse relationship is not apparent.

Beyers: Clinicians at the University of Stellenbosch working in asthma and tuberculosis clinics have the impression that it's rare to find these two diseases coexisting in one patient, even though both diseases are prevalent in the population. Do you have a similar impression?

Bateman: No. We've studied other respiratory diseases in our clinic, and found that tuberculosis is common in patients with several forms of lung disease. I can't comment on asthma.

Kaplan: You showed that children with lower levels of γ-interferon (IFN-γ) had the same levels of interleukin 10. Was this in normal individuals?

Beyers: These studies were done on tuberculin skin test-positive individuals with no history of tuberculosis and with normal chest X-rays. However, the individuals came from the deprived Ravensmead/Uitsig community.

Kaplan: Have you looked in individuals with tuberculosis, and would you get the same response in an antigen-specific assay?

Beyers: No, but in the future we will look at antigen-specific responses using whole blood cultures, as described by Weir et al (1994).

Kaplan: Your data suggest that there's an age where children and young adults are less susceptible to tuberculosis than mature adults. Is this because their immune response is reversed?

Beyers: We can only hypothesize at this stage. It's difficult to reconcile lower IFN-γ levels with the 'protected' time period. The low production of IFN-γ may be limited to children under the age of five years. If children produce less IFN-γ following specific antigen stimulation during the golden age period, we have to re-think our ideas about the way the immune system is handling *M. tuberculosis*. We still cannot be sure that an excessive immune response, including excessive IFN-γ production, is not contributing to immunopathology in adults.

References

Shirakawa T, Enomoto T, Shimazu S, Hopkin JM 1997 The inverse association between tuberculin responses and atopic disorder. Science 275:77–79

Weir RE, Morgan AR, Britton WJ, Butlin CR, Dockrell HM 1994 Development of a whole-blood assay to measure T cell responses to leprosy: a new tool for immunoepidemiological field studies of leprosy and immunity. J Immunol Methods 176:93–101

Yong AJ, Grange JM, Tee RD et al 1989 Total and anti-mycobacterial IgE levels in serum from patients with tuberculosis and leprosy. Tubercle 70:273–279

Analysis of the genome of *Mycobacterium tuberculosis* H37Rv

S. T. Cole and *B. G. Barrell

*Unité de Génétique Moléculaire Bactérienne, Institut Pasteur, 28 Rue du Dr Roux, 75724 Paris Cedex 15, France and *The Sanger Centre, Wellcome Trust Genome Campus, Hinxton CB10 1IL, UK*

Abstract. The powerful combination of genomics and bioinformatics is providing a wealth of information about *Mycobacterium tuberculosis*, the aetiological agent of human tuberculosis, that will facilitate the conception and development of new therapies. The starting point for genome sequencing was the integrated map of the 4.4 Mb circular chromosome of the widely used, virulent reference strain, *M. tuberculosis* H37Rv. Cosmids and bacterial artificial chromosomes were selected from ordered libraries and subjected to systematic shotgun sequence analysis. This approach simplified sequence assembly as the genome is rich in repetitive DNA. In common with most bacteria, >90% of the potential coding capacity is used, and probable or tentative functions could be attributed to >70% of the genes. The potential biological roles of two of the principal driving forces in genome dynamics, insertion sequence elements and polymorphic multigene families are discussed.

1998 Genetics and tuberculosis. Wiley, Chichester (Novartis Foundation Symposium 217) p 160–177

One of the most significant and far-reaching events in the history of molecular biology was undoubtedly the deciphering of the genetic code in the 1960s, and the subsequent demonstration of its ubiquitous use by all forms of life, ranging from relatively simple micro-organisms such as bacteria to complex mammals like humans. This ability to read the message encoded in the genes enabled the predictions of the 'central dogma', the flow of information from DNA, via RNA to proteins, to be tested experimentally. Thanks to striking technological developments in the fields of DNA sequencing and bioinformatics, it is now a relatively straightforward task to obtain the sequences of large numbers of genes rapidly or, in the case of a bacterium like *Mycobacterium tuberculosis*, to sequence the whole genome. The enormous progress made in computer-assisted database searching during the last few years, coupled with the ability to perform reverse genetics, means that the functions of the majority of the genes can now be predicted *in silico* then tested in the laboratory.

Sequencing strategies

Progress in genome sequencing has been so phenomenal that it is probable that the complete genetic repertoires of all of the leading bacterial pathogens will be available before the year 2000. One of the major reasons for using a genomic approach to study a bacterium is its cost-effectiveness, because essentially everything that we need to know about an organism, from its biology to its behaviour, is encrypted in its genome. Thus, a vast body of information can be obtained for a relatively modest, single investment. Two different but equally effective strategies have been employed to obtain bacterial genome sequences, the whole genome shotgun approach (Fleischmann et al 1995), in which the complete genome sequence is reassembled from the sequences of a large number of small fragments, and the systematic sequencing of large insert (20–45kb) clones from ordered cosmid or bacteriophage libraries selected from a sequence ready contig map (Blattner et al 1997). Both methods have their strengths and weaknesses (Table 1). The principal advantages of the whole genome shotgun method, which involves the cloning of small chromosomal fragments (<2kb) directly into sequencing vectors, are its randomness, due to the ability to avoid lethal expression of toxic genes, and, above all, the reduced expense. One major disadvantage of this approach is the fact that useful sequence only becomes available in the final stages of the project. By contrast, the ordered clone-based method regularly generates packages of fully analysed, exploitable information, but it is more costly as the mapping, an essential prerequisite, can be labour intensive and time consuming. A detailed genome map is a powerful tool, however, as it aids in clone selection, enables regions harbouring repetitive DNA to be identified and isolated, and allows progress to be monitored and sequence assembly to be verified independently. An added benefit of this strategy is the availability of large insert clones, immortalized sources of DNA, for use in future genetic and biological experiments.

The *Mycobacterium tuberculosis* H37Rv genome project

An efficient compromise, which has been used in the *M. tuberculosis* H37Rv genome sequencing project, is to combine both approaches, by fully sequencing selected cosmids (Philipp et al 1996a) or bacterial artificial chromosome (BAC) clones containing large inserts (Brosch et al 1998), and by sequencing a large number of clones derived from a whole genome shotgun library that gives two- to threefold genomic coverage. In parallel, end sequencing of the inserts present in several thousand cosmid or BAC clones is undertaken. In this way, one not only generates data rapidly and efficiently but

TABLE 1 Strategies for microbial genome sequencing

Strategy	*Insert size*	*Cloning*	*Mapping*	*Problems*	*Useful sequence*	*Side-products*
Whole genome shotgun	Small (<2 kb)	Simple	Not required	Genome, complexity, repetitive DNA, occasional unclonability	Late	Small insert clones for SAGE
Directed or clone based	Large (>35 kb)	Complex	Labour intensive but valuable	Unclonability, incomplete genome coverage	Early, constant delivery	Small insert clones for SAGE. Large insert clones for biology, genetics, etc.

SAGE, systematic analysis of gene expression.

also benefits from an in-built topological check on shotgun sequence assembly provided by the end sequences. This is particularly important for genomes rich in repetitive DNA, as is true for *M. tuberculosis*. In the final stages of sequencing, the BAC clones in particular serve as templates for gap filling as they generally carry those regions of the genome which are underrepresented or missing from cosmid or plasmid libraries. This was indeed the case with H37Rv as several loci that were apparently unclonable in multicopy cosmids, as *Sau*3AI partial digest fragments, or in small insert libraries containing randomly sheared fragments, were fully covered in the pBeloBac11 bank that carries large inserts generated by partial digestion with *Hin*dIII (Brosch et al 1998). From the ~5000 clones initially isolated a canonical set of 68 BACs has been established that covers essentially the complete genome. This readily manageable collection of stable clones represents a valuable resource for the future, and should find use in applications as diverse as comparative genome mapping and the systematic analysis of gene expression (SAGE).

At the time of writing, the sequencing of the H37Rv genome is entering the final phase. There are currently about 15 sequence contigs, ranging in size from 1.2 to 1540kb, that are all linked to their neighbours by BAC or cosmid clones. The total contig length is 4393kb, which is close to the genome size of ~4400kb estimated for the circular chromosome of H37Rv by pulsed field gel electrophoresis, thus indicating that the remaining gaps are probably small. Intensive efforts are now being made to close these and to analyse and annotate the composite sequence. From our preliminary inspection of completed cosmid sequences it is clear that *M. tuberculosis,* unlike the leprosy bacillus (Cole 1994, Honoré et al 1993), has densely packed coding regions and the genome is likely to comprise ~4000 protein-coding sequences. Genes are identified by a combination of methods involving positional base preference, codon usage, pattern recognition, and similarity to known genes or their products. Database searches have led to the attribution of precise, or putative, functions to over 70% of the genes and, as in other genome sequencing projects, it is a matter of great interest to determine the biological roles of the remaining 30% that have no counterparts in other bacteria. Much insight into the biochemistry, general metabolism and physiology of *M. tuberculosis* has been obtained from our initial analysis of the large body of sequence data, and many new leads for chemotherapy and immunoprophylaxis have been highlighted, but these will not be discussed here. Instead, we will concentrate on two salient features: (1) the possible role of insertion sequences in genome dynamics; and (2) the potential significance of two dispersed repetitive DNA sequences, i.e. the major polymorphic tandem repeat (MPTR; Hermans et al 1992) and the polymorphic GC-rich sequence (PGRS; Poulet & Cole 1995a, Ross et al 1992), that both belong to multigene families encoding acidic proteins.

Genome dynamics

One of the most striking properties of members of the *M. tuberculosis* complex is the remarkable lack of genomic diversity observed at the nucleotide level and the rarity of synonymous or silent (third position) codon changes. This is important in the context of immunity and vaccine development as it means that the majority of the proteins should be exactly the same in all strains and, as a consequence, the potential for antigenic drift is restricted. Based on the systematic sequence analysis of 26 loci in a large number of independent isolates of the tubercle bacillus (Kapur et al 1994, Sreevatsan et al 1997), Musser and his colleagues have concluded that the genome of *M. tuberculosis* is either unusually inert for a bacterium or that the organism is relatively young in evolutionary terms, i.e. <15 000 years old. While this hypothesis is seductive, and consistent with both the period when the domestication of cattle occurred, (i.e. the possible conversion of the host range of the tubercle bacillus from bovines to humans) and the archaeological record (Salo et al 1994), other factors that might contribute to the remarkable lack of genetic diversity are the inefficiency of homologous recombination that has been observed (Kalpana et al 1991) and the long generation time. Alternatively, it is possible that, as a result of long exposure to the potentially mutagenic environment of the macrophage, the DNA repair system may be outstandingly efficient, although the fact that drug-resistant mutants emerge with the same frequency (10^{-6}–10^{-7}) in *M. tuberculosis* as in *Escherichia coli* argues against this (Sreevatsan et al 1997).

During the last five years, a large body of evidence has emerged indicating that the principal source of genomic polymorphism in *M. tuberculosis* is due to the transposition of an endogenous insertion sequence, IS6110, commonly used for typing purposes (Hermans et al 1990, McAdam et al 1990, Otal et al 1991, van Embden et al 1993), or, somewhat less frequently, to alterations of the PGRS or MPTR loci. It was known from mapping studies that 16 copies of IS6110 and six copies of the more stable element, IS1081, resided within the genome of H37Rv (Philipp et al 1996a), but scrutiny of the genomic sequences led to many more insertion sequence elements being discovered. In addition to these, >12 novel insertion sequences have been identified by sequence comparison, as well as a smaller number of repetitive sequences that bear some of the hallmarks of mobile genetic elements (Fig. 1). At least two prophages have been detected and more may be present. As one of the prophages and one of the new insertion sequences are missing from the genomes of recent clinical isolates of *M. tuberculosis* (Mahairas et al 1996, Philipp et al 1996b, S.T. Cole & B.G. Barrell, unpublished data 1997), this could indicate that horizontal transfer of genetic material occurs in nature and that, despite its intracellular niche, the tubercle bacillus might be able to acquire new genes if they were carried by mycobacteriophages or transposable elements. This is also consistent with the findings of a comparative mapping study (Philipp et al 1996b), that insertions/deletions are probably the principal source of genomic diversity within the *M. tuberculosis* complex. The distribution of these elements on the chromosome is particularly informative as there appears to have been a selection against insertions in the quadrant encompassing the origin of replication (Fig. 1). Strikingly, in a recent transposon mutagenesis study a similar bias was observed (Bardarov et al 1997), again suggesting that this chromosomal region may be less permissive to insertional events.

PGRS and MPTR are members of the PE and PPE multigene families

During a number of molecular epidemiological studies performed to investigate the relatedness of strains of the tubercle bacillus two families of repetitive DNA were detected. MPTR (Hermans et al 1992) was believed to consist of numerous

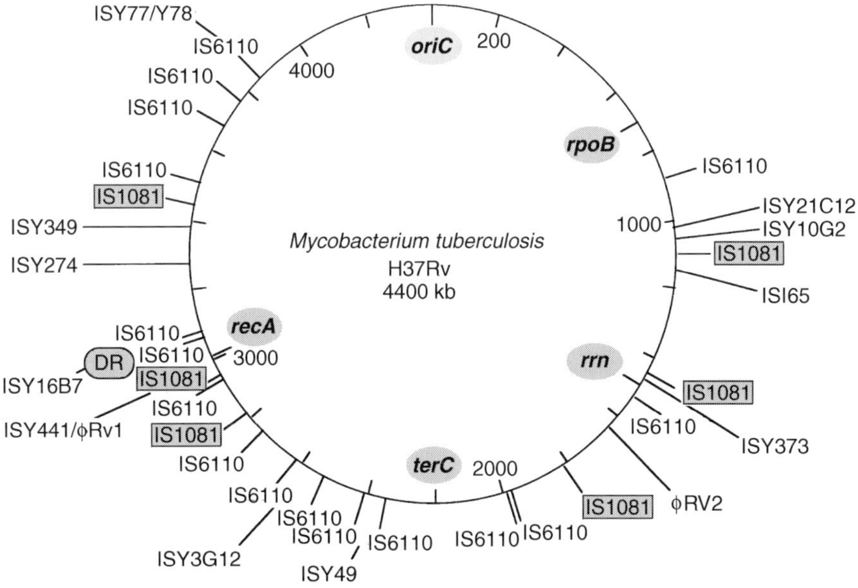

FIG. 1. Distribution of mapped and sequenced insertion sequence elements on the genome of *Mycobacterium tuberculosis* H37Rv. The positions of the previously mapped IS6110 and IS1081 are shown. Based on analysis of about 70% of the genome sequence, new insertion sequence elements and prophages were identified and their approximate location determined with respect to the integrated map (Philipp et al 1996a). Insertion sequence elements are provisionally referred to by the corresponding cosmid name and the prophages designated as ϕRv1 and ϕRv2. The locations of a few genetic markers and the direct repeat (DR) region harbouring the progenital copy of IS6110 are indicated.

tandem copies of the sequence GCCGGTGTTG (or its complement) separated by 5 bp spacers, whereas PGRS (Poulet & Cole 1995a,b) comprised multiple copies of the motif CGGCGGCAA. Initial mapping work with H37Rv suggested that at least 26 loci harboured PGRS elements (Philipp et al 1996a, Poulet & Cole 1995a), whereas MPTR appeared to be even more abundant (Hermans et al 1992). Although it was initially felt that both PGRS and MPTR corresponded to families of dispersed, non-coding repeats, it is now clear that they belong to multigene families encoding proteins that are rich in Gly, Ala and Asn, and to a lesser extent Ser and Thr (data not shown). Multiple sequence alignments performed with the putative PGRS and MPTR proteins revealed that they were members of two larger protein families that are now referred to as the PE and PPE families, respectively, and that these shared a common organization.

Members of the PE protein family all have a highly conserved N-terminal domain of ~110 amino acid residues that is followed by a C-terminal segment which varies in size, sequence and repeat copy number (Fig. 2). The name PE derives from the fact that the motif Pro-Glu (PE) is found in almost all cases at positions 8 and 9 and, based on preliminary analysis of about 70% of the genome sequence, the family is expected to contain ~80 members. Phylogenetic analysis of the degrees of relatedness of the members of the PE family shows them to fall into two groups, the larger of which contains roughly 65% of proteins carrying multiple copies of Gly-Gly-Ala repeats, corresponding to the PGRS motif, or Gly-Gly-Asn repeats, whereas members of the other group share limited sequence similarity in their C-terminal domains (Fig. 3). The predicted molecular weights of the PE proteins, which are all acidic, vary considerably as a few members only contain the ~110 amino acid N-terminal domain while the majority have C-terminal extensions ranging in size from 100 to >500 residues.

Like the PE family, the PPE protein family (Fig. 2) also has a conserved N-terminal domain that comprises ~180 amino acid residues and has a Pro-Pro-Glu (PPE) motif at positions 7 to 9 followed by C-terminal segments that vary considerably in sequence and length. The PPE family of acidic proteins (mean pI = 4.4) is expected to contain >20 members and these fall into at least three groups, one of which constitutes the MPTR class that is characterized by the presence of multiple, tandem copies of the motif Asn-X-Gly-X-Gly-Asn-X-Gly. The second subgroup contains a characteristic, well-conserved motif around position 350 (Gly-X-X-Ser-Val-Pro-X-X-Trp), whereas the other group contains proteins that are unrelated except for the presence of the common 180-residue PPE domain.

At the time of writing, there is little information concerning the significance, subcellular location and biological roles of the PE and PPE protein families which, if expressed together, would represent, numerically, about 2% of the protein species present in *M. tuberculosis*. Genes homologous to PGRS and

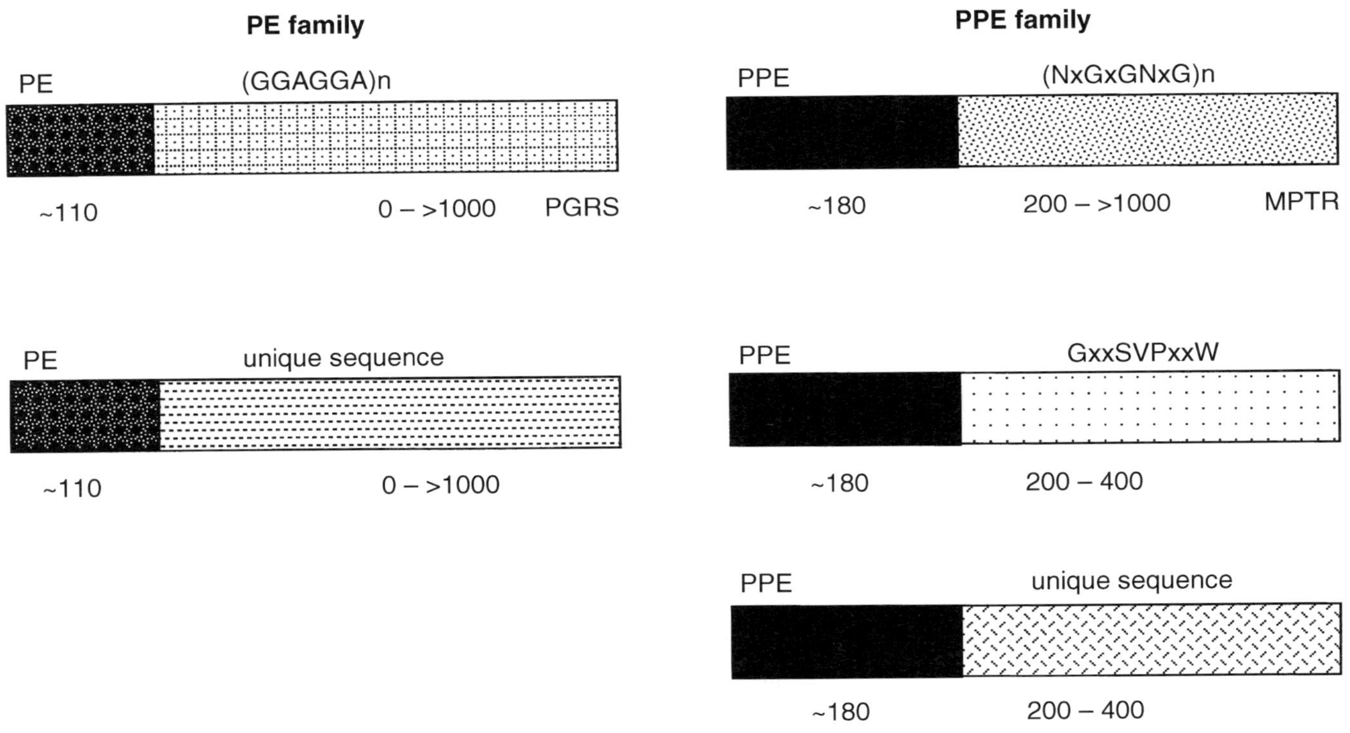

FIG. 2. Schematic summary of the structures of proteins belonging to the PE and PPE families. MPTR, major polymorphic tandem repeat; PGRS, polymorphic GC-rich sequence.

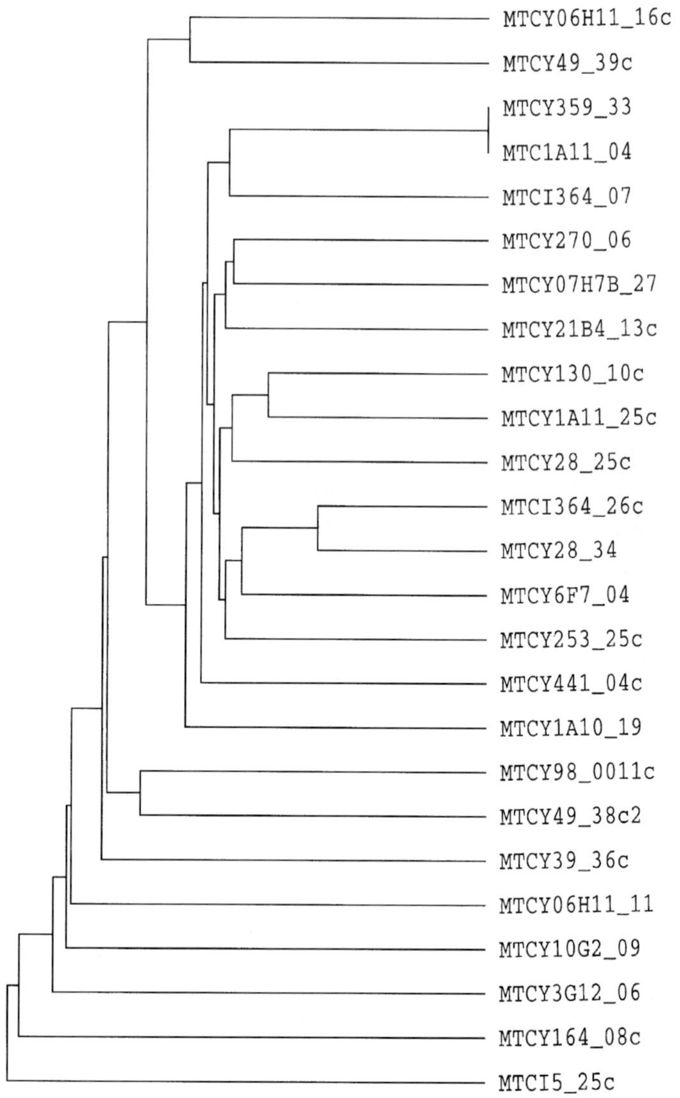

FIG. 3. Phylogenetic tree of selected members of the PE family. The tree was deduced from a multiple sequence alignment and relational analysis and proteins are referred to by the cosmid identifiers used throughout the project. Note the tight clustering of the polymorphic GC-rich sequence (PGRS) members, MTCY06H11.16c through MTCY1A10.19, compared to the other PE proteins, MTCY98.0011c through MTCI5.25c, which have unique sequences in the C-terminal domain.

MPTR-coding sequences have been detected in *Mycobacterium gordonae*, *Mycobacterium kansasii* and *Mycobacterium marinum* (Poulet & Cole 1995a,b, Ross et al 1992), and in *Mycobacterium asiaticum*, *Mycobacterium gastri* and *Mycobacterium szulgai*, respectively (Hermans et al 1992). On examination of the protein database from the extensively sequenced *Mycobacterium leprae*, no PGRS- or MPTR-related polypeptides were detected, but some proteins belonging to the non-MPTR subgroup of the PPE family were found. These include the serine-rich antigen described by Stoker et al (Vega-Lopez et al 1993) who identified the corresponding gene in an expression library using serum from lepromatous leprosy patients. Immunological studies with a recombinant form of the protein showed that sera from the majority of lepromatous and tuberculoid patients recognized this protein, suggesting that it corresponds to a major antigen.

Antigenic variation in *Mycobacterium tuberculosis*?

While it is still too early to attribute biological functions to the PE and PPE families, it is tempting to speculate that they could be important protein antigens that may represent a source of antigenic variation. A number of observations support this contention. It is known that one PGRS member, WHO antigen 22T, a 55 kDa protein present in culture filtrate and cell extracts capable of binding fibronectin (Abou-Zeid et al 1991), is produced during infection and disease, and elicits an antibody response. When sera from 14 different tuberculosis patients were screened for reactivity to a recombinant form of 22T, strongly positive responses were observed in eight cases, suggesting that either individuals mount different immune responses or that this PGRS protein may not be produced by all strains of *M. tuberculosis*. Indirect support in favour of the latter is provided by Southern blotting studies with a PGRS-specific probe as striking differences in the hybridization profiles were observed (Fig. 4).

Given the remarkably low level of nucleotide sequence variation in structural genes (Sreevatsan et al 1997), and hence the high degree of conservation of restriction sites, it is probable that the restriction fragment length polymorphisms observed with PGRS genes, as well as with MPTR genes, reflect physical differences such as repeat length or copy number. This could result from inter- or intragenic recombinational events between the repetitive sequences comprising the corresponding coding sequences or strand slippage during replication, as discussed previously for the PGRS members (Poulet & Cole 1995a,b). It is probably significant in this respect that the genes encoding PE and PPE proteins often occur in clusters on the chromosome of H37Rv.

Antigenic variation among the surface proteins of many intra- and extracellular pathogens is commonly observed, presumably reflecting the selective pressure of the host's immune system. This has been well documented for both eukaryotes

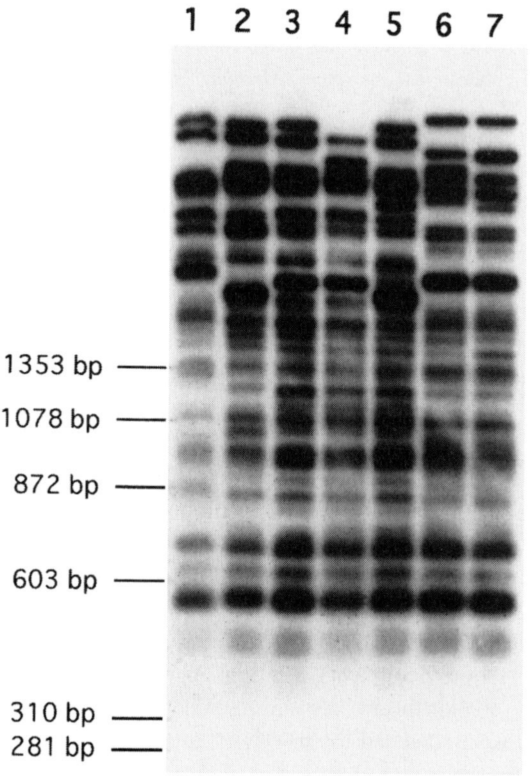

FIG. 4. Restriction fragment length polymorphism analysis of *Mycobacterium tuberculosis* using a polymorphic GC-rich sequence (PGRS)-specific probe hybridized to *Alu*I digestions. Lane 1: *M. tuberculosis* H37Rv. Lane 2: strain 161. Lane 3: strain 133. Lane 4: strain 253. Lane 5: strain 272. Lane 6: strain 164. Lane 7: strain 205. The positions of size markers used to calibrate the gel are shown.

such as trypanosomes, and prokaryotes, well exemplified by *Neisseria gonorrhoeae* or *Haemophilus influenzae*. In one of the best studied cases, the pili of *Neisseria* spp., antigenic variation often involves recombinational events in which previously silent coding sequences are inserted into expression sites (Haas et al 1992, Robertson & Meyer 1992). In *N. gonorrhoeae* and *H. influenzae*, other means of achieving antigenic variation include RecA-independent recombination, or replication errors such as strand slippage, within simple sequence (penta- or tetranucleotide) repeats in genes coding for surface proteins, thereby culminating in the use of alternative reading frames (Hood et al 1996). Recombination between copies of tandemly arranged, repetitive motifs that code for multiple surface-exposed domains of the M-protein of Streptococci is another well characterized mechanism of antigenic variability (Robertson & Meyer 1992). Given the many potential parallels between these systems and the PE/PPE structures, it is important that further studies be performed to investigate the molecular basis of the polymorphism associated with the genes comprising the PPE and PE families. If genetic alterations led to significant antigenic variability they could be important for understanding protective immunity in tuberculosis and explain the varied responses seen in different Bacillus Calmette–Guérin vaccination programmes (Bloom & Fine 1994).

Acknowledgements

Important contributions to the work described here were made by R. Brosch, K. Eiglmeier, T. Garnier, S.V. Gordon, S. Poulet, C. Churcher, D. Harris, K. Badcock, D. Basham, D. Brown, T. Chillingworth, R. Connor, R. Davies, K. Devlin, T. Feltwell, S. Holroyd, S. Gentles, K. Jagels, J. McLean, S. Moule, L. Murphy, K. Oliver, J. Osborne, J. Parkhill, M. Quail, M-A. Rajandream, J. Rogers, S. Rutter, K. Seeger, J. Skelton, R. Squares, S. Squares, J. Sulston, K. Taylor and S. Whitehead. Financial support from the Wellcome Trust and the Association Française Raoul Follereau is gratefully acknowledged.

References

Abou-Zeid C, Garbe T, Lathigra R et al 1991 Genetic and immunological analysis of *Mycobacterium tuberculosis* fibronectin-binding proteins. Infect Immun 59:2712–2718

Bardarov S, Kriakov J, Carriere C et al 1997 Conditionally replicating mycobacteriophages: a system for transposon delivery to *Mycobacterium tuberculosis*. Proc Natl Acad Sci USA 94:10961–10966

Blattner FR, Plunkett IG, Bloch CA et al 1997 The complete genome sequence of *Escherichia coli* K-12. Science 277:1453–1462

Bloom BR, Fine PEM 1994 The BCG experience: implications for future vaccines against tuberculosis. In: Bloom BR (ed) Tuberculosis: pathogenesis, protection, and control. American Society for Microbiology, Washington DC, p 531–557

Brosch R, Gordon SV, Billault A et al 1998 Use of a *Mycobacterium tuberculosis* H37Rv bacterial artificial chromosome (BAC) library for genome mapping sequencing and comparative genomics. Infect Immun 66:2221–2229

Cole ST 1994 The genome of *Mycobacterium leprae*. Int J Lepr 62:122–125

Fleischmann RD, Adams MD, White O et al 1995 Whole genome random sequencing and assembly of *Haemophilus influenzae* Rd. Science 269:496–512

Haas R, Veit S, Meyer TF 1992 Silent pilin genes of *Neisseria gonorrhoeae* MS11 and the occurrence of related hypervariant sequences among other gonococcal isolates. Mol Microbiol 6:197–208

Hermans PWN, van Soolingen D, Dale JW et al 1990 Insertion element IS986 from *Mycobacterium tuberculosis*: a useful tool for diagnosis and epidemiology of tuberculosis. J Clin Microbiol 28:2051–2058

Hermans PWM, van Soolingen D, van Embden JDA 1992 Characterization of a major polymorphic tandem repeat in *Mycobacterium tuberculosis* and its potential use in the epidemiology of *Mycobacterium kansasii* and *Mycobacterium gordonae*. J Bacteriol 174:4157–4165

Honoré N, Bergh S, Chanteau S et al 1993 Nucleotide sequence of the first cosmid from the *Mycobacterium leprae* genome project: structure and function of the Rif-Str regions. Mol Microbiol 7:207–214

Hood DW, Deadman ME, Jennings MP, Bisercic M 1996 DNA repeats identify novel virulence genes in *Haemophilus influenzae*. Proc Natl Acad Sci USA 93:11121–11125

Kalpana GV, Bloom BR, Jacobs WR 1991 Insertional mutagenesis and illegitimate recombination in mycobacteria. Proc Natl Acad Sci USA 88:5433–5437

Kapur V, Whittam TS, Musser J 1994 Is *Mycobacterium tuberculosis* 15,000 years old? J Infect Dis 170:1348–1349

Mahairas GG, Sabo PJ, Hickey MJ, Singh DC, Stover CK 1996 Molecular analysis of genetic differences between *Mycobacterium bovis* BCG and virulent *M. bovis*. J Bacteriol 178:1274–1282

McAdam RA, Hermans PWM, van Soolingen D et al 1990 Characterization of a *Mycobacterium tuberculosis* insertion sequence belonging to the IS3 family. Mol Microbiol 4:1607–1613

Otal I, Martin C, Vincent-Lévy-Frébault V, Thierry D, Gicquel B 1991 Restriction fragment length polymorphism analysis using IS6110 as an epidemiological marker in tuberculosis. J Clin Microbiol 29:1252–1254

Philipp WJ, Poulet S, Eiglmeier K et al 1996a An integrated map of the genome of the tubercle bacillus *Mycobacterium tuberculosis* H37Rv and comparison with *Mycobacterium leprae*. Proc Natl Acad Sci USA 93:3132–3137

Philipp WJ, Nair S, Guglielmi G, Lagranderie M, Gicquel B, Cole ST 1996b Physical mapping of *Mycobacterium bovis* BCG Pasteur reveals differences from the genome map of *Mycobacterium tuberculosis* H37Rv and from *Mycobacterium bovis*. Microbiology 142:3135–3145

Poulet S, Cole ST 1995a Characterisation of the polymorphic GC-rich repetitive sequence (PGRS) present in *Mycobacterium tuberculosis*. Arch Microbiol 163:87–95

Poulet S, Cole ST 1995b Repeated DNA sequences in mycobacteria. Arch Microbiol 163:79–86

Robertson BD, Meyer TF 1992 Genetic variation in pathogenic bacteria. Trends Genet 8:422–427

Ross BC, Raios K, Jackson K, Dwyer B 1992 Molecular cloning of a highly repeated DNA element from *Mycobacterium tuberculosis* and its use as an epidemiological tool. J Clin Microbiol 30:942–946

Salo WL, Aufderheide AC, Buikstra J, Holcomb TA 1994 Identification of *Mycobacterium tuberculosis* DNA in a pre-Columbian Peruvian mummy. Proc Natl Acad Sci USA 91:2091–2094

Sreevatsan S, Pan X, Stockbauer KE et al 1997 Restricted structural gene polymorphism in the *Mycobacterium tuberculosis* complex indicates evolutionarily recent global dissemination. Proc Natl Acad Sci USA 94:9869–9874

van Embden JDA, Cave WM, Crawford JT et al 1993 Strain identification of *Mycobacterium tuberculosis* by DNA fingerprinting: recommendations for a standardized methodology. J Clin Microbiol 31:406–409

Vega-Lopez F, Brooks LA, Dockrell HM et al 1993 Sequence and immunological characterization of a serine-rich antigen from *Mycobacterium leprae*. Infect Immun 61:2145–2153

DISCUSSION

Mizrahi: I would like to begin by applauding you, because this research project is going to change the way all of us are going to work. It's useful to know that the sequencing of the *Mycobacterium tuberculosis* genome is essentially completed. Have you gained any insights into two particularly perplexing aspects of the organism, i.e. its slow growth and dormancy?

Cole: There are many reasons why it grows slowly. The one I like best is the one that Bill Jacobs has tried to tackle experimentally. He has looked at the ribosomal RNA operon, which is close to the terminus of replication rather than near the origin of replication, where it would be found in a fast-growing bacterium. Therefore, it doesn't benefit from gene dosage, which affects genes close to the origin or replication. A number of other genes that are linked to the ribosomal RNA operon are also markers that normally benefit from a gene dosage effect. For example, unlike in other bacteria, the ATP synthase operon is also near the terminus of replication. The issue of dormancy is fairly complicated. The current paradigm is that this involves some form of metabolic shut-down, which is genetically programmed in that there is a series of activators or repressors which turn on corresponding genes involved in each stage of the metabolic shut-down. We know, for instance, that the tubercle bacillus can respire oxidatively, but from the genome sequence we also know that it has a repertoire of genes that enable it to grow in conditions where oxygen is limiting or even absent. I would imagine that there's some kind of metabolic shut-down, where it goes from an oxidative state, to a microaerophilic state and then a fully anaerobic state. Perhaps then a set of genes is switched on to just keep it ticking over. We can identify candidate activators, repressors, etc. from the sequence.

Anderson: It may not actually be dormant, because HIV was thought to be dormant until an understanding of its high rate of turnover was available. You showed hybridization results for six isolates compared with the standard. Were they isolates from different patients, and if so have you looked for repetitive isolates?

Cole: Yes, they were isolates from different patients. I would be interested to examine serial isolates from the same patient, and also to look at Phil Hopewell's isolates, because his laboratory has the best collection of consecutive isolates from well-characterized patients.

Anderson: You mentioned that there was variation in the PE and PPE gene families, although it is difficult to make quantitative statements if you only have six isolates. Have you compared this with another bacterium which has a well-defined set of genes encoding the antigens that are recognized by the immune system?

Cole: First, we have no real evidence that these proteins are major antigens, although they are recognized by the immune system, and patients do have antibodies that are capable of detecting them. In *Haemophilus* and *Neisseria* many of the genes coding for surface antigens are composed of tetranucleotide or petanucleotide repeats. Between given isolates of these species there is variation in the copy number of those repeats, and this in turn leads to reading frame changes and antigenic variation. This has been well investigated in Gram-negative bacteria (Robertson & Meyer 1992, Hood et al 1996). In Gram-positive bacteria, perhaps the best example would be the M protein in *Streptococcus*, which has a repetitive structure. Antigenic variation in the M protein results from recombination events between the sequences coding for the repetitive motifs of the protein. There are a number of interesting parallels.

Brennan: On the one hand, you recalled Jim Musser's work indicating that there was little polymorphism in many structural genes (Kapur et al 1994), then on the other you described this extraordinary polymorphism in the PE and PPE gene families. Could you explain this?

Cole: The problem is that the amount of genome information was extremely limited when Jim started his project. He was working on a set of housekeeping genes, which are extremely well conserved throughout biological kingdoms. Therefore, they may not have been the best genes to look at to detect variation within an organism. The PE and PPE gene families have repetitive structures and are thus more amenable to replication errors or possibly recombinational errors, whereas such errors are less common in genes that have unique sequences. We know, for instance, that homologous recombination is poor in *M. tuberculosis*, and this could be a reason why genetic diversity in housekeeping genes is so limited. Jim maintains that if you select for drug-resistant strains you find them at the same frequency as when you select in *Escherichia coli*, and Jim uses this as an argument against the possibility that *M. tuberculosis* has got an efficient DNA repair system. He thinks that variation can be obtained but it has to be driven by something.

Steyn: The GC content of the *M. tuberculosis* genome is 66–67%. Is it uniformly spread throughout the genome or is it concentrated in certain areas?

Cole: There are bits that are more AT rich than others, and there are bits, for instance the polymorphic GC-rich sequence (PGRS) regions, that are more GC rich. This is why we've been held up finishing off the project. The GC content of the PGRS regions is about 90%, which makes it difficult to sequence for various chemical and biochemical reasons. It also makes it difficult to do a sequence assembly, because the sequence is so simple that you can't find the place where one part overlaps with the next part.

Steyn: In recent times, when would the opportunity for gene transfer have occurred? *M. tuberculosis* is sequestered within macrophages during the disease process and does not come into contact with other micro-organisms, unlike gut bacteria or environmental organisms in the soil.

Cole: That is a good point. One always assumes that the tubercle bacillus is hermetically sealed in its vacuole and it is not going to encounter anything else. I'm not saying that there's a lot of horizontal gene transfer going on, but this must have happened in the past. The new insertion sequences are not present in all isolates; there is some variability associated with them. Therefore, the genome information has given us new epidemiological tools to examine the relatedness between two strains.

van Helden: Do you have any ideas about where the cytochrome P450 genes come from and what they might be doing?

Cole: This is an interesting question. There are at least 10 enzyme systems in *M. tuberculosis* that use cytochrome P450 as a cofactor. This is somewhat surprising because this was thought to be a eukaryotic cofactor. The closest similarities between the mycobacterial P450 enzymes are two enzymes present in filamentous fungi, so one could perhaps speculate that either the fungi obtained them from the mycobacteria or vice versa.

Rook: Sterols and steroid-like compounds are used as signalling molecules in many plants and fungi. Also, *Mycobacterium vaccae* is used in commercial processes for generating steroid precursors for human use. Therefore, I thought it would be fun to give the steroid most abundant in the human, namely dehydroepiandrosterone (DHEA) which is present at more than 10^{-5} M, to *M. tuberculosis*, and to look at whether it generates a range of interesting steroids and whether any of these are biologically active. *M. tuberculosis* certainly generates the active derivative 3β,17β-androstenediol (S. Al Nakhli, J. P. Honour & G. A. W. Rook, unpublished results 1997)

Cole: One of Julian Davies' collaborators, Yossi Av-Gay, when he saw these interesting genes, set out to answer that question, and the answer is that it is capable of degrading it.

Kaplan: Are there any indications that the laboratory isolates of H37Rv are significantly tampered with genetically, at least compared to some of the clinical isolates?

Cole: I can give you a top-down answer and a bottom-up answer. For the top-down answer, if you do a global comparison of the structure of the genome of H37Rv with those of other *M. tuberculosis* isolates you find that they are highly conserved. If you then take the bottom-up approach and compare individual nucleotides, and we've been doing this with the CSU strain, you find that there is hardly any variation at the nucleotide level. The only differences that we've seen are in some of the repetitive sequences, such as insertion sequences, or in the copy number of repeated DNA sequences. For example, in H37Rv there may be four copies of a repeat, and in the CSU strain there may be only three. It seems as though the variation is local rather than global, and that it is as a result of minor replication errors rather than gross events, such as recombination or inversions.

One of the most common tautologies is the concept of 'virulent' *M. tuberculosis*. You don't have to say that it is virulent, because by definition it is virulent. The avirulent strains are those that require an adjective.

Anderson: But it's also the definition of what's similar, because the genes that have been sequenced in chimpanzees and humans are often extremely similar, yet the phenotypes are different.

Blackwell: Are the PE and PPE gene families important in pathogenicity? Do they occur in environmental mycobacteria? And are they variable?

Cole: There are two ways to answer this: at the nucleotide level; and by means of Southern blotting. Southern blotting with repetitive probes is often misleading. If one uses a probe to address which organisms contain the PE and PPE proteins, one finds that they're present in *Mycobacterium kansasii*, *Mycobacterium gastri* and *Mycobacterium marinum*, but not in *M. avium*, *Mycobacterium intracellulare* and all the other 30 or so mycobacteria that people have looked at. I can answer this question in another way. The other organism for which we have a large body of sequence data is *Mycobacterium leprae*, which doesn't have the PGRS or the major polymorphic tandem repeat (MPTR) families. However, it does have a few genes which belong to the family that I called unique. For instance, in *M. leprae* there is a protein called the serine-rich antigen. It is a 45kDa protein that has been worked on by Neil Stoker's group (Vega-Lopez et al 1993). This antigen is recognized by patients with both lepromatous leprosy and tuberculoid leprosy. It belongs to the PPE family, but not the MPTR branch, rather the unique branch. Therefore, some of the members are there, but in *M. leprae* they are rare. We have virtually the complete sequence of *M. leprae*, but we have found that all the abundant PE and PPE proteins are missing, which is interesting in the light of Bacillus Calmette–Guérin protecting against *M. leprae* but not *M. tuberculosis*.

Davies: Given that it's been shown that intracellular pathogens can deliver DNA into the host genome under laboratory circumstances, have any studies looked into the possibility of good matches between the available *M. leprae* or *M. tuberculosis* sequences, and the available human genome sequence?

Cole: The problem with this question is that one has to be rigorous. If you do database comparisons you get hits for things that have been worked on intensively. It doesn't mean to say that it's the best hit. For instance, one finds a number of enzymes that seem more closely related to those present in humans, or other eukaryotes, than the corresponding enzymes in related prokaryotes simply because the enzymes have been worked on more intensively in eukaryotes. It is an exciting idea that there could be exchange of genetic information between the host and the parasite, but we're far from proving it.

Mizrahi: In support of that, we've compared the databases against human expressed sequence tag (EST) databases, and we find that if we do the analysis monthly, the outcomes change because of the recent and continuing influx of prokaryotic genome information. Therefore, hits that were showing up as strong hits are no longer strong hits. The issue of bias is important. One of the criteria we have tried to build in, however, is atypical codon usage in mycobacterial genes. So far, we have found one instance in which there is a coincidence between atypical codon usage in the mycobacterial genome and its closest match in humans.

Colston: I would also guess that if there are homologous eukaryotic and prokaryotic genes, and the mycobacterial genes are more similar to the eukaryotic than the prokaryotic genes, then it may simply be that they have coevolved under similar stresses, rather than that there has been transfer of material.

References

Hood DW, Deadman ME, Jennings MP, Bisercic M 1996 DNA repeats identify novel virulence genes in *Haemophilus influenzae*. Proc Natl Acad Sci USA 93:11121–11125

Kapur V, Whittam TS, Musser J 1994 Is *Mycobacterium tuberculosis* 15,000 years old? J Infect Dis 170:1348–1349

Robertson BD, Meyer TF 1992 Genetic variation in pathogenic bacteria. Trends Genet 8:422–427

Vega-Lopez F, Brooks LA, Dockrell HM et al 1993 Sequence and immunological characterization of a serine-rich antigen from *Mycobacterium leprae*. Infect Immun 61:2145–2153

Bacterial genetics and strain variation

Paul D. van Helden

MRC Centre for Molecular and Cellular Biology, Department of Medical Biochemistry, Faculty of Medicine, University of Stellenbosch, PO Box 19063, Tygerberg, 7505, South Africa

Abstract. An entire genome sequence will provide valuable information, but the genome of only one individual will limit interpretation of that information. Knowledge concerning genome variation in both eukaryotic and prokaryotic organisms such as *Mycobacterium tuberculosis* is likely to yield information of equal value and provide fundamental insights concerning the function of the genome. The variability in the genome between individual strains may be small and well defined, but it may cause large phenotypic changes (e.g. point mutations causing drug resistance). Clinical and epidemiological observations have led to the development of hypotheses, assumptions and models concerning disease dynamics. However, genome variation studied by molecular epidemiology has made new insights possible, which have allowed us to examine prevailing dogmas concerning tuberculosis. Recent results suggest that historical dogmas may well hold true in some communities, but not all. The information gathered from studying strain variation can be used for modelling disease dynamics, prediction of epidemics, policy planning and for monitoring the outcome of new interventions, as well as for gaining insight into the life processes of the organism. However, molecular epidemiology has its own limitations, some of which result from our lack of understanding of genome variation. We need further information in order to understand clonality and evolution of this organism so that our use of molecular tools in epidemiology and drug development may become more relevant and accurate.

1998 Genetics and tuberculosis. Wiley, Chichester (Novartis Foundation Symposium 217) p 178–194

The association of serotype with disease has been used to refine the study of classical epidemiology of infectious disease. Historically, bacterial typing by serotype or phenotype analysis has suggested that bacteria are clonal and that a line of infection can therefore be traced. However, in tuberculosis the tracing of routes of infection by phenotype analysis has not proven to be sensitive or specific and other techniques are therefore required. In this context, genotyping could be theoretically considered the most accurate form of typing of any organism, because it has been shown that strains of bacteria are genomically different, as are individual humans. Although we are likely to gain an enormous amount of knowledge and information from the sequence of any genome, the study of the genome of only one individual will provide only limited information and be restricted in its practical applications. A better understanding of genome variation is likely to yield information of equal value and provide fundamental insights concerning the function of the genome. This applies to both eukaryotic and prokaryotic organisms such as *Mycobacterium tuberculosis*. The variations in the genome may be small and well defined, but they can cause large phenotypic changes (e.g. point mutations resulting in drug resistance). Thus, genome variations may be responsible for many different phenotypes and families of bacteria which may vary from, for example, highly virulent to attenuated strains. The study of the genome is likely to yield information for both drug development in the longer term and for short-term applications, such as the study of the dynamics of the disease.

Clinical and epidemiological observations using the tools available at the time have led to the development of hypotheses, assumptions and models concerning disease dynamics. However, since tuberculosis patients appear to harbour only one bacterial strain and each strain has a unique genotype, the new branch of science named molecular epidemiology has made new insights possible. This technology has allowed us to examine some of the prevailing dogmas concerning this disease and ask new questions. Results of studies in different communities suggest that these dogmas may well hold true in some communities, particularly developed countries with low incidence of disease, but they may not necessarily be valid in all communities. There is therefore a danger in extrapolating findings from one community and geographical area to another. In addition, limitations in traditional epidemiology have been revealed. The information gathered from studying strain variation in a community can also be used for modelling the disease dynamics, predicting epidemics and for planning policy. This technology can also be used to monitor the results of any new intervention on the dynamics of disease in the community being studied. However, molecular epidemiology is not a panacea, it is still in its infancy and has its own limitations, some of which are due to our lack of understanding of genome variation in the bacillus. We need further information in order to understand the clonality, genome variation, ecology and evolution of this organism and therefore improve our use of molecular tools in epidemiology and drug development. Although the use of antibiotics is clearly important for combating infectious disease at present, real breakthroughs are likely to come from a better understanding of ecology, social behaviour and an appreciation for the interaction between host and pathogen as part of the biosphere (Dubos 1960).

In this chapter the overall concept of bacterial variation will be considered in the context of molecular epidemiology and the dynamics of the disease. The measurement of variation will be by genotype analysis using markers detecting polymorphisms.

Experimental procedures

A number of different genotyping methods for *M. tuberculosis* have been developed. These techniques rely on the occurrence of repeated sequences in the genome of the organism. The number, location and variability of flanking sequences and restriction enzyme cutting sites in the genome generate polymorphisms in the genome that can be used for typing. The methods used may be based on PCR alone (Haas et al 1993), on PCR and hybridization, e.g. spoligotyping (De la Salmoniere et al 1997), or on Southern blotting (van Embden et al 1993). A number of probes are currently in use for the application of the latter technology, of which the most commonly used is IS6110 (van Embden et al 1993). A polymorphic GC-rich sequence (PGRS; Burman et al 1997) or a direct repeat region (or oligonucleotide repeat regions; Torrea et al 1995) may also be used. It has been noted, particularly in the case of low copy number strains (less than four copies of IS6110 or direct repeat), that a single probe is inadequate and that additional probes should be used (Warren et al 1996a, Burman et al 1997). Once accurate fingerprint patterns matched to standards have been obtained, it is important that some computerized analytical method be used in order to achieve accurate comparisons within a large sample base (Alland et al 1994, Small et al 1994, Warren et al 1996b). The uses, advantages and disadvantages of the various approaches will not be discussed here, but typing results will be used to discuss concepts.

Results and discussion

IS6110 is a transposon-like element identified in the genome of *M. tuberculosis* that occurs at various locations and at variable copy number. This element is probably most widely used as a polymorphic marker for genotyping and while it has its weaknesses, it will be used as an example to illustrate the concepts concerning genome variability and molecular epidemiology in this discussion. An example where IS6110 was used as a marker in restriction length polymorphism analysis is shown in Fig. 1. This figure shows the salient points of genotype comparison, namely the case where some strains match others (100% similarity index e.g. T2 and R6–R9), others where closely related strains can be seen (e.g. compare U5 with R15, and R23 with R24) and others with no match or no similarity index (compare R30 with R31, R38 and T2). This figure represents strains obtained from a rural community (R) in Mpumalanga, from an urban community (U; Warren et al 1996b) in Ravensmead/Uitsig and from a large drainage area around a metropolitan urban community (T) in the Western Cape Province. No contact occurs or has occurred as far as can be ascertained between the rural and urban communities (approximately 2000 km apart and of quite different economic,

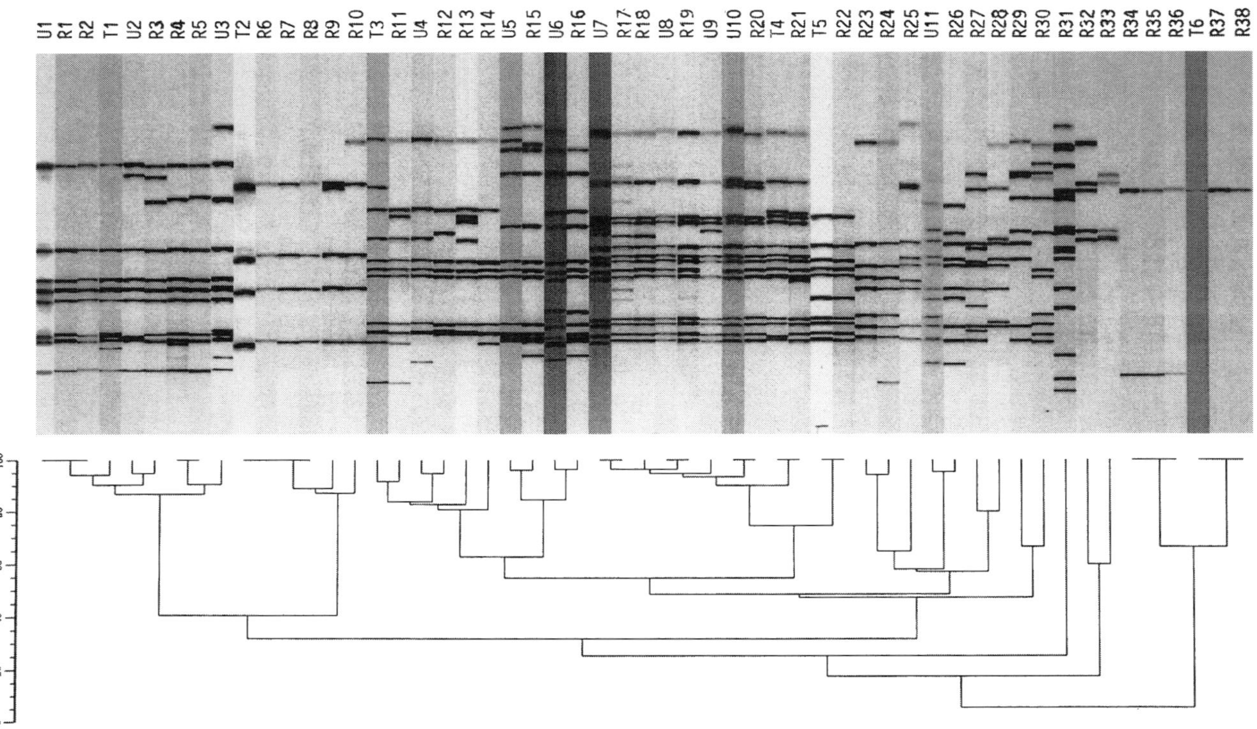

FIG. 1. IS6110-generated DNA fingerprint patterns of strains of *Mycobacterium tuberculosis* from a rural community (R), from an urban community (U) 2000 km distant from R (Warren et al 1996b) and from a large drainage area surrounding the urban community (up to 500 km distant from R; T). Scale shows similarity index. Figure shows identical matches (e.g. R6–R9), similar (clonal variants R23 and R24) and matching strains where no epidemiological relationship is evident (e.g. R17 compared to U8).

ethnic and language groups) or between the rural and town communities. However, genotypically identical strains are seen within these communities.

The recovery of genotypically identical strains from apparently unassociated hosts is evidence for a clonal population structure (O'Rourke & Spratt 1994). Figure 1 also illustrates the occurrence of clonal variants (e.g. R17–R21), where there is in all likelihood a common progenitor. Note that strain families (similarity index >80%) are represented by strains from different communities, thus suggesting the spread of a common ancestor or variants springing from this ancestor. This could perhaps have occurred during the pioneer expansion in the early colonial history of the country. In this context, it is interesting to note that strains which dominate in Asia (van Soolingen et al 1995) have a particular signature pattern that is detectable in many South African strains. This may reflect early colonial trade with Asia or result from the importation of Asian slaves and workers in colonial times. Yet, the majority of strains from the rural community in South Africa do not match any strains from the urban community (the latter is a large database). This is suggestive of at least two routes of infection into the rural community or of sufficient genomic change that a common ancestor can no longer be easily recognized.

It is not possible to study accurately the dynamics of disease in a given area by limited sample selection, and therefore we believe that it is better to focus on a defined group in a given area and attempt full coverage, because sampling rate will influence cluster analysis and the error effects will increase as the sample size or rate decreases. Comprehensive analysis is currently being attempted in a high incidence urban community (Ravensmead/Uitsig; Warren et al 1996b) and results from the strains collected over approximately 3.5 years will be discussed. Identical strains have been grouped into clusters and the results analysed using the formula:

$$\frac{\text{No. of strains in clusters} - \text{No. of clusters.}}{\text{Total number of strains}}$$

The results of this analysis are shown in Table 1. This so-called $(n-1)$ formula (Small et al 1994) is used to estimate recent transmission. Given that the commonly accepted clinical definition of recent transmission in tuberculosis is transmission occurring over 24 months, it can be seen that an estimate of recent transmission in this community is somewhat less than 50% of the total tuberculosis cases. While it can be seen that the extent of clustering rises between six and 42 months, it can also be ascertained that the rise after 12 months is slow. Yet, it can be seen that the majority of strains in this community (70%) have at least one match within 42 months. The majority of the clusters of strains seen in this high incidence community consists of clusters of two or three individuals (81% of the total). However, it is known that IS6110 alone cannot accurately cluster

TABLE 1 Cluster analysis in a high incidence community

Time (months)[a]	*Per cent clustering (n – 1)*[b]	*Per cent clustering (n)*[c]
6	12–30	30
12	41	56
24	48	62
42	53	70

[a]The estimates for six months depend on the window and are variable. The 12-month estimate is the average of three separate years (1993–1995).
[b]Clustering ($n-1$) was calculated according to the formula:

$$\frac{\text{No. of strains in clusters} - \text{No. of clusters}}{\text{Total No. of strains}}$$

[c]Clustering (n) was simply the proportion of strains in clusters.

strains with five or fewer elements of IS6110 and these should therefore be excluded from the analysis. When this is done the percentage of strains clustered within 42 months drops to 46%. It has also been shown in two studies (Warren et al 1996a, Burman et al 1997) that subcluster analysis will almost certainly decrease the estimate of clustering by about 8–10% and therefore a more accurate reflection of recent transmission is probably 38–40%. This low value was not expected in this high incidence community, nor that it would coincide with estimates of clustering from large metropolitan areas in developed countries (Alland et al 1994, Small et al 1994), low incidence developed and developing countries (Hermans et al 1995), and island communities (i.e. French Polynesia; Torrea et al 1996). This phenomenon remains to be satisfactorily explained; however, it must be acknowledged that subgroups of persons in various communities may show vastly different clustering and therefore recent transmission estimates. Examples of this would be prison communities or HIV communities, where clustering is high. It should therefore be realized that estimates quoted are an average for any given community. Cluster estimates are based on 100% identity in pattern matching, which is a further source of error, because if similar strains are epidemiologically linked, these will be excluded (Mazurek et al 1991). More research is required to clarify this risk, because even strains with 100% matching may not be epidemiologically linked (Braden et al 1997, and as indicated in this chapter).

Pattern matching is problematical in fingerprint analysis, and it has been shown that no single probe is sufficiently accurate for epidemiological matching. There is therefore a need to develop a working system that can be universally applied. A future scenario is one where multiplex PCR may be done on any given strain, so that a number of well-known and highly polymorphic sites can be simultaneously amplified to give a consistent, easily assignable pattern. These patterns which would ideally have exactly the same number of fragments in each case, could be easily loaded and compared by spreadsheet analysis. Some potential sites have been identified (e.g. Goyal et al 1994, Vera-Cabrera et al 1997), but this idea has not yet been tested.

Some of the commonly held dogmas which may be challenged by molecular epidemiology are that high incidence implies high transmission rate (previously shown to be incorrect) and that multiple cases in a household are the result of household transmission. Our results would suggest that this occurs in no more than 50% of the households studied, but with a mother–child relationship carrying a far higher transmission risk. Furthermore, the analysis of drug-resistant cases in our study community suggests that the majority of drug-resistant cases are transmitted and not acquired. These results hold considerable significance for therapy and policy planning.

Clearly, although the estimate of the extent of recent transmission may be similar in different communities, there are quite different risk factors driving the transmission. In a diverse area (e.g. a large city) an average estimate is therefore of less value than that obtained in a limited and well-defined study area or community. In planning the control of tuberculosis, an accurate estimate of clustering is likely to be of considerable importance. Should the cluster estimates obtained thus far prove to be correct, i.e. implying that the majority of cases are due to reactivation, then even if we achieve a 100% cure rate of cases, a substantial reservoir of disease is likely to remain present in communities for decades. This is particularly true if no active case finding or contact tracing and prophylaxis of contacts is done. Furthermore, modelling and prediction of the cause of the disease could be extremely valuable for policy makers and planners (Garcia et al 1997), but accuracy is clearly required. Essentially, one needs to calculate the effective reproductive rate (R) of tuberculosis to predict the course of an epidemic, where if $R > 1$ an epidemic will prevail, but if $R < 1$ an epidemic will wane. The calculation of R is theoretically relatively simple, since R is equal to the infection rate multiplied by the duration of infection multiplied by the probability that a contact will become infectious (Blower et al 1996). It is clear that an authentic determination of R depends on the accuracy of the various functions used in this calculation which is in turn dependent on the accuracy and reliability of molecular epidemiology. The calculations that may be made from data acquired in molecular epidemiology studies will enable an impartial assessment of the efficacy of past, current and future control programmes.

It is curious that although most clusters consist of two to four individuals and there is a rapid decline in the number of clusters from two to nine individuals, there are four large clusters identified in our high incidence urban community (Warren et

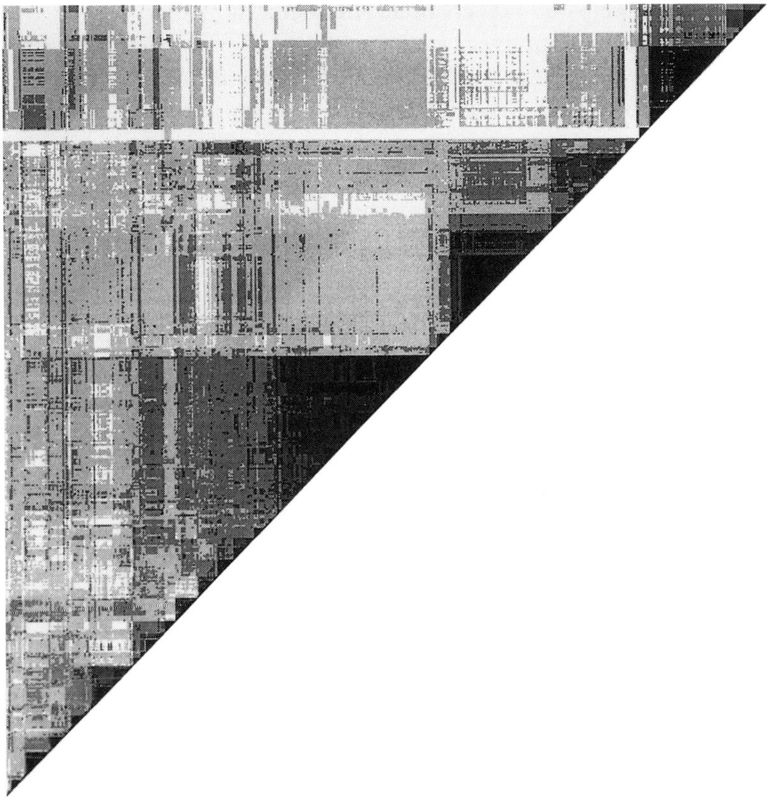

FIG. 2. Similarity matrix of a collection of strains from a well-defined high incidence tuberculosis community (Warren et al 1996b) collected over 3.5 years. Diagonal represents strains of 100% similarity coefficient with corresponding strains. Grey tones represent less than 100% similarity. Note clustering into some dominant families and clear division into two major groups (clear space).

al 1996b, R. Warren, M. Richardson & S. Sampson, unpublished results 1996). These clusters include 11, 12, 15 and 35 individuals, respectively, to date, and although these four clusters represent only about 4% of the number of clusters identified, they include 20% of the strains found in clusters or 15% of the total number of strains. It is possible that these strains represent those where a higher degree of bacterial fitness may be represented (e.g. transmissibility, virulence, pathogenicity). This may be related to strain uptake (Cywes et al 1997), host resistance (Rhoades & Orme 1997) or many as yet unknown factors. Not only are these strains disproportionately represented, but their clonal variants are also present in large numbers. This is represented in Fig. 2, where every strain in the large database is compared to every other strain and a similarity index plotted. The darker shaded triangles represent strains with an increasing degree of similarity, and they suggest that a few families dominate in the community. This diagram may therefore represent clonal expansion of highly successful strains in this relatively stable community. Neo-Darwinism explains that genotypes that have a higher level of fitness will contribute disproportionately to successive generations (Brookfield 1996, Shapiro 1995). Some mutations arise more frequently because of selection and are therefore useful (e.g. drug resistance; Shapiro 1995), but there may also be gradual changes in phenotype, where mutations are cumulative and each may have only a minor but dominant effect. Figure 2 shows another significant feature, namely a clear segregation between two groups of strains. At this stage the significance of these two independent groups remains obscure.

It is interesting that the largest cluster identified contains 21 IS6110 inserts and consists of 35 individuals. The results of copy number analysis correlated with clustering are also suggestive that fitness may relate to copy number, since at higher copy number the likelihood of changes is enhanced. For example, the results presented in Table 2 show that those strains with a higher copy number of IS6110 (17–23 copies) occur significantly more frequently ($p < 0.009$) in clusters than as unique strains.

Our lack of understanding of much of the biology of *M. tuberculosis* is partly a consequence of a dearth of information concerning the ecology of the organism. A population of bacteria may be either panmictic (with random association between loci), clonal or some function between these two extremes. The bacteria may be sexual, but because of epidemic status may appear to be superficially clonal. At present, there is no evidence for panmixia in tuberculosis, but rather clonality, with the possibility that all tuberculosis bacteria may be traced to one progenitor (Sreevatsan et al 1997). In the clones (clonal variants) and families of strains seen in *M. tuberculosis*, there is significant evidence for association between loci (i.e. temporary disequilibrium because of explosive expansion). This still needs to be tested by linkage disequilibrium analysis (non-random association of loci), which should theoretically prove clonality. However, the proof will be complex because

TABLE 2 IS6110 copy number in clustered or unique strains

	Number of strains		
IS6110 copy number	*Clustered (C)*	*Unique (U)*	*Ratio (C : U)*
6–16	211	136	1.55 : 1
17–23	64	20	3.2 : 1

this equilibrium can also arise by genetic drift. Recent evidence suggests that transposition is biased towards certain sites (S. Sampson, R. Warren, M. Richardson & P. van Helden, unpublished results 1997, M. D. Cave, personal communication 1997), which is another factor that will influence the interpretation of results.

Pathogenic bacteria may require some degree of recombination in order to achieve variation (e.g. cell surface epitopes) to protect against (periodic) selection. The reason for the apparently large variation in *M. tuberculosis* still remains obscure, given that much of the genome is quite possibly invariant and highly conserved (Sreevatsan et al 1997). The strain heterogeneity seen in any given geographic area or population group may then be due to heterogeneity in hosts and could be locally biased due to ethnic differences (the relative proportion of susceptible compared to resistant hosts). The degree of heterogeneity could also reflect the stage of the epidemic, since in a low incidence community there will be few hosts and possibly little opportunity for change, whereas in a high incidence community the opportunity for diversification is probably high. If we do not understand the rate of change (diversification), it is possible that our estimate of clustering and understanding of molecular epidemiology will be faulty. In this context, it could be that unique strains (often detected in older patients and thought to be associated with latency) may be unique owing to hypermutability induced by starvation (Yarmolinsky 1995) such as is experienced in a moribund state (Rosenberg 1994).

Thus, for many possible reasons, we have a number of variants that may exist in order to preserve diversity. The concept that there may be phenotypic variations with different degrees of virulence is commonly held (see Fig. 3, model A). However, this model probably reflects an oversimplified situation and the population biology of tuberculosis is possibly more accurately depicted by model B in Fig. 3. A three-dimensional figure would enhance the complexity and almost certainly provide a more accurate depiction of the population biology. In this case, the third dimension could represent host susceptibility or resistance, which will also be variable and dependent on both genetics and environment. By availing itself of the opportunity to exist in different forms, the tuberculosis bacterium can best ensure its survival, because clearly in a highly virulent form there may be less chance for long-term survival. Similarly, with no variation and in a limited population of hosts, a rapidly progressing form may disappear (Rhodes & Anderson 1996).

By operating in different modes, the bacillus is able to ensure its survival by (periodic) explosive expansion (Chevrel-Dellagi et al 1993) in local or susceptible populations, as well as ensuring its long-term survival in the event of the failure of the above strategy. Thus, tuberculosis is likely to be endemic in both large and small (isolated) population groups. The 'rapids' allow for rapid clonal expansion,

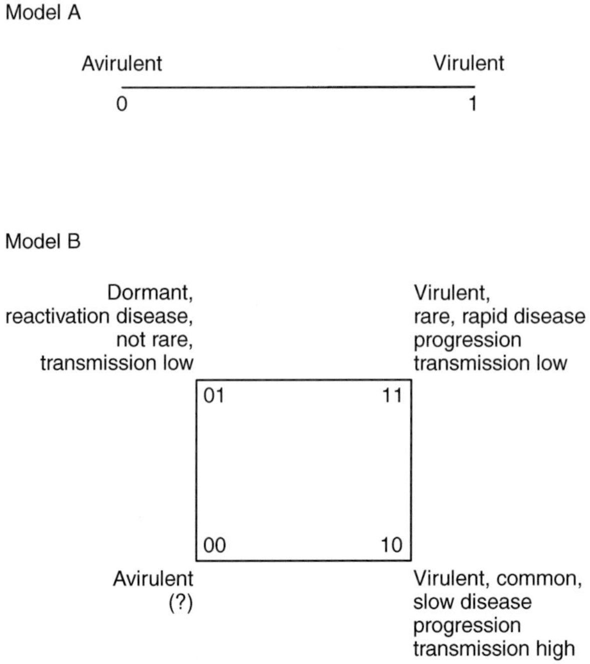

FIG. 3. Conceptual models for strain phenotype.

but the 'persisters' do it more invidiously by possibly interfering with the host immune function, resulting in dormancy. Thus, it is unlikely that the genome of *M. tuberculosis* is a simple structure with individual non-redundant genes coding for a single virulence factor, rather virulence (or other qualitative characteristics) will be a function of degree, probably determined by multiple genes and products. Unfortunately, genetics alone cannot predict the fitness of genotype as yet, but the analysis of genetic variation is an exciting and important area for study.

A population of bacteria is a natural experiment, where during natural recombination (knockout) mutants will probably be generated. Some of these will be lethal and not found in nature. Any natural knockout found in a patient will thus invalidate the knockout gene as a target for chemotherapeutics. Conversely, certain knockouts may dominate (e.g. a large cluster), by providing the organism with a natural advantage. Thus, genes found to be involved in pathology, virulence, transmissibility or other functions can be validated as targets prior to drug design and development.

The synthetic generation of recombinant knockouts will complement this natural experiment, but has the inherent problem that *in vivo* testing will be extremely complex and time consuming. A careful combination of recombinant experimentation together with analysis of strains found in the natural experiment of life may provide a synergistic answer towards discovery of new drugs so urgently needed to combat tuberculosis.

Acknowledgements

I thank the Medical Research Council of South Africa, the University of Stellenbosch, Tygerberg Hospital, Glaxo Wellcome Action TB Initiative and National Institutes of Health grant RO1 A135265–03 for financial support. The European Union contract Biomed1-BMH1-CT93–1614 assisted with computer analysis. I would also like to thank my colleagues for many hours of stimulating discussion and collaboration, particularly R. Warren, J. Hauman, I. Wiid, W. Bourn, T. Victor, N. Beyers, A. van Rie, P. Donald, M. Richardson, S. Sampson and C. Classen. Finally, I thank Eileen van Helden for many forms of assistance, not the least of which is her editorial assistance in completing this chapter.

References

Alland D, Kalkut GE, Moss AR et al 1994 Transmission of tuberculosis in New York City — an analysis by DNA fingerprinting and conventional epidemiological methods. New Engl J Med 330:1710–1716

Blower SM, Small PM, Hopewell PC 1996 Control strategies for tuberculosis epidemics: new models for old problems. Science 273:497–500

Braden CR, Templeton GL, Cave MD et al 1997 Interpretation of restriction fragment length polymorphism analysis of *Mycobacterium tuberculosis* isolates from a state with a large rural population. J Infect Dis 175:1446–1452

Brookfield J 1996 Population genetics. Curr Biol 6:354–356

Burman WJ, Reves RR, Hawkes AP et al 1997 DNA fingerprinting with two probes decreases clustering of *Mycobacterium tuberculosis*. Am J Respir Crit Care Med 155:1140–1146

Chevrel-Dellagi D, Abderrahman A, Haltiti R, Koubaji H, Gicquel B, Dellagi K 1993 Large-scale DNA fingerprinting of *Mycobacterium tuberculosis* strains as a tool for epidemiological studies of tuberculosis. J Clin Microbiol 31:2446–2450

Cywes C, Hoppe HG, Daffé M, Ehlers MRW 1997 Non-opsonic binding of *Mycobacterium tuberculosis* to complement receptor type 3 is mediated by capsular polysaccharides and is strain dependent. Infect Immun 65:4258–4266

De la Salmoniere Y-OG, Li MH, Torrea G, Bunschoten A, van Embden J, Gicquel B 1997 Evaluation of spoligotyping in a study of the transmission of *Mycobacterium tuberculosis*. J Clin Microbiol 35:2210–2214

Dubos R 1960 Mirage of health. Allen & Unwin, London

Garcia A, Maccario J, Richardson S 1997 Modelling the annual risk of tuberculosis infection. Int J Epidemiol 26:190–203

Goyal M, Young D, Zhang Y, Jenkins PA, Shaw RJ 1994 PCR amplification of variable sequences upstream of the katG gene to subdivide strains of *Mycobacterium tuberculosis* complex. J Clin Microbiol 32:3070–3071

Haas WH, Butler WR, Woodley CL, Crawford JT 1993 Mixed-linker polymerase chain reaction: a new method for rapid fingerprinting of isolates of the *Mycobacterium tuberculosis* complex. J Clin Microbiol 31:1293–1298

Hermans PWM, Messadi F, Guebrescabher H et al 1995 Analysis of the population structure of *Mycobacterium tuberculosis* in Ethiopia, Tunisia and the Netherlands: usefulness of DNA typing for global tuberculosis epidemiology. J Infect Dis 171:1504–1513

Mazurek GH, Cave MD, Eisenach KD, Wallace RJ, Bates JH, Crawford JT 1991 Chromosomal DNA fingerprint patterns produced with IS6110 as strain-specific markers for epidemiology study of tuberculosis. J Clin Microbiol 29:2030–2033
O'Rourke M, Spratt BG 1994 Further evidence for the non-clonal population structure of *Neisseria gonorrhoeae*: extensive genetic diversity within isolates of the same electrophoretic type. Microbiology 140:1285–1290
Rhoades ER, Orme IM 1997 Susceptibility of a panel of virulent strains of *Mycobacterium tuberculosis* to reactive nitrogen intermediates. Infection Immunity 65:1189–1195
Rhodes CJ, Anderson RM 1996 A scaling analysis of measles epidemics in a small population. Philos Trans R Soc Lond B Biol Sci 351:1679–1688
Rosenberg SM 1994 In pursuit of a molecular mechanism for adaptive mutation. Genome 37:893–899
Shapiro JA 1995 Adaptive mutation: who's really in the garden? Science 268:373–374
Small PM, Hopewell PC, Singh SP et al 1994 The epidemiology of tuberculosis in San Francisco. N Engl J Med 330:1703–1708
Sreevatsan S, Pan X, Stockbauer KE et al 1997 Restricted structural gene polymorphism in the *M. tuberculosis* complex indicates evolutionarily recent global dissemination. Proc Natl Acad Sci USA 94:9869–9874
Torrea G, Levee G, Grimont P, Martin C, Chanteau S, Gicquel B 1995 Chromosomal DNA fingerprinting analysis using the insertion sequence IS6110 and the repetitive element DR as strain-specific markers for epidemiological study of tuberculosis in French Polynesia. J Clin Microbiol 33:1899–1904
Torrea G, Offredo C, Simonet M, Gicquel B, Berche P, Pierre-Audigier C 1996 Evaluation of tuberculosis transmission in a community by 1 year of systematic typing of *Mycobacterium tuberculosis* clinical isolates. J Clin Microbiol 34:1043–1049
van Embden JDA, Cave MD, Crawford JT et al 1993 Strain identification of *Mycobacterium tuberculosis* by DNA fingerprinting: recommendations for a standardized methodology. J Clin Microbiol 31:406–409
van Soolingen D, Qian L, de Haas PEW et al 1995 Predominance of a single genotype of *M. tuberculosis* in countries of East Asia. J Clin Microbiol 33:3234–3238
Vera-Cabrera L, Howard ST, Laszlo A, Johnson WM 1997 Analysis of genetic polymorphism in the phospholipase region of *Mycobacterium tuberculosis*. J Clin Microbiol 35:1190–1195
Warren R, Richardson M, Sampson S et al 1996a Genotyping of *M. tuberculosis* with additional markers enhances accuracy in epidemiological studies. J Clin Microbiol 34:2219–2224
Warren R, Hauman J, Beyers N et al 1996b Unexpectedly high strain diversity of *Mycobacterium tuberculosis* in a high-incidence community. S Afr Med J 86:45–49
Yarmolinsky MB 1995 Programmed cell death in bacterial populations. Science 267:836–837

DISCUSSION

Davies: If the adenylate cyclase gene of *Streptomyces*, which like mycobacteria is an actinomycete, is disrupted the organism develops an unusual growth defect. It grows to stationary phase, but fails to undergo a particular transition in growth phase normally associated with the switching on of certain metabolic pathways. Does a natural adenylate cyclase knockout of *Mycobacterium tuberculosis* grow normally, and are its properties as a pathogen affected?

van Helden: I can't answer that specific question because we haven't gone back to that strain and looked at it from any of those points of view. The gene encoding adenylate cyclase is a multicopy gene in *M. tuberculosis*, so just one gene knockout may not have any effects.

Mizrahi: There's at least one copy that is eukaryotic like and is not related to the *Streptomyces* gene.

Anderson: I didn't quite follow the argument about the estimated rate of reactivation versus active transmission. Are you totally excluding the possibility that somebody can be infected twice with the same strain?

van Helden: They can probably be infected twice with the same strain. If you analyse two cultures from a person, you don't know whether the person was not sterilized the first time round and then reactivated with the same strain, or whether they were reinfected by the same strain.

Anderson: So, there is uncertainty about recent transmission versus reactivation?

van Helden: Yes. The $n-1$ formula is useful in a low incidence environment but perhaps not in a high incidence environment.

Hopewell: I don't understand the concept of families, as defined by restriction length fragment polymorphism typing, and how they develop, or at least how strains develop identical patterns from non-related sources. All the data suggest divergence in patterns rather than convergence. Unless you have a completely closed community where there are no mutations of the IS6110 patterns, then divergence rather than convergence is more likely, so how do families develop?

van Helden: We don't know enough to be able to say whether convergent or divergent changes occur. We have certain strains that are relatively stable, and we have a few that appear to be unstable, and they could be the ones that drive divergence. I would rather say that the diversity of *M. tuberculosis* has arisen over the last 6000–10 000 years by clonal expansion. However, this diversity could also be explained by multiple sources of import, i.e. from Asia, Europe and America.

Hopewell: About 25% of people with repeated positive cultures over time (i.e. a positive culture more than 90 days after the initial positive) have at least a one band change, i.e. either an addition, deletion or a shift.

van Helden: The amount of change we see in our community is much less than you see in San Francisco. In our collection we have about five strains that demonstrate changes following repeat cultures.

Hopewell: Different strains may behave differently in terms of pattern shifts.

Anderson: I also didn't totally understand your point about fitness and virulence. Fitness is simply reproductive success of the pathogen, in the strictest Darwinian sense. However, I didn't follow the argument about which out of the virulent, avirulent or moderate strains would be the most fit.

van Helden: But all strains are virulent. It's a matter of degree.

Anderson: Virulence can only be defined, in terms of transmission success, by the likelihood of transmission, which can be measured by the average duration of infection and the probability that it's going to be reactivated later on, etc. It's not clear to me whether evolution would drive the pathogen towards more or less virulence.

van Helden: Sreevatsan et al (1997) proposed that the furthest evolved strains are the least virulent, but this has not been tested functionally or biologically.

Steyn: As a general rule, the more virulent organisms, and by that I mean the ones that cause disease, also have to transmit rapidly. You left me with the impression that you had a different relationship between virulence and transmission.

van Helden: I don't want to discuss the definition of virulence. I am using the term loosely. If you could provide me with a better definition I would be grateful.

Steyn: A virulent organism would be one that caused a more severe disease, a more rapid disease progression and a more rapid transmission.

Anderson: The literature on this is large. There is a precise quantitative trade-off between transmissibility and virulence, in other words average survival or duration of infectiousness. Evolution for a variety of pathogens has gone in different directions. Sometimes it drives towards higher virulence, and sometimes to lower, but it is possible stably to maintain a variety of strains that differ in virulence. The molecular epidemiology areas of mycobacteria transmission, and genetic diversity within populations, how it's maintained and how it's changing, is about to open, and our notions about this are going to change dramatically in the coming years.

Ehlers: I don't understand how you can reach any conclusions about reactivation versus recent transmission from your data. You have shown clearly that there are four or five clusters. The members of those clusters may not be identical but they are highly related. This suggests to me that a limited number of strains are being exchanged within this community. It is possible that with each transmission from one host to the next there is a brief new acquisition of a polymorphism. You don't know the rate of change within a high incidence, high transmission community, so I don't think you can say anything about transmission versus reactivation.

van Helden: I agree. We have to study the evolution and obtain a measure of rate of change. If we don't know the rate of change estimates of transmission could be wrong. This may be why we don't understand why in a high incidence community we do not detect a high degree of recent transmission.

Hopewell: This is understandable because most of the evidence suggests that acquisition of a new infection is much more difficult once someone becomes infected. Therefore, in a high incidence community people acquire their infection early on and then develop reactivation disease later.

van Helden: On the other hand, the contact tracing we have done within the clusters shows that we can only determine realistic personal contact for someone who has an identical strain. We have not been able to demonstrate personal contact in people, even within a family, who have minor differences. This is a circular problem, and we can only solve it by doing the molecular biology and sociology together.

Kaplan: You have a community that's separated from the outside world by crossing the street. If you compared this community with the one four streets away, for example, would you lose reactivation? Your estimate of clusters versus non-clusters is lower than it is in reality, simply because, even though you have a large number of strains, it is a small sample within the entire community.

van Helden: We picked that study community because it is bounded on three sides by non-suburban areas (industrial areas and a cemetery). Therefore, crossing the street for this suburb is only one of the four sites. Secondly, the people don't move in and out of the community to any large extent.

Kaplan: But they do come into contact with people across the street.

van Helden: Absolutely. And I don't deny that this is a weak point in the study.

Kaplan: Would the estimate have therefore increased if you had studied a larger area?

van Helden: Possibly, but we don't have the resources to do this.

Fourie: I would like to take this one step further. This study area should not be seen as an island, in terms of behavioural contacts, i.e. with whom people work and socialize. The population dynamics in Cape Town, as I understand, have changed dramatically over the past 50 years, and I believe that you will not understand the diversity in this local area unless you understand the strain diversity within the Eastern Cape, where there has been a high rate of migration by people seeking work. We need to set up a national database, as matter of urgency.

van Helden: I agree that we need a national database. With respect to understanding the Eastern Cape, I am not sure that this would have much impact on the study community because the study community didn't exist prior to about 1947. In the early 1960s, about half of the population were forced to settle in Ravensmead by the Group Areas act. Therefore, we have a young study community comprising an ethnic group that probably had little contact, if any, with people from the Eastern Cape.

Fourie: This confirms my own feeling that you're looking at active transmission and not reactivation disease. You're looking at strains being introduced all the time, and you don't know where they're coming from.

Fine: I find your interpretation of the apparently low proportion of reactivation disease in this study community to be intuitively unreasonable. National databases may help, but some basic descriptive epidemiology would be more helpful. I would guess that you would see some interesting trends if you broke down the community simply by age, by residence time and by migration history.

van Helden: An anthropologist is currently looking into the history of the area, so we are trying to get a handle on this information.

Bateman: I would like to address the question of reinfection in a patient who already has tuberculosis. There are now examples of multidrug resistance arising in patients who are admitted to hospital with sensitive *M. tuberculosis*, and who pick up other strains in the hospitals. It's not difficult to imagine that the same situation occurs in the community. I suspect that the full picture will not be obtained from setting up a national database. The answers will come from looking at the microenvironment: at person-to-person transmission within limited geographic, domestic or work-place environments.

Hopewell: In the study by Small et al (1993), those patients who went to hospital and acquired a drug-resistant organism, went to hospital because they had end-stage AIDS, and so were severely immunocompromised.

Bateman: I know of three other cases, one from the Western Cape and two from another province, where multidrug-resistant strains were acquired in hospital. I don't know if any of these patients had AIDS.

References

Small PM, Shafer RW, Hopewell PC et al 1993 Exogenous reinfection with multidrug-resistant *Mycobacterium tuberculosis* in patients with advanced HIV infection. N Engl J Med 238:1137–1144

Sreevatsan S, Pan X, Stockbauer KE et al 1997 Restricted structural gene polymorphism in the *M. tuberculosis* complex indicates evolutionarily recent global dissemination. Proc Natl Acad Sci USA 94:9869–9874

Antibiotic resistance in mycobacteria

Julian Davies

Department of Microbiology and Immunology, University of British Columbia, #300-6170 University Boulevard, Vancouver, BC Canada V6T 1Z3

Abstract. Multiple drug resistance in mycobacteria compromises the use of antibiotics. Although the genetic and biochemical bases of antibiotic resistance in mycobacteria are largely understood, a number of questions remain to be addressed. This chapter discusses the potential roles of hypermutability and compensatory mutations in establishing stable resistant phenotypes in the pathogenic mycobacteria.

1998 Genetics and tuberculosis. Wiley, Chichester (Novartis Foundation Symposium 217) p 195–208

Ever since the introduction of effective antimycobacterial agents in the 1940s, successful therapy has been frustrated by the inevitable development of strains resistant to those agents. In recent years high rates of mortality due to infection by mycobacteria have been reported, especially in immunosuppressed patients (Nolan 1997). The increased frequency of multidrug-resistant *Mycobacterium tuberculosis* and the identification of *Mycobacterium bovis* strains resistant to up to 10 different antimicrobial agents in AIDS sufferers in Spain and other parts of Europe threaten continued antibiotic use (Blázquez et al 1997).

The appearance of antibiotic resistance during the course of treatment of *M. tuberculosis* infections has been well documented. Streptomycin, the first effective antibiotic available for tuberculosis therapy, was discovered in 1944 and introduced into general practice in 1947; within a year reports of streptomycin-resistant strains of *M. tuberculosis* were frequent. The same scenario has unfortunately been repeated for every antibiotic since introduced for tuberculosis, and indeed for any bacterial infection in humans.

Mechanisms and development of antibiotic resistance

Several reviews of the mode of action of antimycobacterial drugs and their biochemical mechanisms of resistance have been published recently (Blanchard 1996, Cole 1994, Musser 1995, Webb & Davies 1998). This information is summarized in Table 1 and Fig. 1. Briefly, for most antibiotics available for the

TABLE 1 Genetics of resistance to common antimycobacterial antibiotics

Drug	*Target*	*Altered gene*
Isonicotinic hydrazide	Mycolic acid synthesis	*inhA*, *katG*, *aphC*
Ethambutol	Arabinosyltransferase	*embCAB*
Pyrazinamide	?	*pncA*
Streptomycin	Translation	*rpsL*, *rrnA*
Manamycin	Translation	*rrnA*
Viomycin	Translation	*rrnA* + *rrnB*
Clarithromycin	Translation	*rrnB*, ?
Rifamycin (rifampin)	Transcription	*rpoB*, ?
Ciprofloxacin	Replication	*gyrA*, *parC*?, efflux

? indicates that uncharacterized mechanisms of resistance exist.

treatment of mycobacterial infections, there are several mechanisms of resistance that have been well characterized on a genetic and biochemical basis. For some agents whose mode of action remains incompletely characterized (such as pyrazinamide), the detailed resistance mechanisms have proved difficult to assign. A number of efflux pump systems have been identified in mycobacteria, and the role of some level of increased drug efflux in the overall development of antibiotic resistance appears to be important (Takiff et al 1996). It is possible that an initial low level resistance due to increased drug efflux during therapy provides a selective advantage (survival), which then increases the possibility of the appearance of mutations leading to resistance by other mechanisms, such as target modification.

In this brief review I would like to discuss and speculate on several aspects of antibiotic resistance and its phenotypic expression in pathogenic mycobacteria which set them apart from other bacteria. The current upswing in multidrug-resistant infections by Gram-negative and Gram-positive bacteria in hospitals and the community is due in the majority of cases to the acquisition of plasmid-determined antibiotic resistance genes. Resistance due to chromosomal mutation is the exception (although certain 'exceptions' are highly significant, such as methicillin-resistant staphylococci, and antibiotic-resistant meningococci and pneumococci). By contrast, antibiotic resistance in clinical isolates of mycobacteria is due almost exclusively to the mutational alteration of chromosomal genes. However, as described later, antibiotic resistance in clinical situations may be a more complex genotype than was once realized.

There are a few instances in which acquisition of an antibiotic resistance gene from an exogenous source has apparently occurred in mycobacteria. Martin et al

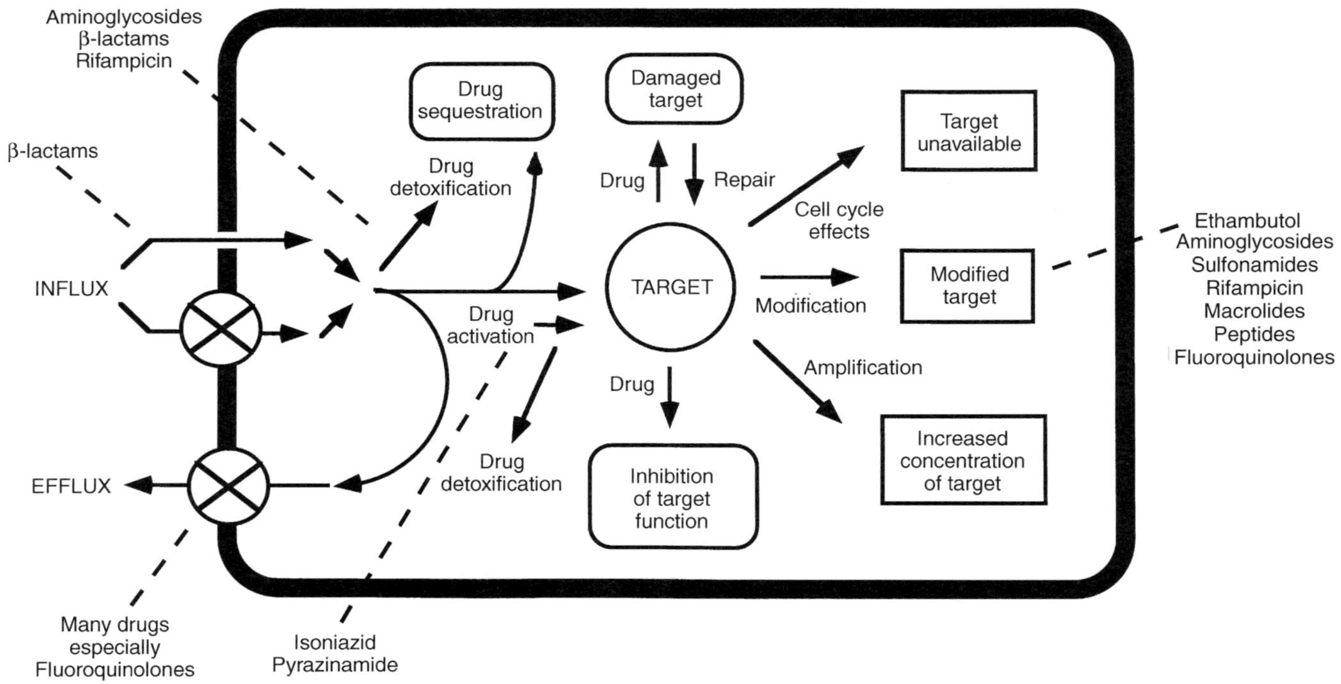

FIG. 1. The cartoon represents the cellular mechanisms of resistance to antibiotics at various critical stages in antibiotic action.

(1990) have identified a defective type I integron that is highly resistant to sulfonamide drugs in the chromosome of *Mycobacterium fortuitum*. A composite transposon Tn610 carries (between two IS6100 elements) two genes similar to those in the type I antibiotic resistance integrons typically found in Gram-negative bacteria: a truncated integrase gene (*intI1*) and a gene for sulfonamide resistance (dihydropteroate synthase, *sul3*). Since the IS6100 sequences are related to the widely dispersed IS6, especially as found in Gram-negative bacteria, it is highly likely that Tn610 acquisition by mycobacteria originated in a member of the Enterobacteriaceae. The interspecific vector involved has not been identified, but we can speculate that the transfer was probably mediated by bacteriophages. In addition, to date integrons and their associated antibiotic resistance gene cassettes have been identified in Enterobacteriaceae (Recchia & Hall 1995).

Pang et al (1994) may have uncovered another example of resistance gene transfer into mycobacteria. While examining several instances of coexistence of streptomycetes and mycobacteria in human infections they found that the genes for tetracycline resistance, normally associated with a tetracycline-producing streptomycete, were shared by both genera. Other examples of gene acquisition and mobile genetic elements in the mycobacteria have been noted. Plasmids have been identified in a number of mycobacterial species but they do not appear to be involved in clinically significant antibiotic resistance (Picardeau & Vincent 1997). Although the available evidence indicates that the mechanisms of antibiotic resistance are almost exclusively due to mutation, one cannot eliminate the contingency that horizontal resistance gene transfer could occur in the mycobacteria.

This prompts the question: what is different about mycobacteria? In the first place it is possible that transferable resistance has not occurred (yet) in mycobacteria because the antibiotics employed in treatment are not broad-spectrum agents and, with the exception of streptomycin, have not been used on a worldwide scale for non-human applications, as animal growth promoters, for example. Thus, prudent antibiotic usage for mycobacterial infections may have inadvertently reduced the selection pressure for resistance.

There are other distinct genetic and physiological characteristics that set mycobacteria apart. Mycobacterial infections are:

(a) due to species with slow growth rates;
(b) intracellular pathogens that successfully parasitize host cells;
(c) caused by large populations of pathogenic organisms in a specific host environment (e.g. in the lung);
(d) able to pass through a non-growing (dormant) phase and be reactivated many years later; and

(e) often intrinsically resistant to a number of antibiotics (Jarlier & Nikaido 1994).

These factors lead to chronic infection and corresponding difficulty in treatment. The necessity to maintain appropriate drug levels over many months of treatment predisposes to the development of resistance, and the inherent toxicity of most anti-mycobacterial agents contributes to the difficulty of achieving therapeutic success.

The genetic constitution of mycobacteria is also a factor in the development of resistance. In particular, while most bacterial pathogens, such *as Escherichia coli, Salmonella typhimurium, Staphylococcus aureus,* etc., have multiple copies of the rRNA genes (up to seven; Keener & Nomura 1996), mycobacteria such as *M. tuberculosis* possess only single copies of 16S and 23S rRNA genes (Böttger 1994, Gonzalez-y-Merchand et al 1996, Suzuki et al 1987). As a result, a single rDNA alteration in mycobacteria generates a dominant phenotype. Thus, resistance to an antibiotic such as streptomycin, which inhibits protein biosynthesis by binding to the 30S ribosomal subunit (Cundliffe 1981), most often results from mutations in the gene for ribosomal protein S12 (*rpsL*) that lead to high level streptomycin resistance in many different bacterial genera (Cundliffe 1979). Some *rpsL* mutations even determine a streptomycin-dependent phenotype, and such have been found for *M. tuberculosis* (Honoré et al 1995). In most bacterial genera mutation in an rRNA gene has no phenotype, although the over-expression of a mutated (altered) 16S rRNA gene (*rrnA*) has been shown to confer streptomycin resistance (Morgan et al 1988). However, in *M. tuberculosis*, and other slow-growing mycobacteria possessing a single copy of the *rrnA* gene, mutations conferring streptomycin resistance are dominant and are expressed phenotypically (Finken et al 1993, Meier et al 1996). The same is true for translation inhibitors that bind to the 50S subunit, such as clarithromycin. Mutation in either an r-protein (L22) or 18S rRNA (*rrnB*) can lead to high level resistance (Böttger 1994, Meier et al 1994, 1996).

Hypermutability

At this point it is worth considering the generation of antibiotic resistance mutations in mycobacteria. For many resistance alleles, the frequency of spontaneous mutation in laboratory studies is relatively low, especially with reference to functions involved in replication (*gyrA*), transcription (*rpoB*) or translation (*rpsL, rrnA, rrnB*), which take place generally at rates of 10^{-7} per generation (or less) in bacteria. Do mutations conferring antibiotic resistance occur at a more significant rate *in vivo*? While studies with *M. tuberculosis* infection provide no evidence for such an increase, recent findings with other bacterial pathogens may be significant. LeClerc et al (1996) found that human pathogenic *Salmonella enterica* and *E. coli* exhibited an unexpectedly high incidence of the mutator phenotypes commonly associated with methyl-directed mismatch repair. It was suggested that there exists a relationship between hypermutability and pathogenicity; such genetic variability may provide an advantage in survival of the pathogen and the establishment of successful infection (Taddei et al 1997a). An increased mutation rate likely contributes to the potential of pathogens to develop antibiotic resistance in conjunction with 'fitness' mutations (see later). Could this be the case for a pathogen such as *M. tuberculosis*? Given that it is an obligate intracellular pathogen which is exposed to the DNA-damaging effect of active oxygen species, hypermutability could be an important factor in *M. tuberculosis* physiology in favouring adaptation to this unstable environment, especially in the case of non-dividing cells (dormancy). The *mutT* locus present in most bacteria encodes a hydrolase that removes the mutagenic base 8-oxo-dGTP (Taddei et al 1997b); are bacterial pathogens more likely to be deficient in this function? Studies of potential mutator genes leading to the hypermutability *in vivo* of bacterial pathogens such as *M. tuberculosis* are called for.

Compensatory mutations

There is another factor related to the development of significant antibiotic resistance in bacteria that has not been discussed with reference to mycobacteria. When components of essential macromolecular components (such as ribosomes) are altered by mutation, as in antibiotic-resistant strains, the mutants are frequently growth defective, which can lead to reduction in the pathogenicity of the antibiotic-resistant strains as compared to the drug-sensitive wild-type. This may be more apparent in the case of an intracellular pathogen that is exposed to severe growth-limiting stress in the host. Recent research with Gram-negative pathogens has revealed the possibility of additional factors associated with the acquisition of successful antibiotic resistance genotypes; these factors warrant study in mycobacteria as well.

Several researchers (Lenski 1997, Levin et al 1997, Schrag & Perrot 1996) have studied the effects of the acquisition of an antibiotic resistance plasmid, or a mutation conferring antibiotic resistance, on the ability (fitness) of bacteria to survive in competition with wild-type organisms or to tolerate a stressful environment (pathogenesis). It was found that the change from sensitive to resistant had negative effects on fitness; in other words, the antibiotic-resistant mutants were enfeebled. However, during prolonged growth of antibiotic-resistant strains, variants arose spontaneously that were as fit as the wild-type strains without a change in the resistant phenotype. This restoration of fitness was due to compensatory mutations occurring intra- or extragenically with reference to the original resistance allele. Acquisition of such mutations led to resistant strains that could compete successfully with wild-type strains for growth in a hostile environment (Table 2). Even more striking are experiments by Björkman et al (1998) which have analysed the appearance of virulent streptomycin-resistant *S. typhimurium* in mouse infection studies. While the majority of infecting organisms did not survive, rapid selection of fast-growing (surviving) strains took place; these strains had acquired additional mutations (sometimes intracodon!) compensating the growth defect without altering the resistant phenotype. In some cases, two compensatory mutations were identified. These analyses were extended to mutants resistant to rifampicin (*rpoB*) and nalidixic acid (*gyrA*) in *S. typhimurium*, and similar multiple mutations were identified. The general conclusion is that all antibiotic-resistant microbial pathogens are probably multiple mutants: that is, the initial mutation to generate the resistance phenotype, plus the compensating mutation(s) to maintain virulence. There may be more to resistance than meets the eye!

It is reasonable to ask if the same situation applies to mycobacteria; do streptomycin-resistant and rifampicin-resistant strains of *M. tuberculosis* isolated during the course of infection possess both target modification and associated compensation? Is this likely to be true for all cases of antibiotic resistance in mycobacterial pathogens? In an analysis of *rpoB* mutations conferring rifamycin resistance (Cole 1994), it was noted that double mutations and deletions were quite common in *M. tuberculosis* (see Fig. 2.). Are these the result of compensation? Early studies with *E. coli* showed that there is antagonism between *rpsL* and *rpoB* mutations (Chakrabarti & Gorini 1975, 1977); yet coexisting mutations of this kind are common in multidrug-resistant *M. tuberculosis*. Are additional compensatory changes needed to re-establish the fitness of such strains? Is the primary resistance mutation growth limiting, and does the secondary mutation restore the normal growth characteristics of the organism? Or, on the contrary, since *M. tuberculosis* grows so sluggishly at the best of times, there may be no need for compensatory mutations in this genus (Cole 1994).

TABLE 2 Overcoming the 'cost' of resistance: restoring fitness

Drug	*Resistance*	*Compensation*
Streptomycin	*rpsL*	*rpsL*, rpsD, rpsE,?*
Rifampicin	*rpoB*	*rpoB*,?*
Ciprofloxacin	*gyrA*	*gyrA *,?*

*represents second-site mutations; ? represents additional unidentified mutations.

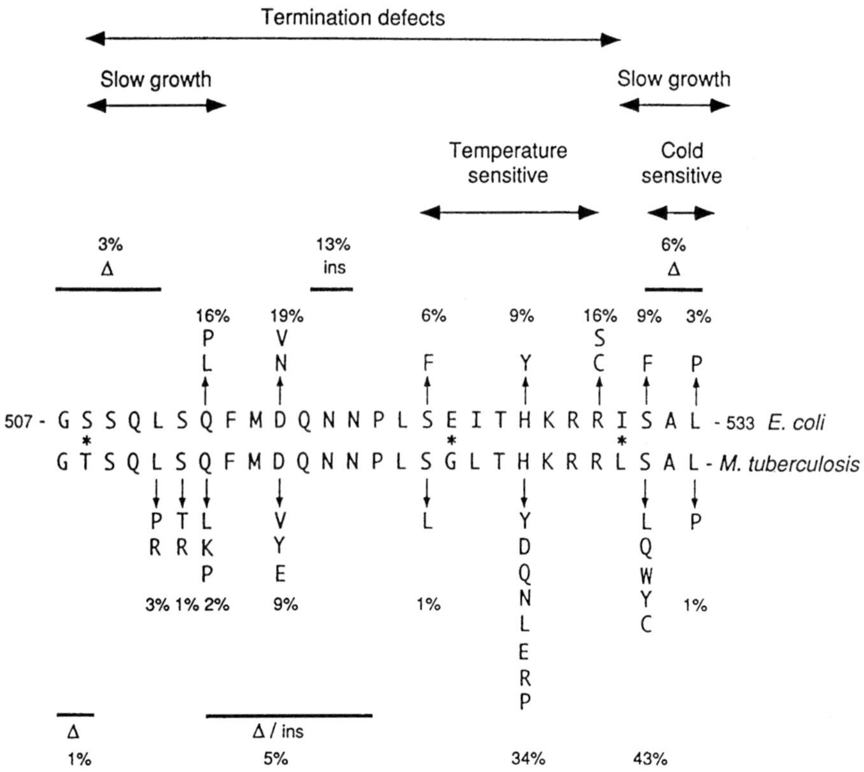

FIG. 2. The multiplicity of mutations leading to rifampicin resistance in the gene for the beta subunit of RNA polymerase (*rpoB*) in *Escherichia coli* (above sequence) and *Mycobacterium tuberculosis* (below sequence). For details see Cole (1994).

In the studies mentioned above it should be noted that compensatory mutations, when isolated from their cognate resistance mutation, often reduce survival fitness of the host. Comparable studies with mycobacteria need to be carried out. What biochemical mechanisms may be involved in determining fitness? Do compensatory mutations simply prevent mutant reversion or suppress the adverse effect(s) of the structural mutant? Do the compensatory mutations modulate interaction between pathogen and host? Finally, it will be of considerable interest to examine the genotype of streptomycin- or clarithromycin-resistant isolates that have altered rRNA genes (*rrnA* or *rrnB*) to see what type of compensatory change, if any, is associated with this type of mutation.

The development of resistance to antibiotics by bacteria is inescapable. Unhappily, the resistance mechanisms and the means by which they are established in human mycobacterial infections are still not well understood. The presence of compensatory mutations may complicate the situation, since the fitness 'cost' of resistance is likely to be a significant factor in determining both survival and virulence. Combinations of drugs and fully maintained courses of therapy are the most obvious and successful approaches to the treatment of susceptible strains at the present time. For multidrug-resistant strains, the only available treatment appears to be the use of cocktails of different classes of antibiotics, which must be rational combinations based on good biochemical evidence of drug action. However, novel antibiotics are needed, and if what I have proposed is correct, the use of anti-mutators might be justified, although molecules with this property of lowering mutation rates have not been well studied and have never been employed in medicine.

When all is said and done, it seems that the development of an effective vaccine is the ultimate solution to the problem of tuberculosis.

Acknowledgements

I wish to thank D. Davies for invaluable assistance in preparing this manuscript, and the Canadian Bacterial Diseases Network and the National Science and Engineering Council of Canada for their support of my research leading to these cogitations.

References

Björkman J, Hughes D, Andersson DI 1998 Virulence of antibiotic resistant *Salmonella typhimurium*. Proc Natl Acad Sci USA 95:3949–3953

Blanchard JS 1996 Molecular mechanisms of drug resistance in *Mycobacterium tuberculosis*. Annu Rev Biochem 65:215–239

Blázquez J, Espinosa de Los Monteros LE, Samper S et al 1997 Genetic characterization of multidrug-resistant *Mycobacterium bovis* strains from a hospital outbreak involving human immunodeficiency virus-positive patients. J Clin Microbiol 35:1390–1393

Böttger EC 1994 Resistance to drugs targeting protein synthesis in mycobacteria. Trends Microbiol 2:416–421

Chakrabarti SL, Gorini L 1975 A link between streptomycin and rifampicin mutation. Proc Natl Acad Sci USA 72:2084–2087

Chakrabarti SL, Gorini L 1977 Interaction between mutations of ribosomes and RNA polymerase: a pair of *strA* and *rif* mutants individually temperature insensitive but temperature sensitive in combination. Proc Natl Acad Sci USA 74:1157–1161

Cole ST 1994 *Mycobacterium tuberculosis*: drug-resistance mechanisms. Trends Microbiol 2:411–415

Cundliffe E 1979 Antibiotics and prokaryotic ribosomes: action, interaction, and resistance. In: Chambliss G, Craven GR, Davies J, Davis K, Kahan L, Nomura M (eds) Ribosomes. Structure, function and genetics. University Park Press, Baltimore, p 555–581

Cundliffe E 1981 Antibiotic inhibitors of ribosome function. In: Gale EF, Cundliffe E, Reynolds PE, Richmond MH, Waring MJ (eds) The molecular basis of antibiotic action. John Wiley, Chichester, p 402–547

Finken M, Kirschner P, Meier A, Wrede A, Böttger EC 1993 Molecular basis of streptomycin resistance in *Mycobacterium tuberculosis*: alterations of the ribosomal protein S12 gene and point mutations within a functional 16S ribosomal RNA pseudoknot. Mol Microbiol 9:1239–1246

Gonzalez-y-Merchand JA, Colston MJ, Cox RA 1996 The rRNA operons of *Mycobacterium smegmatis* and *Mycobacterium tuberculosis*: comparison of promoter elements and of neighbouring upstream genes. Microbiology 142:667–674

Honoré N, Marchal G, Cole ST 1995 Novel mutation in 16S rRNA associated with streptomycin dependence in *Mycobacterium tuberculosis*. Antimicrob Agents Chemother 39:769–770

Jarlier V, Nikaido H 1994 Mycobacterial cell wall: structure and role in natural resistance to antibiotics. FEMS Microbiol Lett 123:11–18

Keener J, Nomura M 1996 Regulation of ribosome synthesis. In: Neidhardt FC (ed) *Escherichia coli* and *Salmonella*, vol 2. ASM Press, Washington, DC, p 1417–1431

LeClerc JE, Li B, Payne WL, Cebula TA 1996 High mutation frequencies among *Escherichia coli* and *Salmonella* pathogens. Science 274:1208–1211

Lenski RE 1997 The cost of antibiotic resistance—from the perspective of a bacterium. In: Antibiotic resistance: origins, evolution, selection and spread. Wiley, Chichester (Ciba Found Symp 207), p 131–151

Levin BR, Lipsitch M, Perrot V et al 1997 The population genetics of antibiotic resistance. Clin Infect Dis 24 (suppl 1):S9–S16

Martin C, Timm J, Rauzier J, Gomez-Lus R, Davies J, Gicquel B 1990 Transposition of an antibiotic resistance element in mycobacteria. Nature 345:739–743

Meier A, Kirschner P, Springer B et al 1994 Identification of mutations in 23S rRNA gene of clarithromycin-resistant *Mycobacterium intracellulare*. Antimicrob Agents Chemother 38:381–384

Meier A, Sander P, Schaper K-J, Scholz M, Böttger EC 1996 Correlation of molecular resistance mechanisms and phenotypic resistance levels in streptomycin-resistant *Mycobacterium tuberculosis*. Antimicrob Agents Chemother 40:2452–2454

Morgan EA, Gregory ST, Sigmund CD, Borden A 1988 Antibiotic resistance mutations in *Escherichia coli* ribosomal RNA genes and their uses. In: Tuite MF, Picard M, Bolotin-Fukuhara M (eds) Genetics of translation, new approaches. Springer-Verlag, Berlin, p 43–53

Musser JM 1995 Antimicrobial agent resistance in mycobacteria: molecular genetic insights. Clin Microbiol Rev 8:469–514

Nolan CM 1997 Nosocomial multidrug-resistant tuberculosis: global spread of the third epidemic. J Infect Dis 176:748–751

Pang Y, Brown BA, Steingrube VA, Wallace RJ, Roberts MC 1994 Tetracycline resistance determinants in *Mycobacterium* and *Streptomyces* species. Antimicrob Agents Chemother 38:1408–1412

Picardeau M, Vincent V 1997 Characterization of large linear plasmids in mycobacteria. J Bacteriol 179:2753–2756

Recchia GD, Hall RM 1995 Gene cassettes: a new class of mobile element. Microbiology 141:3015–3027

Schrag SJ, Perrot V 1996 Reducing antibiotic resistance. Nature 381:120–121

Suzuki Y, Yoshinaga K, Ono Y, Nagata A, Yamada T 1987 Organization of rRNA genes in *Mycobacterium bovis* BCG. J Bacteriol 169:839–843

Taddei F, Matic I, Godelle B, Radman M 1997a To be a mutator; or how pathogenic and commercial bacteria can evolve rapidly. Trends Microbiol 5:427–428

Taddei F, Hayakawa H, Bouton M-F et al 1997b Counteraction by MutT protein of transcriptional errors caused by oxidative damage. Science 278:128–130

Takiff HE, Cimino M, Musso MC et al 1996 Efflux pump of the proton antiporter family confers low-level fluoroquinolone resistance in *Mycobacterium smegmatis*. Proc Natl Acad Sci USA 93:362–366

Webb V, Davies J 1998 Antibiotics and antibiotic resistance in mycobacteria. In: Ratledge C, Dale JW (eds) Mycobacteria: molecular biology and virulence. Chapman and Hall, London, in press

DISCUSSION

Steyn: The *katG/ahpC* mutations are a natural examples of compensatory mutations. Is it possible that adaptive mutations occur in mycobacteria, as has been proposed for other micro-organisms (Foster 1993)?

Davies: Yes. There is variability amongst hosts and also, but perhaps to a lesser extent, amongst strains. A critical functional interplay is necessary to maintain the pathogenicity of multidrug-resistant strains, and such mutations (*katG*) may be important in this process.

Anderson: You raised two interesting questions about the emergence of drug resistance and its persistence. Population genetic methods are going to be extremely important in answering those questions, but it is a field that is somewhat neglected by those working in the drug resistance field. Why are you surprised that resistance emerges quickly? The rate of emergence of resistance, whether it's to an antibiotic or to an insecticide, is not simply a function of the mutation rate, which can be small. The rates of emergence and fixation of resistance, or the stability of a balanced polymorphism, are dependent on the net selective pressure (i.e. the total population size of the pathogen exposed to selection), its rate of turnover and the per capita mutation rate. I would have thought that bacteria, even slow-growing mycobacteria, have enormous potential to fix resistance quickly.

My second point relates to your question about how mycobacteria transmit with a fitness disadvantage. Even the simplest population genetics framework states that irrespective of the fitness disadvantage polymorphisms are likely to appear, allowing both susceptible and resistant strains to coexist. The balance of that polymorphism, in other words the frequency of the resistant alleles versus the susceptible alleles, will depend on the intensity of the selective pressure. Unfortunately, for most resistant problems we are good at recording frequency of resistance, but we're often bad at quantifying the selective pressure, which is the volume of drug usage in a particular community and its change over time. So my question to you is, in the areas where multidrug resistance is important for mycobacteria, are there any communities in which you have precise drug volume usage data?

Davies: Such information should be available, but all we have at the moment is general information, in the sense that since 1950 more than a million metric tons of antibiotics have been released on the surface of the earth, so essentially we are bathed in a dilute solution of antibiotics. We could probably get figures for specific antibiotic use now, but we cannot obtain reliable information for the 1950s, for example. The experiments you suggest should have been done had all the necessary information been collected.

Anderson: It's starting to be collected routinely for some of the antibiotic areas, such as penicillin resistance in bacteria.

Davies: A complete and accurate analysis of drug use, clinical practice, etc. will only take place when the next new superdrug is introduced. I am not trying to defend the early bacterial geneticists and I'm not embracing their notions, because one of the things they clearly didn't take into account was the actual bacterial load. If you have got 10^{10} organisms in a lung then you're going to get a resistant strain.

Kaplan: It's not fair to say that they didn't know it was going to happen. Rene Dubos predicted before streptomycin was introduced that drug resistance would evolve within a few months, and he was right (see Moberg 1996). So, there were those who thought about such problems, rather than celebrating the success of antibiotics.

Davies: If you look at all the drugs that have been introduced, with one exception, resistance was established quickly. The big surprise was the appearance of multidrug resistance in Japan, which everybody said would not happen. But then it was realized that multiple copies of different resistance genes could be present together on a plasmid. In terms of the appearance of double and triple mutants, I still find it difficult to understand how such strains arise because it is difficult to isolate a double streptomycin–rifamycin mutant in the laboratory unless chemical mutagenesis is employed.

Colston: You can if you select one then the other, which is what is happening in the field.

Davies: I agree, but it may still be difficult to obtain certain combinations. As I mention in my chapter, resistance to rifamycin plus streptomycin is interesting because the two mutations are antagonistic.

Anderson: I didn't realize there were 10^{10} bacteria, because HIV is always cited as 10^{11} viral particles/patient per day. With an RNA virus every mutation across the entire genome is possible every day, so it's hardly surprising that resistance occurs. But even for 10^{10} bacteria, in which there is a much slower mutation rate and a much lower replication rate, a large net mutation rate (rather than a per capita rate) still occurs.

Cole: I was intrigued by your idea of compensatory mutations; for instance, the restoration of lost function in the case of streptomycin-resistant *Mycobacterium tuberculosis* strains. I've always found it rather staggering that although streptomycin has not been used extensively for more than 20 years in European countries, the frequency with which one encounters primary streptomycin resistance is incredibly high. I always felt that the standard *rpsL* mutations that occur should have been counter-selected with time, but presumably there was a compensatory mutation which was rapidly acquired and fixed the *rpsL* allele in the community in the absence of selective pressure from the antibiotic.

Davies: That's an interesting point because in strains that have compensatory mutations, it is possible that if the streptomycin resistance mutation is removed then the strain becomes less fit. Therefore, the two mutations together are fixed, but a single mutation is deleterious to the cell. This has been shown by Levin's laboratory (Lipsitch & Levin 1997). Lenski (1997) also demonstrated the same phenomenon with plasmid-determined resistance. He showed that bacteria containing a plasmid will not compete very well with a wild-type cell, but eventually mutations will accumulate in the plasmid-containing cell which results in that cell out-competing the wild-type. In some cases those mutations make the cell grow faster than wild-type, even in the absence of the plasmid. Although there are likely to be different phenomena associated with plasmid-determined versus chromosomal mutations to resistance, there's no doubt that all of the mutations which have been identified as compensating a chromosomal resistance gene are deleterious. It's not quite clear how they work. Do they just fix the mutation in some way?

Anderson: Lenski suggested, on the basis of theory, that the rate of increase to a particular frequency of drug resistance is much more rapid than the rate of decay once the selective pressure is removed. The only detailed epidemiological example of this phenomenon is from Iceland (Austin et al 1998), which reduced the use of penicillin to treat pneumococcal infections for a number of years. The frequency of resistance then decreased by roughly 13%. Therefore, in that particular case the resistant form was less fit than the susceptible. Are there any other examples of this?

Davies: Not that I'm aware of. However, penicillin resistance in pneumococcus has not disappeared entirely in Iceland. I agree that people are now much more careful about the way it's used. In Scandinavia, with the exception of Norway, penicillin use is well controlled and there are relatively low levels of resistance. Spain recently instituted a national policy of trying to reduce antibiotic usage in hospitals, and levels of resistance decreased substantially. The problem is that the resistance genes are always present in the population, so once you start using the drug again resistance increases rapidly. As Stewart Cole mentioned earlier, streptomycin is not used in some countries, but bacterial resistance to the drug is still common.

Kaplan: If I'm not mistaken there was a relatively high level of isoniazid resistance in South Africa, which was then reduced. The levels are now much lower.

Davies: I seem to remember that mutation to isoniazid resistance may be deleterious to the response of the pathogen to the immunological defence of the host cells. Is this true?

Kaplan: Isoniazid resistance, at least in our hands, does not affect the replication of organisms in monocytes *in vitro*, but it does affect the sensitivity of the bacilli to reactive oxygen metabolites. Therefore, in the host, where they could well be exposed to successive oxidative bursts, they may be compromised.

van Helden: In our study community we see transmission of drug-resistant strains in patients before they are diagnosed, although I can't say whether these patients are only infected by that one strain and that it out-grows a sensitive strain because we have no data on that.

Davies: It's a difficult situation to analyse because you can't go back to the original wild-type strain and ask what happens when antibiotics were introduced therapeutically on a large scale.

References

Austin DJ, Kristinsson KG, Anderson RM 1998 The relationship between the volume of antimicrobial consumption in human communities and the frequency of resistance, in press

Foster PL 1993 Adaptive mutation: the uses of adversity. Ann Rev Microbiol 47:467–504

Lenski RE 1997 The cost of antibiotic resistance—from the perspective of a bacterium. In: Antibiotic resistance: origins, evolution, selection and spread. Wiley, Chichester (Ciba Found Symp 207), p 131–151

Lipsitch M, Levin BR The within-host population dynamics of antibacterial chemotherapy: conditions for the evolution of resistance. In: Antibiotic resistance: origins, evolution, selection and spread. Wiley, Chichester (Ciba Found Symp 207), p 112–130

Moberg CL 1996 René Dubos: a harbinger of microbial resistance to antibiotics. Microb Drug Resist 2:282–297

Mechanisms for isoniazid action and resistance

Lynn Miesel, Denise A. Rozwarski†, James C. Sacchettini* and William R. Jacobs, Jr.

*Department of Microbiology and Immunology, Howard Hughes Medical Institute, Albert Einstein College of Medicine of Yeshiva University, 1300 Morris Park Avenue, Bronx, New York 10461, *Department of Biochemistry and Biophysics, Texas A&M University, College Station, TX 77843, †Department of Biochemistry, Albert Einstein College of Medicine, Bronx, New York 10461, USA*

Abstract. Isoniazid is the most widely used antituberculosis drug. Genetic studies in *Mycobacterium smegmatis* identified the *inhA*-encoded, NADH-dependent enoyl acyl carrier protein reductase as the primary target for this drug. A reactive form of isoniazid inhibits InhA by reacting with the NAD(H) cofactor bound to the enzyme active site forming a covalent adduct (isonicotinic acyl NADH) that is apt to bind with high affinity. Resistance can occur by increased expression of InhA or by mutations that lower the enzyme's affinity to NADH. Both of these resistance mechanisms are observed in 30% of clinical tuberculosis isolates. Mutation in *katG*, which encodes catalase peroxidase, is the most common source for resistance. Another mechanism for isoniazid resistance, in *M. smegmatis*, occurs by defects in NADH dehydrogenase (Ndh) of the respiratory chain. Genetic data indicated that *ndh* mutations confer resistance by lowering the rate of NADH oxidation and increasing the intracellular NADH/NAD$^+$ ratio. An increased amount of NADH may prevent formation of isonicotinic acyl NADH or may promote displacement of the isonicotinic acyl NADH from InhA. While our studies have identified this mechanism in *M. smegmatis*, results reported in early literature lead us to believe that it can occur in *Mycobacterium tuberculosis*.

1998 Genetics and tuberculosis. Wiley, Chichester (Novartis Foundation Symposium 217) p 209–221

Isoniazid is a highly effective antituberculosis agent that has been the core of tuberculosis chemotherapy and prophylaxis since the early 1950s (Robitzek & Selikoff 1952, Ad hoc committee 1995). However, isoniazid resistance is becoming more common, such that as much as 30% of tuberculosis isolates are resistant in many large cities (Cohn et al 1997). In our effort to develop better tuberculosis drugs, we are trying to understand the mechanisms of isoniazid action and resistance.

When Gardner Middlebrook isolated the first isoniazid-resistant mutants of *Mycobacterium tuberculosis*, he observed that most of the mutants had lost catalase peroxidase activity (Middlebrook 1954). Loss of the *katG*-encoded catalase peroxidase activity is sufficient to confer resistance and is the most commonly found resistance mechanism (Heym et al 1994, Morris et al 1995, Musser et al 1996, Rouse et al 1996, Telenti et al 1997, Zhang et al 1992). KatG converts the isoniazid prodrug to a reactive species that attacks an essential target (Winder 1960, Johnsson & Schultz 1994). Results of recent structural studies indicate that the reactive species may be an isonicotinic acyl radical or anion (discussed below; Rozwarski et al 1998).

Early biochemical studies showed that mycolic acid synthesis is the primary pathway inhibited by isoniazid. Mycolic acids are the long branched β-hydroxy fatty acids (C_{70}–C_{80}) that comprise a major structural component of the mycobacterial cell envelope (reviewed in Winder 1982). Studies by Takayama et al (1972) showed that the inhibition of mycolic acid synthesis occurs at low concentrations comparable to the bactericidal concentration and precedes the onset of cell death.

Identification of InhA

The enzymatic target of isoniazid was identified from a genetic study with *Mycobacterium smegmatis* in which isoniazid-resistant mutants that are co-resistant to a structurally related drug, ethionamide, were isolated (Fig. 1; Banerjee et al 1994). This genetic approach was based upon the idea that isoniazid and ethionamide inhibit a common target enzyme such that resistance to both drugs can only occur by an alteration in that enzyme target. A *M. smegmatis* mutant resistant to both drugs was found to have a mutation in *inhA* which caused a serine to alanine substitution at position 94 in the InhA enzyme. This mutation was also found in an isoniazid-resistant isolate of *Mycobacterium bovis*, indicating that the target is conserved in members of the *M. tuberculosis* complex and that resistance can occur by the same mechanisms (Wilson et al 1995). Allelic exchange studies in *M. smegmatis* showed that the *inhA* substitution is

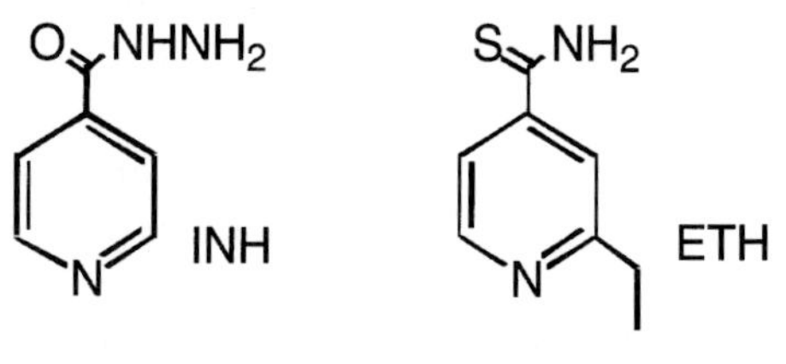

FIG. 1. Structure of isoniazid (INH) and ethionamide (ETH).

responsible for the isoniazid and ethionamide resistance. Increased expression of wild-type *inhA* also conferred isoniazid and ethionamide resistance to *M. smegmatis*. These findings led Banerjee and colleagues to conclude that isoniazid resistance can occur by either increasing the amount of the InhA enzyme or by altering the enzyme target so that it is not inhibited by the active drug form (Banerjee et al 1994).

Sequence analysis of the *inhA* locus in clinical isolates of *M. tuberculosis* identified *inhA* mutations in about 30% of the isoniazid-resistant strains but not in the isoniazid-sensitive strains (Heym et al 1994, Morris et al 1995, Musser et al 1996, Ristow et al 1995, Telenti et al 1997). Most of these mutations are located in the *inhA* promoter and several are within the *inhA* structural gene (L. Basso, personal communication 1998). Many of these *inhA* mutant strains of *M. tuberculosis* are also resistant to ethionamide (low level resistance), which is consistent with traits of the *M. smegmatis inhA* resistance mutants (Banerjee et al 1994, Heym et al 1994).

Biochemical evidence that InhA is the target for isoniazid

Enzymatic characterization of recombinant purified InhA enzyme showed that InhA is involved in fatty acid elongation and/or mycolic acid synthesis. It is an enoyl-ACP reductase that catalyses the NADH-dependent reduction of the double bond at position two of the fatty acid which is linked to the acyl carrier protein (ACP) part of the type II fatty acid synthesis system (Quémard et al 1995). Kinetic analysis of InhA indicated that it preferentially reduces long fatty acyl substrates (at least 16 carbon atoms) consistent with the proposed role for InhA in synthesizing very long-chain fatty acids in mycolic acid synthesis.

Isoniazid inhibits the InhA enzyme by a mechanism that requires NADH, Mn^{2+} and oxygen (Basso et al 1996, Johnsson et al 1995, Zabinski & Blanchard 1997). Inhibition was demonstrated using an *in vitro* reaction system with purified InhA enzyme; the combination of isoniazid, NADH and Mn^{2+} caused the loss of InhA activity (Fig. 2). KatG accelerates the inhibition process perhaps by oxidizing Mn^{2+} to form Mn^{3+} (Magliozzo & Marcinkeviciene 1997). Structural studies in which InhA crystals were formed in the presence of isoniazid, NADH and $Mn^{2+(3+)}$ clearly showed that isoniazid was covalently attached to the nicotinamide ring of NADH at the site of the hydride exchange (isonicotinic acyl-NADH; Fig. 3; Rozwarski et al 1998). The isonicotinic acyl-NADH was bound to InhA in the NADH-binding site, forming a close association with amino acids and a structural arrangement that is apt to have a higher affinity than NADH bound to InhA. Rozwarski et al (1988) proposed that the isonicotinic acyl-NADH forms by the addition of an isonicotinic acyl radical to an NAD· radical or

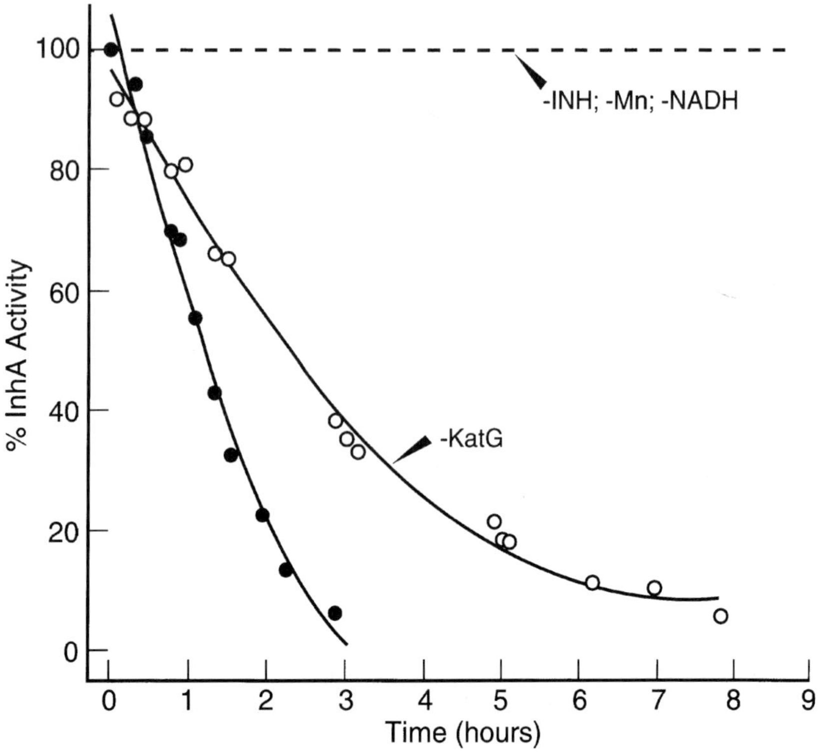

FIG. 2. Inhibition of InhA by isoniazid. InhA (3 μM) was incubated in 100 mM phosphate buffer, pH 7.5 at 25°. Solid lines: mixtures included NADH (100 μM), isoniazid (100 μM) and $MnCl_2$ (1 μM) with KatG (1 μM; filled circles) or without KatG (open circles). The dashed line represents results from control experiments in which mixtures lacking $MnCl_2$, NADH or isoniazid showed no inhibition of InhA (Zabinski & Blanchard 1997, Basso et al 1996). Adapted from Zabinski & Blanchard (1997) and reprinted with permission (Copyright 1997 American Chemical Society).

by an isonicotinic acyl anion to an NAD^+ (Fig. 3). Isonicotinic acyl-NADH forms within the InhA active site.

A mechanism for isoniazid resistance

Resistance to isoniazid can occur by mutations that reduce the affinity of InhA enzyme for the NADH cofactor (Quémard et al 1995). Sequence analysis of *inhA* in resistant clinical isolates found that most of the amino acid substitutions are located within the enzyme's NADH-binding site (Fig. 4; L. Basso, personal communication 1998). These mutations cause lower affinity to NADH, indicated

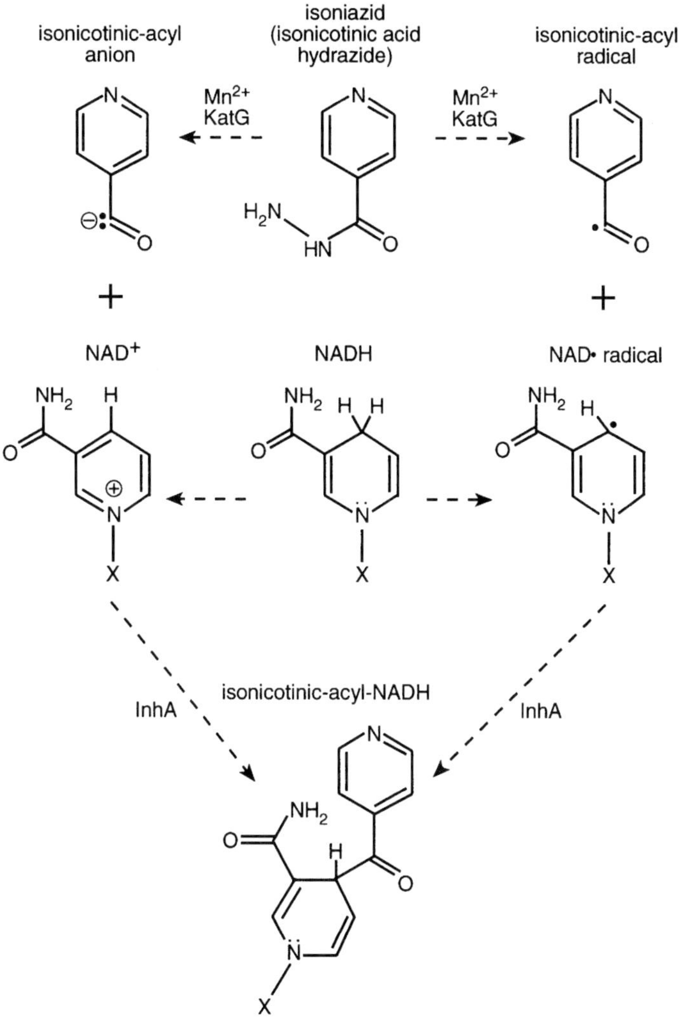

FIG. 3. The proposed pathway for formation of the isonicotinic acyl-NADH inhibitor of InhA (Rozwarski et al 1998). According to this model, isonicotinic acyl anion or isonicotinic acyl radical covalently attaches to a form of NADH (NAD^+ or NAD·) in the active site of InhA. The free radical pathway is the favoured mechanism because isoniazid-dependent inhibition of InhA occurs at a faster rate in the presence of NADH than with NAD^+ (Johnsson et al 1995, Rozwarski et al 1998). Rozwarski proposed that Mn^{3+} ions induce formation of isonicotinic acyl radicals or isonicotinic anions and that KatG facilitates this activation by catalysing the oxidation of Mn^{2+} to Mn^{3+} (Rozwarski et al 1998, Zabinski & Blanchard 1997, Magliozzo & Marcinkeviciene 1997). Reprinted from Rozwarski et al (1998) with permission (Copyright 1998 American Association for the Advancement of Science).

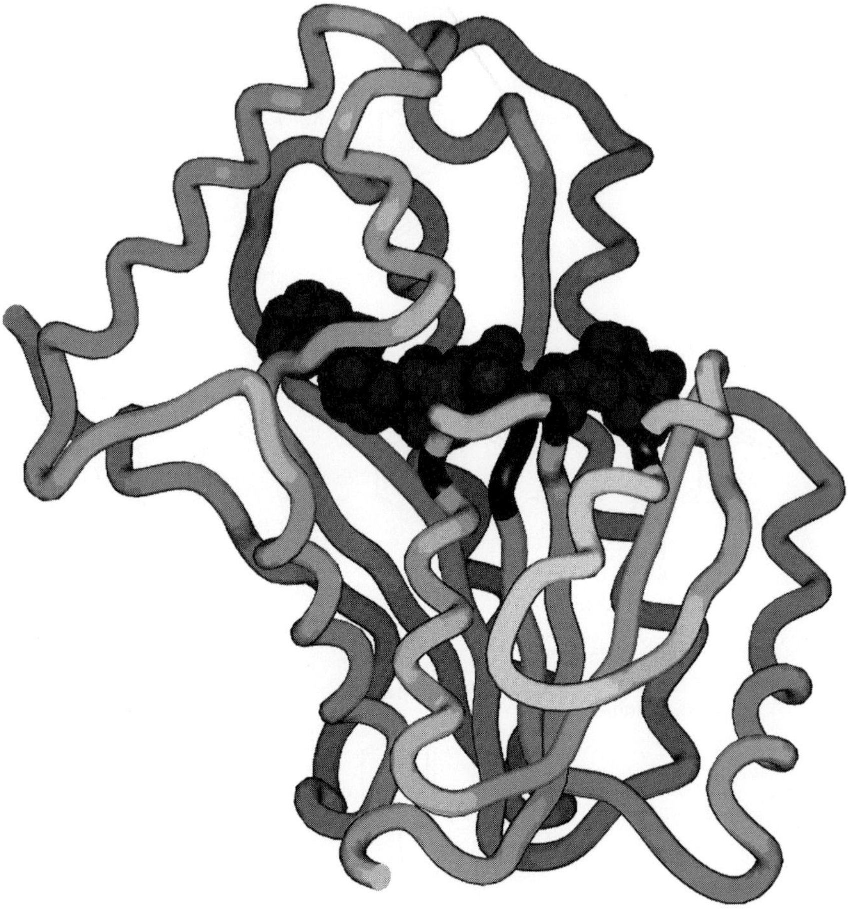

FIG. 4. Positions of substitutions in InhA that are correlated with resistance in clinical tuberculosis isolates. Shown in lighter grey is the alpha-carbon backbone of a single subunit of InhA. The subunit is a single domain, in which the central core contains a Rossmann fold. Shown in darker grey is a CPK model of the bound isonicotinic-acyl-NADH inhibitor, sitting on top of the shelf created by the Rossmann fold. Shown in black are portions of the alpha-carbon backbone which correspond to the locations of clinical isolate mutations. Interestingly, all of these locations are near the isonicotinic-acyl-NADH inhibitor.

by an increased dissociation constant (at least a 10-fold increase) while other enzymatic parameters (e.g. V_{max}) are not significantly altered. These substitutions may confer resistance by lowering the enzyme's affinity to isonicotinic acyl-NADH or by altering the kinetic mechanism of the enzyme (L. Basso, personal communication 1998; Rozwarski et al 1998).

Defects in NADH oxidation confer isoniazid resistance

Temperature-sensitive (*ts*) mutants that have thermolabile InhA defects would be particularly useful for defining the function of InhA in fatty acid elongation and/or mycolic acid synthesis. We have tried to obtain an *inhA ts* mutant by isolating isoniazid-, ethionamide-resistant mutants at 30°C that are unable to grow at 42°C, even in the absence of drugs (Miesel et al 1998; Fig. 5). About half of all isoniazid-resistant mutants isolated at 30°C have these traits. Allelic exchange studies showed that these mutants do not have *inhA* mutations. Most of the mutants require amino acid supplements for growth at 30°C (some are specifically serine/glycine auxotrophs; Fig. 5); this auxotrophic trait is not expected for *inhA* mutants.

The resistance of the *ts* mutants is caused by defects in NADH dehydrogenase (Ndh), which catalyses the first step in the electron transport chain: NADH oxidation and quinone (Q) reduction ($NADH + Q \rightarrow NAD^+ + QH_2$) (Miesel et al 1998). Assays of NADH dehydrogenase activity indicated that the mutants have reduced activity which can be restored by expression of the wild-type enzyme (Table 1). The mutant phenotypes are also corrected by expression of the wild-type *ndh* gene. Sequence analysis identified substitutions in *ndh* in all the *ts* isoniazid-resistant mutants (Fig. 5).

Genetic data indicated that all the phenotypes of the *ndh* mutants (isoniazid-resistant, ethionamide-resistant, *ts* and auxotrophy) are caused by defects in NADH oxidation (Miesel et al 1998). Expression of NADH-dependent malate dehydrogenase (Mdh) corrects all of the *ndh* mutant phenotypes, causing thermoresistance, prototrophy, and sensitivity to isoniazid and ethionamide. Mdh catalyses the NADH-dependent interconversion of oxaloacetate and malate ($NADH + \text{oxaloacetate} \rightarrow NAD^+ + \text{malate}$) that could provide an additional NADH oxidation system that operates independently from the respiratory chain. Enzyme assays showed that Mdh efficiently catalyses NADH oxidation using oxaloacetate as an electron acceptor (Table 1). *mdh* was recovered from the *M. tuberculosis* complex. *M. smegmatis* does not express this Mdh enzyme (Prasada Reddy et al 1975).

We have proposed that the *ndh* defect in the *M. smegmatis* mutants lowers the rate of NADH oxidation, which would cause an increase in the $NADH/NAD^+$ ratio (Miesel et al 1998). This imbalance causes the multiple phenotypes: isoniazid and ethionamide resistance; thermosensitivity; and auxotrophy. Mdh expression would correct the phenotypes by oxidizing NADH. The genetic data indicate that Ndh is not a target for isoniazid; instead, the Ndh must potentiate isoniazid and ethionamide by oxidizing NADH. The isoniazid resistance in the *ndh* mutants could be due to the increased NADH concentration, which may promote dislocation of the isonicotinic acyl-NADH from InhA. Another possibility is

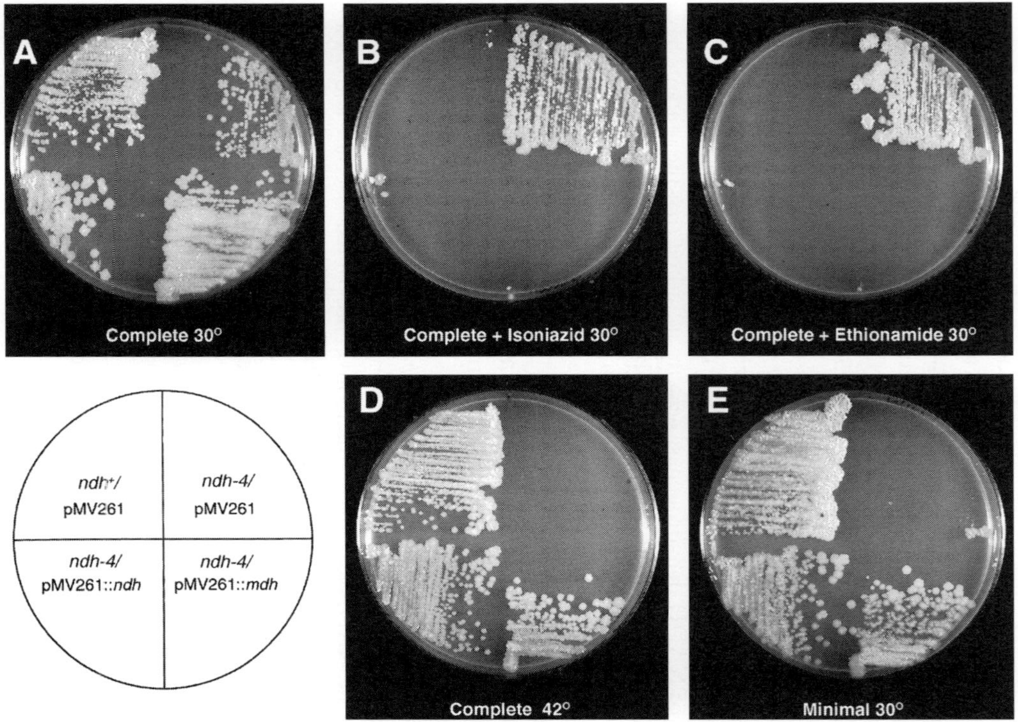

FIG. 5. Phenotypes of a representative *ndh* mutant (*ndh-4*) of *Mycobacterium smegmatis* that is resistant to isoniazid and ethionamide; and complementation with the *ndh* and *mdh* genes (Miesel et al 1998). The isoniazid-sensitive parent strain and the *ndh-4* mutant carry pMV261, which confers kanamycin resistance (top half of each plate). The lower half of each plate shows the complementation by the *Mycobacterium tuberculosis ndh* gene expressed from pMV261 (left) and complementation by the *Mycobacterium bovis* Bacillus Calmette–Guérin *mdh* gene (right). Isoniazid and ethionamide were used at 50 μg/ml, and kanamycin (10 μg/ml) was included into all medium to select for maintenance of plasmids.

that NADH may competitively inhibit KatG-mediated oxidation reactions that activate isoniazid.

Results reported in early literature by Middlebrook & Cohn (1953) suggest that isoniazid resistance in *M. tuberculosis* can occur by metabolic defects which may be similar to the *ndh* defects found in *M. smegmatis*. These investigators attempted to culture isoniazid-resistant tubercle bacilli from patients treated with isoniazid monotherapy and found that eight out of 21 patients produced tubercle bacilli that have growth defects. Of these eight isolates, five were auxotrophs, and three were not cultureable in rich or minimal medium. Perhaps these resistant tubercle bacilli had NADH oxidation defects that prevented their growth on certain types of medium.

Intermediary metabolism influences isoniazid sensitivity

Our genetic data indicate that the NADH oxidation systems influence isoniazid sensitivity. The Mdh enzyme of the *M. tuberculosis* complex is highly active at oxidizing NADH, and expression of this enzyme confers increased isoniazid sensitivity to *M. smegmatis*, a species that otherwise does not have this enzyme (Table 1). We suggest that Mdh enzyme may contribute to the extreme isoniazid sensitivity of *M. tuberculosis*, which is about 100-fold more sensitive than *M. smegmatis*. Although others have suggested that the exquisite isoniazid sensitivity of *M. tuberculosis* is due to a species-specific isoniazid target (Mdluli et al 1996), differences in metabolic functions (the KatG enzyme, peroxide concentrations and the NADH oxidation systems) may fully explain the different sensitivities of *M. tuberculosis* and *M. smegmatis*.

TABLE 1 Correlation between isoniazid sensitivity and NADH oxidation by Ndh or Mdh in *Mycobacterium smegmatis*

Strain genotype[a]	*Isoniazid sensitivity (MIC μg/ml)*	*Ndh activity*[b]	*Mdh activity*[c]
ndh+/–	5	40	2
ndh-4/–	>100	1	10
ndh-4/*ndh* (*Mycobacterium tuberculosis*)	2.5	150	1
ndh-4/*mdh* (*Mycobacterium bovis*)	1	2	10 000

[a]*ndh* chromosome allele/plasmid gene.
[b]Activity is the rate of NADH oxidation (mol/min per mg × 10^7) in the presence of menadione as an electron acceptor. For experimental conditions see Miesel et al (1998).
[c]Rate of NADH oxidation (mol/min per mg × 10^7) in the presence of oxaloacetate as an electron acceptor.
MIC, minimal inhibitory concentration.

Another target for isoniazid?

Mutations in *inhA* were found in 30% of resistant *M. tuberculosis* isolates and were never found in sensitive isolates (Heym et al 1994, Morris et al 1995, Musser et al 1996, Ristow et al 1995, Telenti et al 1997). This indicates a strong selective advantage for the *inhA* mutations only in the presence of the drug and provides compelling evidence that *inhA* is a primary target for isoniazid. Mdluli et al (1996) argue that *M. tuberculosis* has another target for isoniazid. Indeed, the presence of multiple targets may explain why isoniazid is such an effective drug.

The goal: new drug targets

The purpose of these studies is ultimately to develop better tuberculosis drugs. Our studies of isoniazid resistance identified two enzymes, Ndh and InhA, that are likely to be essential in *M. tuberculosis* and are apt to be good drug targets. InhA is a terrific candidate for new drug development since it is a known target for isoniazid. The Ndh enzyme is not essential in the facultative anaerobes such as *Escherichia coli* because these organisms have multiple pathways for NADH oxidation. In contrast, the thermosensitive lethality of the *M. smegmatis ndh* mutants indicates that Ndh is required for viability and is the main enzyme responsible for NADH oxidation (Miesel et al 1998).

Acknowledgements

Financial support was provided by the NIH grant AI36849. We thank L. Basso, J. Musser and J. Blanchard for providing data prior to publication.

References

Ad hoc committee of the scientific assembly on microbiology, tuberculosis and pulmonary infections 1995 Treatment of tuberculosis and tuberculosis infection in adults and children. Clin Infect Dis 21:9–27

Banerjee A, Dubnau E, Quémard A et al 1994 *inhA*, a gene encoding a target for isoniazid and ethionamide in *Mycobacterium tuberculosis*. Science 263:227–230

Basso LA, Zheng R, Blanchard JS 1996 Kinetics of inactivation of WT and C243S mutant of *Mycobacterium tuberculosis* enoyl reductase by activated isoniazid. J Am Chem Soc 118:11301–11302

Cohn DL, Bustreo F, Raviglione MC 1997 Drug-resistant tuberculosis: review of the worldwide situation and the WHO/IUATLD global surveillance project. Clin Infect Dis 24 (suppl 1):S121–S130

Heym B, Honoré N, Truffot-Pernot C et al 1994 Implications of multidrug resistance for the future of short-course chemotherapy of tuberculosis: a molecular study. Lancet 344:293–298

Johnsson K, Schultz PG 1994 Mechanistic studies of the oxidation of isoniazid by the catalase peroxidase from *Mycobacterium tuberculosis*. J Am Chem Soc 116:7425–7426

Johnsson K, King DS, Shultz PG 1995 Studies on the mechanism of action of isoniazid and ethionamide in the chemotherapy of tuberculosis. J Am Chem Soc 117: 5009–5010

Magliozzo RS, Marcinkeviciene JA 1997 The role of Mn(II)-peroxidase activity of mycobacterial catalase–peroxidase in activation of the antibiotic isoniazid. J Biol Chem 272:8867–8870

Mdluli K, Sherman DR, Hickey MJ et al 1996 Biochemical and genetic data suggest that InhA is not a primary target for activated isoniazid in *Mycobacterium tuberculosis*. J Infect Dis 174:1085–1090

Middlebrook G 1954 Isoniazid resistance and catalase activity of tubercle bacilli. Am Rev Tuberc 65:471–472

Middlebrook G, Cohn ML 1953 Some observations on the pathogenicity of isoniazid-resistant variants of tubercle bacilli. Science 118:297–299

Miesel L, Weisbrod TR, Marcinkeviciene JA, Bittman R, Jacobs WR 1998 NADH dehydrogenase defects confer isoniazid resistance and conditional lethality in *Mycobacterium smegmatis*. J Bacteriol 180:2459–2467

Morris S, Bai GH, Suffys P, Portillo-Gomez L, Fairchok M, Rouse D 1995 Molecular mechanisms of multiple drug resistance in clinical isolates of *Mycobacterium tuberculosis*. J Infect Dis 171:954–960

Musser JM, Kapur V, Williams DL, Kreiswirth BN, van Soolingen D, van Embden JDA 1996 Characterization of the catalase–peroxidase gene (*katG*) and *inhA* locus in isoniazid-resistant and susceptible strains of *Mycobacterium tuberculosis* by automated DNA sequencing: restricted array of mutations associated with drug resistance. J Infect Dis 173:196–202

Prasada Reddy TL, Murthy RS, Venkitasubramanian TA 1975 Variations in the pathways of malate oxidation and phosphorylation in different species of mycobacteria. Biochim Biophys Acta 376:210–218

Quémard A, Sacchettini JC, Dessen A et al 1995 Enzymatic characterization of the target for isoniazid in *Mycobacterium tuberculosis*. Biochemistry 34:8235–8241

Ristow M, Möhlig M, Rifai M, Schatz H, Feldmann K, Pfeiffer A 1995 New isoniazid/ethionamide resistance gene and screening for multidrug-resistant *Mycobacterium tuberculosis* strains. Lancet 346:502–503

Robitzek EH, Selikoff IJ 1952 Hydrazine derivative of isonicotinic acid (rimifon, marsilid) in the treatment of active progressive caseous-pneumonic tuberculosis. Am Rev Tuberc 65:402–428

Rouse DA, DeVito JA, Li Z, Byer H, Morris SL 1996 Site-directed mutagenesis of the *katG* gene of *Mycobacterium tuberculosis*: effects on catalase peroxidase activities and isoniazid resistance. Mol Microbiol 22:583–592

Rozwarski DA, Grant GA, Barton DHR, Jacobs WR, Sacchettini JC 1998 Modification of the NADH of the isoniazid target (InhA) from *Mycobacterium tuberculosis*. Science 279:98–102

Takayama K, Wang L, David HL 1972 Effect of isoniazid on the *in vivo* mycolic acid synthesis, cell growth, and viability of *Mycobacterium tuberculosis*. Antimicrob Agents Chemother 2:29–35

Telenti A, Honoré N, Bernasconi C et al 1997 Genotypic assessment of isoniazid and rifampin resistance in *Mycobacterium tuberculosis*: a blind study at reference laboratory level. J Clin Microbiol 35:719–723

Wilson TM, deLisle GW, Collins DM 1995 Effect of *inhA* and *katG* on isoniazid resistance and virulence of *Mycobacterium bovis*. Mol Microbiol 15:1009–1015

Winder F 1960 Catalase and peroxidase in mycobacteria. Am Rev Respir Dis 81:68–78

Winder FG 1982 Mode of action of the antimycobacterial agents and associated aspects of the molecular biology of the mycobacteria. In: Ratledge C, Stanford J (eds) The biology of the mycobacteria. Academic Press, New York, p 353–438

Zabinski RF, Blanchard JS 1997 The requirement for manganese and oxygen in the isoniazid-dependent inactivation of *Mycobacterium tuberculosis* enoyl reductase. J Am Chem Soc 119:2331–2332

Zhang Y, Heym B, Allen B, Young D, Cole S 1992 The catalase-peroxidase gene and isoniazid resistance of *Mycobacterium tuberculosis*. Nature 358:591–593

DISCUSSION

Colston: You have been looking at drugs such as ethambutol and isoniazid that have similar targets, and now you're saying that by using this target, which you identified by using isoniazid, you can look for other drugs. Aren't you going to run into problems with cross-resistance? Isn't it the general principle to look for novel targets rather than concentrate on known targets of existing drugs?

Miesel: We believe that inhibitors of InhA, a target for isoniazid, can be used for developing new drugs without having problems of cross resistance. Inhibitors that do not require chemical activation and that bind differently to the InhA active site may be effective against isoniazid-resistant strains.

Davies: Have you selected for revertants of your temperature-sensitive strains and demonstrated whether they are now sensitive to isoniazid?

Miesel: I have taken the *ndh* mutants that are isoniazid resistant and temperature sensitive and selected for temperature-resistant revertants. Most of these revertants are sensitive to isoniazid, although a small fraction are isoniazid resistant. These isoniazid-resistant revertants may have a partial restoration of NADH dehydrogenase activity that is sufficient to restore growth at higher temperatures. A small defect in NADH dehydrogenase activity is sufficient to confer isoniazid resistance.

Davies: Would this be at a second site?

Miesel: Some of these strains may have mutations in other genes that suppress the defect of *ndh* mutations. Second-site suppressor mutations may activate other NADH oxidation systems (e.g. fermentation systems) or increase the expression of the defective NADH dehydrogenase enzyme.

Davies: Other members of the actinomycetes and streptomycetes have InhA analogues, and yet they're not sensitive to isoniazid. Is your argument that these organisms are insensitive to isoniazid because isoniazid cannot get in or because these organisms have other backup systems for balancing the NAD/NADH system?

Miesel: The actinomycetes and most other bacteria also have enoyl acyl carrier protein (ACP) reductases that are similar to InhA. These enzymes are involved in making the fatty acids that comprise the phospholipids of the cytoplasmic membrane. This differs from InhA, which we believe elongates long fatty acids ($>C_{16}$) as part of a mycolic acid synthesis pathway.

The question is, why are mycobacteria sensitive to isoniazid when other bacteria are not? One possibility is that the activated form of isoniazid specifically inhibits InhA and may fail to inhibit the enoyl reductases involved in phospholipid synthesis. Alternatively, isoniazid may not be activated in other bacteria due to differences in the catalase peroxidase enzyme or to other metabolic differences such as the $NADH/NAD^+$ ratio.

Duncan: I would like your comments on the work that Cliff Barry presented at the ASM Conference 'Tuberculosis: past, present and future' held in Copper Mountain, Colorado in July 1997. His group showed that at concentrations close to the isoniazid MIC, a β-ketoacyl synthase and an acyl carrier protein are cross-linked by isoniazid, thus shutting down mycolate synthesis and resulting in bacterial growth inhibition.

Miesel: Based on biochemical evidence, Cliff Barry proposes that isoniazid covalently attaches to the phosphopantetheine moiety of the ACP and that the isoniazid ACP inhibits fatty acid elongation by forming a complex between ACP and β-ketoacyl ACP synthase.

Isoniazid may have targets other than InhA, and this may explain why isoniazid is such an effective drug. However, the possibility of other targets does not weaken the evidence that InhA is a primary target for isoniazid. Mutations in the promoter region of *inhA* are frequently found in isoniazid-resistant tuberculosis isolates; amino acid substitutions are also found. These mutations must confer a selective advantage during isoniazid treatment because they are never found in isoniazid-sensitive strains. We find that in *Mycobacterium smegmatis*, *inhA* mutations cause low level isoniazid resistance: about a 10-fold increase in the minimal inhibitory concentration. The sensitivity in these strains may be due to inhibition of other targets.

Brennan: There is a structural relationship between isoniazid and nicotinamide, and there is a compensatory effect of malate dehydrogenase and NADH. In view of that, it is difficult to compare your results with those of Cliff Barry. He is saying that isoniazid causes the accumulation of a saturated 24·fatty-acyl ACP in *M. tuberculosis*, and that the target is a Δ-4 desaturase; in other words, an enzyme that creates an unsaturated group. The free hydrogen could then go on to NAD. This seems to be the reverse of what you're saying. One alternative is that the target is InhA, but isoniazid causes an accumulation of the C·24 fatty-acyl ACP downstream. His evidence is based solely on the accumulation of a product, and I'm inferring that the target is the subsequent step, without any direct evidence for a NADH requirement.

Young: I would like to take the opportunity here of broadening this discussion into a general discussion by asking Patrick Brennan to talk about genetics and the biosynthesis of cell wall molecules.

General discussion II

The biosynthesis of cell wall molecules

Brennan: Present-day understanding of the architecture of the mycobacterial cell envelope arises from electron microscopic studies that demonstrate alternating zones of electron density and transparency reminiscent of Gram-negative cells (Brennan & Draper 1994). This evidence is in accord with chemical models of a substantial lipid permeability barrier, a staggered, pseudo-outer membrane contributed in one-half by the *ca.* C_{50} meromycolate and *ca.* C_{25} α-chains of mycolic acids and, in the other, by an assortment of free lipids with medium-chain (C_{30}, e.g. the mycocerosates) and short-chain fatty acids. This concept has received credence from recent permeability, X-ray diffraction, calorimetry and biochemical studies (Brennan & Nikaido 1995). Thus, the outer half of this lipid barrier and the variable glycocalyx zone beyond it (a capsule?) contain the many extracellular polysaccharides and peculiar species-specific glycolipids of mycobacteria. The inner half is composed of the mycolic acids anchored to the cell wall core, the mycolylarabinogalactan peptidoglycan (mAGP) complex. Lipoarabinomannan (LAM), either Man-capped or uncapped (AraLAM), important immunomodulators, intercalate this network.

The insoluble cell wall matrix is defined as a cross-linked peptidoglycan linked to arabinogalactan, esterified at the distal ends to the mycolic acids. Historically, peptidoglycan is thought to consist of alternating units of N-acetylglucosamine (GlcNAc) and a modified muramic acid (Mur). The tetrapeptide side chains consist of L-alanyl-D-isoglutaminyl-meso-diaminopimelyl-D-alanine (L-Ala-D-Glu-A_2pm-D-Ala) with the Glu being further amidated. This type of peptidoglycan is one of the most common found in bacteria. However, mycobacterial peptidoglycan differs in two ways: some or all of the Mur residues are N-acylated with glycolic acid (MurNGly), and the cross-links include those between two A_2pm residues and between A_2pm and D-Ala.

We have shown that the arabinogalactan polysaccharide is unique not only in its elemental sugars, but, unlike most bacterial polysaccharides, it lacks repeating units, composed instead of a few distinct structural motifs. Partial depolymerization of the per-*O*-alkylated arabinogalactan and analysis of the generated oligomers by gas chromatography-mass spectrometry and fast atom-mass spectrometry established (Daffe et al 1990) that: (i) Ara and Gal are in the furanose form; (ii) the non-reducing termini of arabinan consist of the hexasugar motif [β-D-Ara*f*-(1→2)-α-D-Ara*f*]$_2$-3,5-α-D-Ara*f*-(1→5)-α-D-Ara*f*; (iii) the majority of the arabinan chains consist of 5-linked α-D-Ara*f* with branching introduced by 3,5-α-D-Ara*f*; (iv) the arabinan chains are attached to C-5 of some of the 6-linked Gal*f*, and there are 2–3 such arabinan chains; (v) the galactan consists of linear alternating 5- and 6-linked β-D-Gal*f*; (vi) the galactan region of arabinogalactan is linked to the C-6 of some of the MurNGly residues of peptidoglycan via a special diglycosyl–P bridge, α-L-Rha*p*-(1→3)-D-GlcNAc-(1→P); (vii) the mycolic acids are located in clusters of four on the terminal hexa-arabinofuranoside, but only two-thirds of these are mycolated (McNeil et al 1991). More recently, we obtained oligosaccharide fragments containing up to 26 residues from which molecular weights and alkylation patterns were determined by fast atom bombardment-mass spectrometry (FAB-MS). The extended non-reducing ends of the arabinan were shown to consist of a tricosarabinoside (23 mer), with three such units attached to the galactan unit (Besra et al 1995). The galactan was also isolated and was found to consist of 23 Gal residues of the repeating linear structure, [β-D-Gal*f*-(1→5)-β-D-Gal*f*-(1→6)]$_n$, devoid of branching, thereby demonstrating that the points of attachment of the arabinan chains are close to the reducing end of galactan, itself linked to peptidoglycan via the linker disaccharide-P.

The linkage unit represents the 'bull's eye', the 'Achilles' Heel', in terms of drug development, since the whole of the mycolylarabinogalactan complex is attached to peptidoglycan via this unit. In addition, the Ara*f* and Gal*f* residues of arabinogalactan provide attractive targets due to their xenobiotic status in humans. The Gal*f* originates in UDP-Gal*f* by way of UDP-Gal*p* and the respective mutase (Nassu et al 1996); the Ara*f* residues may be formed via the non-oxidatative pentose shunt, probably as P-Ara-P-P *via* P-Ribose-P-P (Scherman et al 1996); and Rha is generated by way of a novel unclustered version of the *rfb* operon/Rfb A-C. The first step in biosynthesis of the complex involves formation of the linkage unit. Membranes from *Mycobacteriumsmegmatis* and *Mycobacterium tuberculosis* catalysed the incorporation of radioactivity from UDP-[^{14}C]GlcNAc into two glycolipids (GL1 and GL2), which were mild-acid labile and mild-alkali stable, features consistent with polyprenol-based glycolipids (Mikušová et al 1996). When tunicamycin was added, a dramatic inhibition of incorporation was observed, suggesting that the initial step in the synthesis of mycobacterial cell wall arabinogalactan involves formation of a polyprenol-P-P-GlcNAc unit (GL1). Incorporation of [^{14}C]Rha from dTDP-[^{14}C]Rha took place only into GL2, suggesting that GL2 was a polyprenol-P-P-GlcNAc-Rha, which was then confirmed chemically. Addition of a cell wall enzyme preparation resulted in the emergence of more polar glycolipids, GL3 and GL4. The inclusion of UDP-[^{14}C]Gal*p* resulted in exclusive labeling of GL3 (trisaccharide) and GL4 (tetrasaccharide), indicating growth of the galactan chain on the polyprenol-P-P-GlcNAc-Rha unit. Glycosyl linkage analysis of [^{14}C]Gal-labelled glycolipids 3 and 4 showed t-Gal*f* and 5-linked Gal*f*, consistent with the proposed structure of arabinogalactan and suggesting the presence of the mutase within the preparations converting UDP-Gal*p* to UDP-Gal*f*. More recently, we successfully solubilized the polymerized product resulting from labelling with either UDP-[^{14}C]GlcNAc or UDP-[^{14}C]Gal*p*. Analysis showed the emergence of even higher intermediates, GL5, 6, and so on, eventually resulting in a polymer which possessed the characteristics of mild acid-lability, mild-alkali stability, solubility in an extremely polar organic solvent and exclusion from Bio-Gel P-100, suggesting a highly polymerized lipid-linked version of GL1–4. The polymer was found to consist of 35–50 residues, and glycosyl linkage analysis produced t-Gal*f*, 5-Gal*f*, 6-Gal*f*, and 5,6-Gal*f*, demonstrating that it contained the full galactan with some branching, probably due to the presence of a few Ara*f* residues.

Another major development was the discovery of C_{50}-P-Ara*f* (Wolucka et al 1994) and chemical synthesis of C_{50}-P-[^{14}C]Ara*f* (Lee et al 1995), and the consequent development of a basic arabinosyl transfer assay. The [^{14}C]Ara*f* was incorporated into wall material; it was boosted by co-incubation with a variety of sugar nucleotides; it was linear with respect to membrane protein and acceptor cell wall concentration; and the β-anomer only was active. Previously, ethambutol (a widely used antituberculosis drug) was shown to inhibit arabinan synthesis. We found that ethambutol inhibited arabinosyl transferase activity in the C_{50}-P-Ara*f* assay to a residual activity of 40% at 50 μg/mL, suggesting that C_{50}-P-Ara*f* is a substrate to more than one arabinosyl transferase for arabinogalactan and LAM biosynthesis and that the primary mode of action of ethambutol is as an inhibitor of some of these arabinosyl transferases. We also demonstrated that *in vivo* ethambutol primarily inhibited the synthesis of the arabinan of arabinogalactan, while inhibition of the corresponding arabinan components of LAM occurred later, suggesting a secondary target and separate pathways for arabinogalactan and LAM biosynthesis. Moreover, while the synthesis of these arabinans was normal in an ethambutol-resistant isogenic strain, the addition of ethambutol to the resistant strain resulted in partial inhibition of the synthesis of the arabinan of LAM, resulting in the emergence of novel, truncated forms of LAM, indicating various degrees of susceptibility to ethambutol of the resistant gene(s) encoding an array of extramembranous arabinosyl transferases (Mikušová et al 1995). Through the use of target overexpression by a plasmid vector as a selection tool, we have now cloned the *Mycobacterium avium* emb region, which contains three open reading frames — *emb*R, *emb*A and *emb*B — that render the otherwise susceptible *M.smegmatis* resistant to ethambutol. The C_{50}-P-Ara*f* assay demonstrated that *emb*A and *emb*B are associated with high level ethambutol-resistant arabinosyl transferase activity, and *emb*R appears to modulate their *in vitro* level of activity (Belanger et al 1996). Thus, it would appear that *emb*A and *emb*B encode the drug target of ethambutol, and represent the putative arabinosyl transferases associated with arabinogalactan synthesis.

Arising from these fascinating results on the differential effects of ethambutol on the synthesis of the arabinan of arabinogalactan and LAM (and in light of the biological importance of LAM), we conducted an intensive and innovative study of the early steps in LAM/lipomannan (LM) biosynthesis (pre-arabinan attachment). The outcome of this work was the demonstration that the phosphatidylinositol mannosides were the precursors of the linear mannan portion of LM/LAM, and the donor of mannose was not GDP-Man but C_{50}-P-Man. The product of these transformations is linear LM, and, apparently, GDP-Man is the source of the branched Man*p* residues of mature LM/LAM. The arabinan of LAM apparently arises in a fashion similar to the arabinan of arabinogalactan.

Ehlers: The mycobacterial wall interests me because of its potential interaction with host cells. The outer layer of the organism is a highly dynamic structure, and the preponderance of certain types of molecules there is surprising, in view of the long-held view that organism is extremely lipid rich and is essentially a waxed particle, which to a large extent is true. However, the stratification of the wall and distinction between different areas is even more extreme than has been elegantly shown over many years by Patrick Brennan and many others. The role of outer polysaccharides and glycolipids in interacting with host cell receptors on first encountering a cell, and later in promoting survival, may be more extensive than we appreciate. I was also fascinated by what David Russell showed yesterday, i.e. the enormous release of mycobacterial glycolipids, which not only may be distributed through the entire cell, but may also be taken up by surrounding tissues. There's an enormous potential for understanding immunopathology and uncovering novel drug targets. The regulation of this dynamic wall, especially the outer layers, may represent an area where there is a lot of variation between strains. We have some evidence for this from *in vitro* studies with clinical isolates (Cywes et al 1997), and we have found that the situation is much more dynamic than we first appreciated. We have to get away from the idea that the organism is static, and that everything interesting happens to the host.

Young: Presumably, the growth phase of the organism is also a variable.

Russell: Release of cell wall lipids seems to be a dynamic, energy-requiring process because it is much less noticeable in killed bugs.

Kaplan: There are two issues here. One is the fundamental understanding of an organism that is interesting, complicated, sophisticated and defies many simple approaches; and the other is the disease. Can we view these as two opposing and separate issues today? I would be the last person to argue against the intellectual value of any kind of pursuit of knowledge — be it at the level of the organism, biosynthetic pathways, or genetic organization and regulation. These approaches are unquestionably valuable. However, we have to consider the physiology of the disease, and we cannot pretend that a colony on agar is the disease, or that a colony on the surface of the lung is the disease. We are dealing with a chronic granulomatous disease that happens to be caused by an organism which lives and replicates inside macrophages. We have to think about the host, and we have to work ultimately on tuberculosis and not *M. tuberculosis*.

Ehlers: I'm not questioning the importance of the host. I am just saying that not long ago people viewed *M. tuberculosis* as an inert particle which was not virulent, and that the immunopathology of tuberculosis could be ascribed solely to an exuberant and inappropriate host reaction, which was more inappropriate in some individuals than in others.

Colston: Clearly, the organism is capable of making these sort of changes, but the changes are likely to be driven by the host. For example, the thing that drives flu virus to changes its haemagglutinin molecule is the host response to the haemagglutinin.

Ress: I would also like to add that perhaps the most useful purpose the macrophage serves is when it undergoes apoptosis in the process of dealing with *M. tuberculosis*, because this is when the viability of *M. tuberculosis* is reduced. The problem occurs when the organism survives within the macrophage in a dormant phase.

References

Belanger AE, Besra GS, Ford ME et al 1996 The *embAB* genes of *Mycobacterium avium* encode an arbinosyl transferase involved in cell wall arabinan biosynthesis that is the target for the antimycobacterial drug ethambutol. Proc Natl Acad Sci USA 93:11919–11924

Besra GS, Khoo K-H, McNeil MR, Dell A, Morris HR, Brennan PJ 1995 A new interpretation of the structure of the mycolyl-arabinogalactan complex of *Mycobacterium tuberculosis* as revealed through characterization of oligoglycosylalditol fragments by fast atom bombardment mass spectrometry and ^{1}H nuclear magnetic resonance spectroscopy. Biochemistry 34:4257–4266

Brennan PJ, Draper P 1994 Ultrastructure of *Mycobacterium tuberculosis*, In: Bloom BR (ed) Tuberculosis: pathogenesis, protection and control. American Society of Microbiology, Washington, D.C., p 271–284

Brennan PJ, Nikaido H 1995 The envelope of mycobacteria. Annu Rev Biochem 64:29–63

Cywes C, Hoppe HC, Daffe M, Ehlers MRW 1997 Nonopsonic binding of *Mycobacterim tuberculosis* to complement receptor type 3 is mediated by capsular polysaccharides and is strain dependent. Infect Immun 65:4258–4266

Daffe M, Brennan PJ, McNeil MR 1990 Predominant structural features of the cell wall arabinogalactan of *Mycobacterium tuberculosis* as revealed through the characterization of oligoglycosyl alditol fragments by gas-chromatography and by ^{1}H and ^{13}C-NMR analyses. J Biol Chem 265:6734–6743

Lee RE, Mikušová K, Brennan PJ, Besra GS 1995 Synthesis of the mycobacterial arabinose donor *β*-D-arabinofuranosyl-1-monophosphoryldecaprenol, development of a basic

arabinosyl-transferase assay, and identification of ethambutol as an arabinosyl transferase inhibitor. J Am Chem Soc 117:11829–11832
McNeil M, Daffe M, Brennan PJ 1991 Location of the mycolyl ester substituent in the cell walls of mycobacteria. J Biol Chem 266:13217–13223
Mikušová K, Slayden RA, Besra GS, Brennan PJ 1995 Biogenesis of the mycobacterial cell wall and the site of action of ethambutol. Antimicrob Agents Chemother 39:2482–2489
Mikušová K, Mikus M, Besra GS, Hancock I, Brennan PJ 1996 Biosynthesis of the linkage region of the mycobacterial cell wall. J Biol Chem 271:7820–7828
Nassu PM, Martin SC, Brown RE et al 1996 Galactofuranose biosynthesis in *Escherichia coli* K12: identification and cloning of UDP-galactopyranose mutase. J Bacteriol 178:1047–1052
Scherman M, Kalbe-Bournonville L, Bush D, Xin Y, Deng L, McNeil M 1996 Polyprenylphosphate-pentoses in mycobacteria are synthesized from 5-phosphoribose pyrophosphate. J Biol Chem 271:29652–29658
Wolucka BA, McNeil MR, de Hoffmann E, Chojnacki T, Brennan PJ 1994 Recognition of the lipid intermediate for arabinogalactan/arabinomannan biosynthesis and its relation to the mode of action of ethambutol on mycobacteria. J Biol Chem 269:23328–23335

The impact of genomics on the search for novel tuberculosis drugs

Ken Duncan

Glaxo Wellcome Research and Development, Medicines Research Centre, Gunnels Wood Road, Stevenage, Hertfordshire SG1 2NY, UK

Abstract. The emergence of multidrug-resistant strains of *Mycobacterium tuberculosis* has highlighted the need for new drugs to treat tuberculosis. Drugs that either shorten the overall duration of therapy or that simplify the regimen would significantly improve compliance and hence reduce treatment failure rates. The drug development process begins with identification and validation of specific targets. These may be relevant for inhibiting growth of the bacterium *in vitro*, and hence yield novel bactericidal agents, or they may be required at other stages of growth, such as survival in host macrophages. With the availability of the complete genome sequence of *M. tuberculosis*, the primary sequence of every drug target in the pathogen is known. A combination of approaches is being employed to exploit the information contained in the genome and thereafter to identify lead compounds that may yield new drugs.

1998 Genetics and tuberculosis. Wiley, Chichester (Novartis Foundation Symposium 217) p 228–238

The chemotherapeutic era of tuberculosis treatment began in 1944 with the discovery of streptomycin by Selman Waksman and his colleagues. Isoniazid, the agent with the most potent activity known against the tubercle bacillus, *Mycobacterium tuberculosis*, was introduced in 1952. The finding that the broad-spectrum antibacterial agent rifamycin has activity against *M. tuberculosis* revolutionized treatment of tuberculosis with the subsequent development of the short-course (albeit six-month) multidrug regimen in use today. This regimen can achieve cure rates of >95% when used correctly. Despite the wide availability of these relatively inexpensive drugs, the number of tuberculosis sufferers continues to rise worldwide, prompting the World Health Organization (WHO) to declare tuberculosis 'a global emergency' in 1993.

The drawbacks of the multidrug regimen are obvious. The number of tablets, their toxic side-effects and the long duration of the therapy lead to poor compliance. Inevitably, this results in a significant rate of treatment failure and, worse, selection of resistant organisms. In a recent global survey, drug-resistant tuberculosis was found in every country that reported data, with a median level of 10.4% primary resistance, and 'hot spots' where resistance as high as 41% was identified (World Health Organization 1997a). Patients with multidrug-resistant tuberculosis are difficult to treat and continue to infect others with the resistant bacteria. In order to overcome the limitations in current therapy, the WHO recommends the implementation of a programme called directly observed therapy, short-course, or DOTS (World Health Organization 1997b). New chemotherapeutic agents with activity against multidrug-resistant tuberculosis and drugs that can provide a shorter and simpler regimen are needed.

Since the 1960s, there has been relatively little progress in tuberculosis drug development. Semi-synthetic rifamycin derivatives such as rifabutin and rifapentine (Baohong et al 1993) have not yet achieved widespread clinical use. The experimental benzoxazinorifamycin, KRM-1648 (Fig. 1), also shows promise (Saito et al 1991, Klemens et al 1994a), but such agents represent incremental steps in therapy improvement. Although they have some advantages over parent rifampicin, such as a longer half-life in humans which may permit intermittent chemotherapy, there are disadvantages of this approach. In particular, overcoming existing resistance mechanisms to the entire class of compounds can be an insurmountable challenge.

New broad-spectrum antibacterial agents that have particularly good activity against *M. tuberculosis* are also being used in the fight against tuberculosis (Fig. 1). The quinolone levofloxacin (Klemens et al 1994b) is already being used in the clinic and the oxazolidinone U-100480 has good *in vitro* and *in vivo* activity (Barbachyn et al 1996).

Researchers at PathoGenesis Corporation have described a series of nitroimidazopyrans, exemplified by PA-824 (Fig. 1), with potent selective antimycobacterial activity. They have no cross-resistance with other antibiotics and work via a novel, as yet uncharacterized mechanism (W. R. Baker, E. L. Keeler, S. Cai, J. A. Towell, D. R. Pastor, J. N. Morgenroth, S. W. Anderson & T. M. Arain, unpublished paper, Interscience Conference on Antimicrobial Agents and Chemotherapy, 15–18 September 1996).

Immunotherapeutic approaches are described in detail elsewhere in this symposium (this volume: Johnson et al 1998, Rook & Hernandez-Pando 1998).

Over 95% of tuberculosis sufferers are in the developing world. The drugs available today can achieve high cure rates. *M. tuberculosis* is a slow-growing, airborne pathogen that requires specialized handling facilities and the available models of infection are lengthy and difficult. Together, these factors make developing a new tuberculosis drug a daunting challenge. In this chapter the factors that influence the direction of a tuberculosis drug development programme are discussed. The availability of the *M. tuberculosis* genome sequence adds a new dimension to our knowledge of this important pathogen, and provides

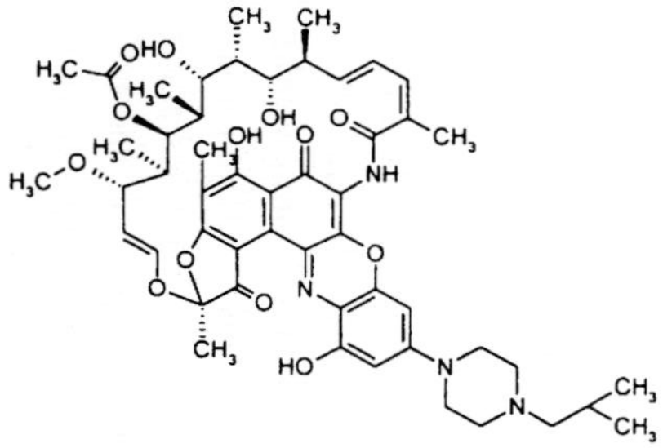

KRM-1648

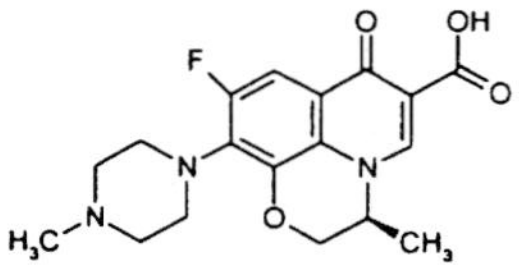

Levofloxacin

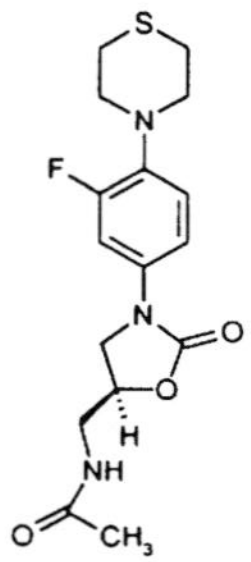

U-100480

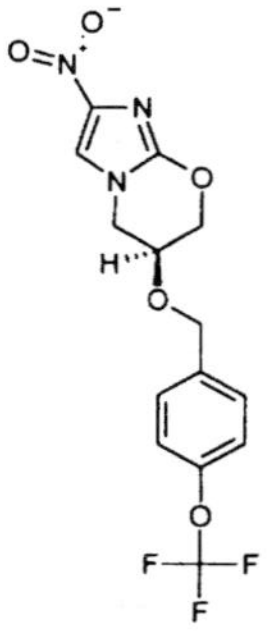

PA-824

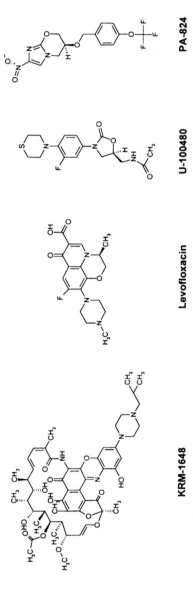

FIG. 1. Structure of potential new tuberculosis drugs.

greater opportunity than ever before for the rapid identification and validation of novel drug targets.

New drug profile

The first step in the tuberculosis drug development process is to consider the desired activity profile and properties we are aiming for, as this will determine the direction of the research programme. The 'ultimate' tuberculosis drug would possess both rapid bactericidal activity and sterilizing activity, thus killing all *M. tuberculosis* populations, and the spectrum of activity would include all multidrug-resistant tuberculosis isolates. The drug must be orally bioavailable, have low toxicity, good tissue distribution to maximize activity against intracellular organisms, a long elimination half-life so that intermittent chemotherapy may be considered and, last but not least, it must be inexpensive to produce given the constraints on pricing that exist. Clearly, it will be difficult to build all these properties into a single molecule and compromises will be made. Critically, any new drug must offer a significant advantage over the drugs already available, such as activity against multidrug-resistant tuberculosis, better tolerability, intermittent dosing or shorter overall therapy duration.

It is most unlikely that *M. tuberculosis* has a single Achilles' Heel, and that more than one agent will always be needed to bring about a complete cure. The most achievable goal is to generate a novel bactericidal agent. Such an agent may be evaluated by the minimal inhibitory concentration (MIC) for *M. tuberculosis* growth in culture, and in short-term acute *in vivo* models of infection. Clinical testing in humans is straightforward, with early bactericidal activity (Mitchison 1996) being a relatively good predictor of final clinical efficacy. On the other hand, disadvantages such as emergence of resistance, fuelled by ineffective or inappropriate combinations, or reservation of the new drug for use in cases where treatment is already failing would make it uneconomic to produce such a drug, unless some other property such as a long half-life would provide an advantage.

Targeting the persisting organisms offers the tantalizing prospect of reducing the duration of therapy. However, specific molecular targets are not so obvious. Significant research effort is required to validate such targets and new *in vitro* and *in vivo* models must be developed if such agents are to be fully evaluated. It will be more difficult to optimize the properties of any agent that has no MIC in culture, and surrogate markers of sterilization must be identified. Clinical trials will be lengthy and require large patient numbers, increasing development costs.

We may therefore envisage intervening at several steps in the disease process. The targets may be visualized on a spectrum, ranging from those functions essential for survival of the bacterium in culture through host–pathogen interaction to host immunity (Fig. 2). The effort required to identify and validate targets varies considerably, with proportionally greater effort being expended on targets that have the highest novelty.

Whole-cell screening

The simplest way to find new tuberculosis drugs is to screen for whole-cell antimycobacterial activity, much in the way that streptomycin and isoniazid were discovered. Realizing that it was impractical to do this with *M. tuberculosis* itself, researchers at Glaxo Wellcome employed a rapid-growing, non-pathogenic surrogate, *Mycobacterium aurum A+* (Chung et al 1995). Using this organism as a host for the *Vibrio harveyi luxAB* genes, we developed a simple bioluminescence assay to screen about 10 000 compounds/day synthesized in bead-based combinatorial libraries (G. A. C. Chung, P. Andrew, S. Polger, J. Silen, C. DeLuca-Flaherty & K. Duncan, unpublished work 1997). Other high throughput assays have been described that monitor viability with bioluminescence (Cooksey et al 1993, Arain et al 1996) or using Alamar Blue (Collins & Franzblau 1997).

Although several novel structural templates with activity against *M. tuberculosis* have been discovered this way, there are significant disadvantages to this approach. Many targets will be masked from the samples being tested, for example if they are intracellular, or are not expressed under the conditions used in the assay. In addition, it is often hard to improve upon the properties of a whole-cell active lead compound when there is no knowledge of the target. It is generally thought to be more effective to screen for agents that modulate the activity of a specific

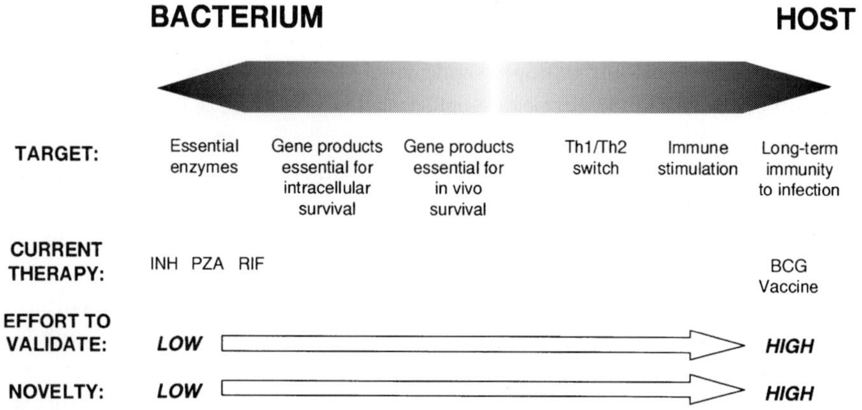

FIG. 2. Targets for novel tuberculosis drugs. BCG, Bacillus Calmette–Guérin; INH, isoniazid; PZA, pyrazinamide; RIF, rifampicin; Th, T helper.

target *in vitro*, and then to further modify any leads obtained to incorporate or improve upon the whole-cell activity. For this approach to be successful, significant effort must be applied to identifying, characterizing and validating appropriate targets. Targets identified by genetic means are not always suitable for screening; often the function of a gene product may not be known or an enzyme's substrate may not be available.

Ideally, one would wish to either hit multiple targets with a single new agent or hit a target that has a role in more than one growth phase. The advantage of the former is that it is highly unlikely that resistance will develop by a single mutation in the gene encoding the target, since the alternative target(s) would still be inhibited. Such a scenario can be envisaged for inhibition of the multiple arabinosyl transferases that recognize the substrate decaprenol-phosphoarabinose (Mikušová et al 1995, Lee et al 1995) or the three proteins that catalyse mycolyl transfer (Belisle et al 1997).

Target identification

Novel drug targets may be identified in several ways. The classical approach is to study a process, identify a protein target and purify sufficient material to obtain sequence information, then clone the gene and use reverse genetics to confirm its essentiality. It is known that disruption of cell wall synthesis is lethal to the bacterium, and several steps in this process are being analysed in this way to reveal targets. In some cases, for example arabinan biosynthesis (Mikušová et al 1995) and linkage region biosynthesis (Mikušová et al 1996), this has only progressed as far as developing a crude assay system. In other cases, enzyme function has been correlated with a single gene product, e.g. UDP-galactopyranose mutase (Weston et al 1997), or several gene products, e.g. mycolyl transferase (Belisle et al 1997).

Alternatively, determining the mode of action of today's drugs identifies targets. For example, the *inhA* gene product was identified as a target for isoniazid (Banerjee et al 1994) and subsequently shown to possess enoyl reductase activity (Quémard et al 1995), a step in fatty acid metabolism. Biochemical studies have indicated that ethambutol acts by inhibiting arabinosyl transferase (Lee et al 1995) and further genetic studies indicated that the *emb* genes encode the target (Belanger et al 1996, Telenti et al 1997).

Progress in identifying essential genes in *M. tuberculosis* has been hampered by the lack of well-characterized mutants. The recent description of a set of temperature-sensitive mutants of *Mycobacterium smegmatis* (A. Belanger, J.C. Porter, G. Hatfull, unpublished paper, ASM Conference on Tuberculosis: past, present and future, 8–12 July 1997) is a step towards defining specific targets in *M. tuberculosis*. Moreover, libraries of *M. tuberculosis* mutants have been generated by randomly inserting a transposon into the chromosome, using plasmid (Pelicic et al 1997) and mycobacteriophage (Bardarov et al 1997) delivery systems. Although it is not possible to recover mutants generated when the transposon interrupts a gene essential for replication, we may nevertheless deduce which gene products are essential for vegetative growth by sequencing the insertion points and by referring to the genome sequence (with the proviso that there may be polar or other effects that will complicate interpretation).

Virulence gene targets

The pathogen must employ a range of virulence mechanisms that enable it to establish an infection and survive within the host. Genes encoding virulence factors may only be expressed at certain phases of infection, and may not be required for growth on agar plates. Alternative strategies are employed to find virulence genes, in both macrophage and whole-animal models. These include IVET (*in vivo* expression technology; Mahan et al 1993), *in vivo*-expressed promoter trapping (F. da Silva-Tatley & M. R. W. Ehlers, unpublished work 1997), signature-tagged mutagenesis (Hensel et al 1995), RNA differential display (Fislage et al 1997), RNA arbitrarily primed PCR (McClelland et al 1995) and proteomics (see below). Although such methods are useful in their own right, they are much more powerful when the whole genome sequence is available. Only a relatively short DNA sequence need be determined in order to locate the complete gene and, furthermore, the gene may be placed into context, important when neighbouring genes may have a related function or may be coexpressed.

The 'genomic approach'

By having the complete sequence of the tuberculosis pathogen at our fingertips, it is self-evident that the sequence of every single target is known. The challenge is to identify which of the ~4000 open reading frames is essential under any given set of growth conditions. The first step is to build a metabolic map, by comparing the mycobacterial genes present with those genes of known function isolated from other organisms. This may provide a clue to the way that *M. tuberculosis* survives and grows. For example, the presence of genes associated with anaerobiosis may suggest a strategy for killing the bacterium in the sterilizing phase. Inevitably, many open reading frames in the genome sequence will have no counterpart in other bacteria, and hence it is not possible to assign a function to the gene product. The employment of high throughput screening to identify antagonists is one way to validate such targets.

Comparison of the *M. tuberculosis* sequence with that of other bacteria reveals both what is in and what is not — genes uniquely found in mycobacteria represent particularly attractive targets. The unusual situation of having the complete genome sequence of two *M. tuberculosis* strains, namely the laboratory strain H37Rv (http://www.sanger.ac.uk/Projects/M_tuberculosis) and a recent clinical isolate (http://www.tigr.org), may be of benefit if we assume that the clinical strain has greater virulence in humans. Differences in the intergenic regions, where gene promoters are located, may suggest clues to pathogenesis.

Proteome analysis

A complementary approach is to study the total protein complement, or 'proteome', of the bacterium at different stages of the growth cycle. Proteins are isolated then separated by electrophoresis in two dimensions, firstly by isoelectric focusing and thereafter by molecular mass, to yield a characteristic pattern of spots after staining (Urquhart et al 1997). The individual protein in each spot may be isolated and identified by accurately determining its mass in a spectrometer either directly, or following trypsin digestion and sizing of the resulting fragments. Reference to a database of predicted protein molecular weights generated from the mycobacterial genome sequence would tell which protein is present in the spot. If the protein appears not to be in the predicted set, techniques such as tandem nanoelectrospray mass spectrometry yield sequence data from which the protein may be identified. This technique is particularly useful for detecting and analysing post-translational modifications such as glycosylation (M. Ward, W. Blackstock, M.-P. Gares, D. B. Young & C. Abou-Zeid, unpublished work 1997).

In summary, the availability of the genome sequence of *M. tuberculosis* is providing many benefits in the search for new drugs to treat tuberculosis. The efficiency and speed with which targets may be identified increases the likelihood that novel potent leads will be found, which may then be developed into the next generation of antituberculosis drugs. Furthermore, new approaches may be followed, thereby increasing the diversity of targets that are available for study.

Acknowledgements

I acknowledge the contribution many people have made to the strategy outlined in this article, including colleagues at Glaxo Wellcome, Gavin Chung, Martin Everett, Karen Kempsell, Pauline Lukey, Ruth McAdam, Steve Martin and Paul Smith, and many academic collaborators participating in the Action TB Initiative.

References

Arain TM, Resconi AE, Hickey MJ, Stover CK 1996 Bioluminescence screening in *invitro* (Bio-*Siv*) assays for high-volume antimycobacterial drug discovery. Antimicrob Agents Chemother 40:1536–1541

Banerjee, A, Dubnau E, Quémard A et al 1994 *inhA*, a gene encoding a target for isoniazid and ethionamide in *Mycobacterium tuberculosis*. Science 263:227–230

Baohong J, Truffot-Pernot C, LaCroix C et al 1993 Effectiveness of rifampin, rifabutin, and rifapentine for preventive therapy of tuberculosis in mice. Am Rev Respir Dis 148:1541–1546

Barbachyn MR, Hutchinson DK, Brickner SJ et al 1996 Identification of a novel oxazolidinone (U-100480) with potent antimycobacterial activity. J Med Chem 39:680–685

Bardarov S, Kriakov J, Carriere C et al 1997 Conditionally replicating mycobacteriophages: a system for transposon delivery to *Mycobacterium tuberculosis*. Proc Natl Acad Sci USA 94:10961–10966

Belanger AE, Besra GS, Ford ME et al 1996 The *embAB* genes of *Mycobacterium avium* encode an arabinosyl transferase involved in cell wall arabinan biosynthesis that is the target for the antimycobacterial drug ethambutol. Proc Natl Acad Sci USA 93:11919–11924

Belisle JT, Vissa VD, Sievert T, Takayama K, Brennan PJ, Besra GS 1997 Role of the major antigen of *Mycobacterium tuberculosis* in cell wall biogenesis. Science 276:1420–1422

Chung GAC, Aktar Z, Jackson S, Duncan K 1995 High-throughput screen for detecting antimycobacterial agents. Antimicrob Agents Chemother 39:2235–2238

Collins LA, Franzblau SG 1997 Microplate alamar blue assay versus BACTEC 460 system for high-throughput screening of compounds against *Mycobacterium tuberculosis* and *Mycobacterium avium*. Antimicrob Agents Chemother 41:1004–1009

Cooksey RC, Crawford JT, Jacobs WR, Shinnick TM 1993 A rapid method for screening antimicrobial agents for activities against a strain of *Mycobacterium tuberculosis* expressing firefly luciferase. Antimicrob Agents Chemother 37:1348–1352

Fislage R, Berceanu M, Humboldt Y, Wendt M, Oberender H 1997 Primer design for a prokaryotic differential display RT-PCR. Nucleic Acids Res 25:1830–1835

Hensel M, Shea JE, Gleeson C, Jones MD, Dalton E, Holden DW 1995 Simultaneous identification of bacterial virulence genes by negative selection. Science 269:400–403

Johnson B, Bekker L-G, Ress S, Kaplan G 1988 Recombinant interleukin 2 adjunctive therapy in multidrug-resistant tuberculosis. In: Genetics and tuberculosis. Wiley, Chichester (Novartis Found Symp 217) p 99–111

Klemens SP, Grossi MA, Cynamon MH 1994a Activity of KRM-1648, a new benzoxazinorifamycin, against *Mycobacterium tuberculosis* in a murine model. Antimicrob Agents Chemother 38:2245–2248

Klemens SP, Sharpe CA, Rogge MC, Cynamon MH 1994b Activity of levofloxacin in a murine model of tuberculosis. Antimicrob Agents Chemother 38:1476–1479

Lee RE, Mikušová K, Brennan PJ, Besra GS 1995 Synthesis of the mycobacterial arabinose donor β-D-arabinofuranosyl-1-monophosphoryldecaprenol, development of a basic arabinosyl-transferase assay, and identification of ethambutol as an arabinosyl transferase inhibitor. J Am Chem Soc 117:11829–11832

Mahan MJ, Slauch JM, Mekalanos JJ 1993 Selection of bacterial virulence genes that are specifically induced in host tissues. Science 259:686–688

McClelland M, Mathieu-Daude F, Welsh J 1995 RNA fingerprinting and differential display using arbitrarily primed PCR. Trends Genet 11:242–246

Mikušová K, Slayden RA, Besra GS, Brennan PJ 1995 Biogenesis of the mycobacterial cell wall and the site of action of ethambutol. Antimicrob Agents Chemother 39:2484–2489

Mikušová K, Mikus M, Besra GS, Hancock I, Brennan PJ 1996 Biosynthesis of the linkage region of the mycobacterial cell wall. J Biol Chem 271:7820–7828

Mitchison DA 1996 Modern methods for assessing the drugs used in the chemotherapy of mycobacterial disease. J Appl Bacteriol 81:72S–80S

Pelicic V, Jackson M, Reyrat J-M, Jacobs WR, Gicquel B, Guilhot C 1997 Efficient allelic exchange and transposon mutagenesis in *Mycobacterium tuberculosis*. Proc Natl Acad Sci USA 94:10955–10960

Quémard A, Sacchettini JC, Dessen A et al 1995 Enzymatic characterisation of the target for isoniazid in *Mycobacterium tuberculosis*. Biochemistry 34:8235–8241

Rook GAW, Hernandez-Pando R 1998 Immunological and endocrinological characteristics of tuberculosis that provide opportunities for immunotherapeutic intervention. In: Genetics and tuberculosis. Wiley, Chichester (Novartis Found Symp 217) p 73–98

Saito H, Tomioka T, Amitani R et al 1991 *In vitro* antimycobacterial activities of newly synthesised benzoxazinorifamycins. Antimicrob Agents Chemother 35:542–549

Telenti A, Philipp WJ, Sreevatsan S et al 1997 The *emb* operon, a gene cluster of *Mycobacterium tuberculosis* involved in resistance to ethambutol. Nat Med 3:567–570

Urquhart BL, Atsalos TE, Roach D et al 1997 'Proteomic contigs' of *Mycobacterium tuberculosis* and *Mycobacterium bovis* (BCG) using novel immobilised pH gradients. Electrophoresis 18:1384–1392

Weston A, Stern RJ, Lee RE et al 1997 Biosynthetic origin of mycobacterial cell wall galactofuranosoyl residues. Tubercle Lung Dis 78:123–131

World Health Organization 1997a Anti-tuberculosis drug resistance in the world. World Health Organization, Geneva

World Health Organization 1997b WHO Report on the Tuberculosis Epidemic, 1997. World Health Organization, Geneva

DISCUSSION

Davies: You mentioned studies on the *Mycobacterium smegmatis* genome. There were many scientists who wanted to have this genome sequenced in the beginning, and it's now clear what a mistake it was not to do so. If you realize what the genome sequence of *Escherichia coli* K12 is telling us about *E. coli* and *Salmonella* pathogenesis, it's clear that comparisons between such organisms can enable the identification of non-hybridizing DNA fragments that may have roles in pathogenesis. Now a hypothetical question. Let's suppose there were a few members of a politically important group in an industrialized country, for example the British Parliament, that came down with multidrug-resistant tuberculosis tomorrow, what cocktail of available orphan drugs would be recommended to treat them?

Duncan: Any new therapy for multidrug-resistant tuberculosis comes under the orphan drug status under the Food and Drug Administration's rules. I can't answer what sort of cocktail would be used, but maybe the clinicians here could. I can only reiterate the point that you can't aim to treat only patients with multidrug-resistant infections, you have to aim to treat everybody with a new therapy.

Bateman: The World Health Organization recently released guidelines for multidrug resistance which provide both standardized and individualized treatment schedules. Both contain combinations of an aminoquinalone (e.g. ofloxacin or ciprofloxacin), pyrazinamide, ethionamide, aminoglycoside (kanamycin or amikacin) and ethambutol, but not isoniazid. This is the standard treatment against which the efficacy of most drugs are being measured today. Factors such as toxicity, cost and resistance are considered when deciding about the use of drugs with borderline efficacy and on the choice of alternatives in the same class (e.g. whether to use levafloxacin or ofloxacin).

Davies: Are there no potential orphan drugs that one would be able to throw into the cocktail straight away?

Bateman: No. Most of the other drugs have questionable efficacies. Some that we currently use (e.g. dapsone and cycloserine) have questionable or low efficacies, but are used as part of a multidrug regimen in the hope that they strengthen the regimen.

Russell: Cells have active mechanisms for getting rid of some of these drugs, and many of these drugs are being used without any thought as to how they are partitioned and sequestered within the cell. For example, an organic anion transporter pumps fluoroquinolone out of the cytosol into the endosomal network, and then it is trafficked out of the cell. However, if probenicid is administered together with fluoroquinolone, the release of the drug from the cell is blocked.

Hopewell: The other category of agent that may be useful is the β-lactam/β-lactamase combination (amoxicillin/clavulanic acid). Not that I would want to advertise this as first-line therapy, but we've treated a series of multidrug-resistant patients with a combination of amikacin and amoxicillin/clavulanic acid intravenously by a peripherally inserted central venous catheter that stays in place for months. We have had several fairly dramatic successes in difficult patients, one of whom had been smear positive for 12 years.

Steyn: Could you comment about the future of immunotherapy. Will a drug ever be designed that will produce effective immunomodulation?

Duncan: There is definitely a place for immunotherapy, but we must first understand the mechanism and then design clinical trials in a rational way. The key point is to design something that is relatively inexpensive and available to everyone.

DNA vaccines against tuberculosis

Jeffrey B. Ulmer[1], Donna L. Montgomery, Aimin Tang, Lan Zhu, R. Randall Deck, Corrille DeWitt, Olivier Denis*, Ian Orme†, Jean Content* and Kris Huygen*

*Department of Virus and Cell Biology, Merck Research Laboratories, WP-16–3 West Point, PA 19486, USA, *Institut Pasteur, 642 rue Engelard, B-1180, Brussels, Belgium and †Department of Microbiology, Colorado State University, Fort Collins, CO 80523, USA*

Abstract. DNA plasmids encoding *Mycobacterium tuberculosis* antigen 85 (Ag85) were tested as vaccines in animal models. Ag85 DNA induced relevant immune responses (i.e. T helper (Th) cells, Th1 cytokines and cytotoxic T lymphocytes) and was protective in mouse and guinea pig models of mycobacterial disease. Therefore, DNA vaccination holds promise as an effective means of preventing tuberculosis in humans. Furthermore, this technique is amenable to identifying the protective antigens of *M. tuberculosis*.

1998 Genetics and tuberculosis. Wiley, Chichester (Novartis Foundation Symposium 217) p 239–253

Tuberculosis causes more deaths (~3 million per year) than any other disease caused by a single pathogen, despite the widespread availability of a tuberculosis vaccine. The lack of overall effectiveness of this vaccine, which is a live attenuated form of *Mycobacterium bovis* termed Bacillus Calmette–Guérin (BCG), makes it imperative that a more effective vaccine be developed. To this end, research is active in several areas, including experimental tuberculosis vaccines based on subunit proteins (Andersen 1994), live attenuated *M. tuberculosis* (Guleria et al 1996), recombinant BCG-expressing cytokines (Murray et al 1996), non-proteinaceous components of mycobacteria (Sieling et al 1995) and DNA vaccines (Huygen et al 1996, Tascon et al 1996).

DNA vaccines consist of an *Escherichia coli*-derived plasmid containing an eukaryotic expression cassette and a gene encoding a foreign antigen that, when inoculated into an animal, results in expression of the antigen *in situ*. DNA vectors have been used to induce immune responses and protective immunity in many experimental models of infectious and non-infectious disease (for review see Donnelly et al 1997). DNA vaccines are commonly administered by intramuscular

[1]Present address: Vaccines Research, Chiron Corporation, 4560 Horton Street, Emeryville, CA 94608, USA

(i.m.) injection, whereupon muscle cells (and possibly others) become transfected. This expression provides a source of antigen for induction of antibodies and T cell responses, including cytotoxic T lymphocytes (CTLs). This ability to induce a broad-based immune response may set DNA vaccines apart from certain other types of vaccines (such as subunit vaccines), which do not in general induce CTLs. For tuberculosis, it is generally regarded that immunity requires both cytokine-secreting T helper (Th) cells and CTL responses (Orme et al 1993). Hence, the application of DNA vaccination to tuberculosis was a logical extension of the technology. This chapter discusses early-stage preclinical studies that demonstrate proof of concept in animal models for this approach to tuberculosis vaccine development.

Materials and methods

DNA constructs

The Ag85A gene from *M. tuberculosis* was cloned into the *Bgl*II site of the V1Jns.tPA vector (Shiver et al 1995), as previously described (Huygen et al 1996). The V1Jns vector, which does not contain a gene insert, was used as a control. To determine relative expression levels of Ag85A *in vitro*, we transiently transfected rhabdomyosarcoma cells (CCL-136) with 10 μg Ag85 DNA together with 10 μg of V1JCAT (chloramphenical acetyltransferase) for transfection normalization according to CellPhect kit instructions (Pharmacia, Piscataway, NJ). Expression of Ag85 was visualized by immunoblot on Immobilon-P membranes (Millipore, Bedford, MA) as detailed elsewhere (Huygen et al 1996, Montgomery et al 1997).

For generation of stably transfected cells expressing Ag85, P815-HTR were transfected using Lipofectamine (according to Gibco [Grand Island, NY] instructions) with Ag85A DNA together with a plasmid for drug resistance to G418 (pcDNA3, Invitrogen [San Diego, CA]). Clonal isolates of drug-selected cells were prepared by limiting dilution in 96-well plates and analysed for Ag85 expression.

Measurement of antibody responses

BALB/c mice (six- to eight-week old females) purchased from Charles River Laboratories (Raleigh, NC) were anesthesized by intraperitoneal injection of ketamine/xylazine (100 mg/kg and 10 mg/kg, respectively), and injected i.m. once in both quadriceps for a total of 5 or 100 μg of Ag85A DNA using a 0.3 cc insulin syringe (Becton-Dickinson, Franklin Lakes, NJ). As negative controls, mice were uninjected or injected with control plasmid DNA (not containing a gene insert) in saline.

Anti-Ag85 antibodies were measured by an ELISA. Maxi-sorb 96-well plates (Nunc, Rochester, NY) were coated for 18 hr at 4°C with BCG culture filtrate at (5 μg total protein/ml) in phosphate-buffered saline (PBS). Serum samples or peroxidase-conjugated rabbit anti-mouse IgG (Zymed, South San Francisco, CA) were diluted in PBS containing Tween 20 (0.05%, v/v) and bovine serum albumin (1%, w/v), and incubated sequentially for 2 h at room temperature with extensive washing between each incubation. Plates were read spectrophotometrically at 490 nm after colour development with 0.1 M citrate buffer (pH 4.5) containing hydrogen peroxide (0.012%, v/v) and O-phenylenediamine (1 mg/ml).

Measurement of T cell responses

For lymphoproliferation, single-cell suspensions of spleen cells from DNA-vaccinated animals were depleted of erythrocytes in ACK Lysis Buffer (Gibco, Grand Island, NY) and stimulated with BCG culture filtrate or purified, native Ag85 (2 μg/ml) *in vitro* in round-bottomed microwell plates at 5×10^5 cells/ml in RPMI-1640 medium, supplemented with HEPES, glutamine, 10% fetal calf serum and 50 μM 2-mercaptoethanol. Cells were cultured for 5 days, and [^{3}H]thymidine was added at 1 μCi/well during the last 24 h. Cells were harvested onto glass fibre filter mats using a Titertek Cell harvester (Huntsville, AL) and radioactivity was measured in a Liquid Scintillation Counter (Betaplate™, Wallac, Gaithersburg, MD).

For CTLs, spleen cells from DNA-vaccinated BALB/c mice were restimulated for seven days *in vitro* with interleukin (IL)-2 (10 U/ml) and stably transfected P815 cells expressing Ag85, or with IL-2 (10 U/ml) and peptide (amino acids 61–80 of Ag85A) at 10 μg/ml. Target cells included P815-HTR mastocytoma cells stably expressing Ag85A or P815 cells untreated or pulsed with peptide, and used at effector : target ratios ranging up to 100 : 1. Details of the assay are given elsewhere (Ulmer et al 1993).

Vector construction and protein expression

To test whether vaccination with a DNA plasmid encoding a *M. tuberculosis* antigen could result in induction of immune responses, we constructed a vector containing the gene for antigen 85A (Ag85A). The vector backbone, termed V1Jns, contains the immediate early promoter and enhancer from cytomegalovirus (with intron A) and the transcription terminator from bovine growth hormone (Montgomery et al 1993, Shiver et al 1995), and had previously been shown to give high levels of expression for several different antigens and reporter proteins (not shown). The native mycobacterial signal sequence from Ag85A was replaced with the eukaryotic signal sequence from tissue-specific plasminogen activator protein, in order to facilitate expression and intracellular trafficking in eukaryotic cells. This construct resulted in expression and secretion of Ag85A in transiently transfected rhabdomyosarcoma cells *in vitro* (Huygen et al 1996, Montgomery et al 1997).

Assessment of immunogenicity in mice

As with many other DNA vaccines tested in animals, Ag85A DNA induced both humoral and cellular immune responses. Anti-Ag85 antibodies, as measured by ELISA, were detected six weeks after a single injection of 5 or 100 μg Ag85A DNA (Fig. 1). Endpoint antibody titres rose substantially with time and were in excess of 1/100 000 at 21 weeks after one injection of 100 μg DNA. The profile of immunoglobulin subclasses induced showed a predominance of IgG2a antibodies, with lower but significant levels of IgG1 and IgG2b (not shown), suggesting that a Th1-type response was preferentially induced.

T cell responses were measured in several ways. Spleen cells from Ag85A DNA-vaccinated mice were restimulated *in vitro* with specific antigen, in the form of either purified Ag85 protein or total culture filtrate proteins from BCG grown *in vitro*. In both instances, specific lymphoproliferation was induced upon restimulation. As with antibody responses, lymphoproliferation appeared to increase in intensity with time after injection. In mice inoculated with either 5 or 100 μg doses of Ag85A DNA (given once), proliferative responses increased between eight and 21 weeks post-injection (Fig. 2). This proliferation was accompanied by secretion of high levels of Th1-type cytokines, including γ-interferon (IFN-γ), the cytokine with

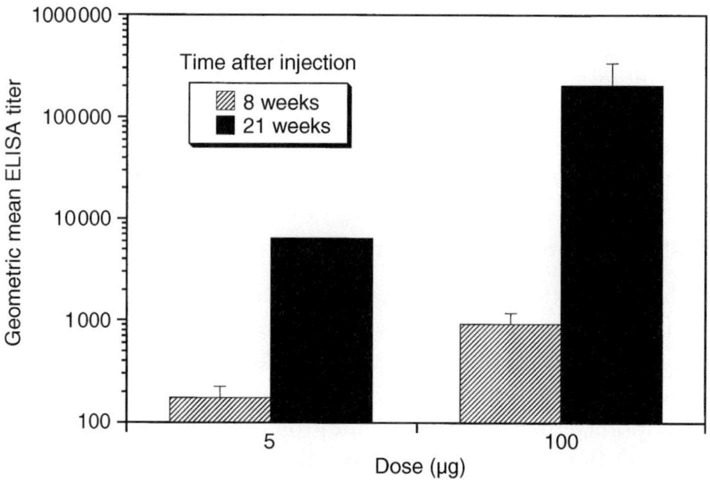

FIG. 1. Induction of anti-Ag85 antibodies by vaccination with Ag85A DNA. BALB/c mice were injected with 5 μg or 100 μg Ag85A DNA and serum samples were collected at eight (hatched bars) and 21 weeks (solid bars). Anti-Ag85 antibodies were measured by ELISA and data shown represent geometric mean ELISA titre $\pm$ S.E.M. for groups of 10 mice.

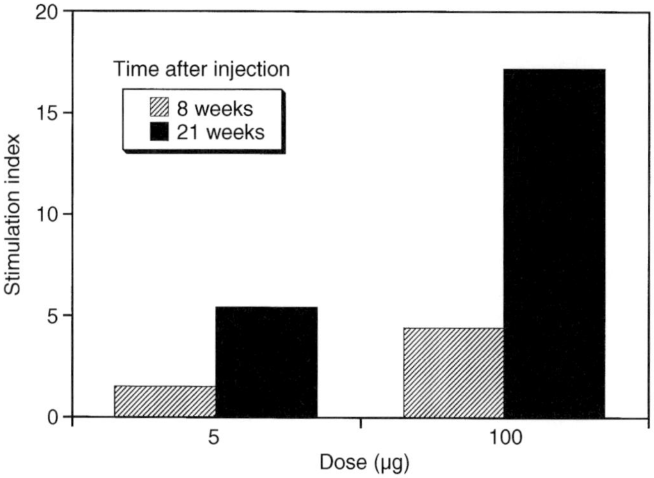

FIG. 2. Induction of lymphoproliferative responses by vaccination with Ag85A DNA. BALB/c mice were injected with 5 μg or 100 μg Ag85A DNA and spleen cells were prepared for antigen restimulation at eight (hatched bars) and 21 weeks (solid bars). Lymphocyte proliferation is shown as stimulation index for pools of three spleens per group.

potent anti-mycobacterial properties. IFN-γ levels in the culture supernatants approached 6 ng/ml, similar to those levels achieved by non-specific activation by the mitogen concanavalin A (not shown). Other cytokines secreted included IL-2, granulocyte macrophage colony-stimulating factor (GM-CSF) and tumour necrosis-α (TNF-α), but little or no IL-4, IL-6 and IL-10 (Huygen et al 1996). Therefore, as suggested by the IgG subtype profile of anti-Ag85 antibodies, Ag85A DNA induced a strong Th1-type response.

Potential induction of CTL responses by Ag85A DNA was assessed in two ways. First, spleen cells were restimulated *in vitro* with complete Ag85 protein then tested for lytic activity against stably transfected target cells expressing Ag85 (Huygen et al 1996). Second, a specific H-2^d-restricted CTL epitope was identified in Ag85A contained within the 20-mer peptide corresponding to amino acids 61–80 (Denis et al 1998). Peptide restimulation of spleen cells (from mice injected with 100 μg Ag85A DNA one year previously) activated Ag85A-specific effector cells capable of lysing target cells expressing Ag85A protein or pulsed with specific peptide, but not untreated cells or cells pulsed with a different Ag85A peptide (Fig. 3).

Assessment of protective efficacy

To measure the ability of Ag85A DNA to confer protective immunity against mycobacterial challenge, we used several animal models. In mice Ag85A DNA

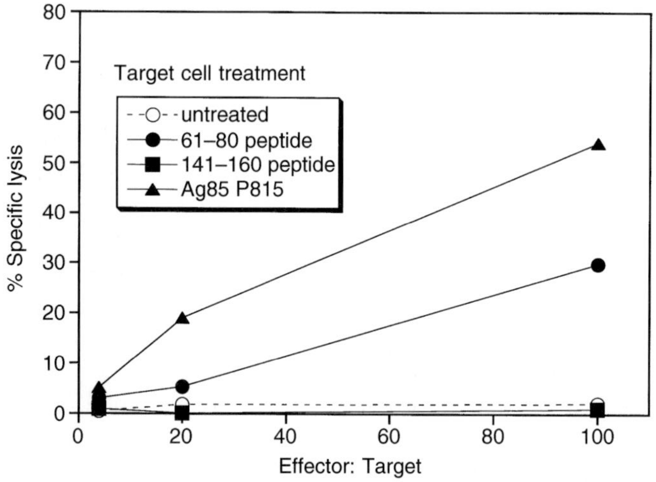

FIG. 3. Induction of cytotoxic T lymphocyte responses by vaccination with Ag85A DNA. BALB/c mice were injected with 100 μg Ag85A DNA and spleen cells were prepared for antigen restimulation at 1 year after injection. Pools of three spleens per group were restimulated with peptide (corresponding to amino acids 61–80 of Ag85A) and tested for lytic activity against P815 target cells that were untreated (open circles), pulsed with 61–80 peptide (solid circles), pulsed with 141–160 peptide (solid squares) or transfected with Ag85A (solid triangles). Data shown are per cent specific lysis for effector:target ratios of 4 : 1, 20 : 1 and 100 : 1.

was protective against systemic challenge with a high dose of BCG, as well as against a low dose aerosol challenge with *M. tuberculosis* (Huygen et al 1996). The level of protection achieved was comparable to that induced by BCG vaccine. In addition, mice were protected from a low dose aerosol challenge with a highly virulent human clinical isolate of *M. tuberculosis* that grows to high levels in the lungs of unprotected mice ($\sim 10^7$ colony-forming units). Finally, guinea pigs were protected from a low dose aerosol challenge with *M. tuberculosis*, as indicated by improved lung pathology and by decreased morbidity (weight loss) and mortality compared to controls in a long-term study (31 weeks; Baldwin et al 1998). Therefore, by several criteria, Ag85A DNA was shown to confer protective immunity against mycobacterial disease in animal models.

Prospects for tuberculosis vaccine development

The data presented here and elsewhere (Huygen et al 1996) demonstrate Ag85A DNA vaccination against tuberculosis in animal models. DNA encoding hsp65 and the 38kDa antigen of *M. tuberculosis* have also been shown to confer protection (Tascon et al 1996, Zhu et al 1997). These data suggest that there may be several protective antigens in *M. tuberculosis* and that a combination of DNA plasmids encoding discrete antigens should be investigated. But which antigens? There are likely to be several thousand proteins expressed by *M. tuberculosis*, thereby greatly complicating the determination of those that are protective against disease. One potential means of identifying these proteins is by taking advantage of the DNA vaccine technology itself. Johnston and colleagues recently showed that pools of many thousand different plasmids each containing a fragment of the genome of a pathogen (in this case *Mycoplasma pulmonis*) could be used to vaccinate and protect mice from challenge (Barry et al 1995). By extension, if these libraries could be successively fractionated to yield the protective plasmids in the mixture, then the protective antigens they encode could be identified. Alternatively, one could take a more directed approach by using open reading frames, as identified using nucleotide sequence information, to test a much more defined set of plasmids. If one wanted to target specific types of *M. tuberculosis* antigens to be tested in this way, high on the list would be those proteins secreted from mycobacteria. These could be identified using bioinformatics by the presence of recognizable signal sequences or functionally using genetic approaches (Lim et al 1995).

In summary, DNA vaccination can be used to express mycobacterial antigens *in situ*, to induce immune responses relevant to protection from mycobacterial disease (i.e. Th1 cells and CTLs) and to confer protection in animal models of tuberculosis. The next step in tuberculosis vaccine development lies in the identification of the protective antigens of *M. tuberculosis*, a process which may be aided by the DNA vaccine technology itself. As a result, a combination of DNA plasmids encoding these antigens can be tested for its potential as an improved tuberculosis vaccine.

References

Andersen P 1994 Effective vaccination of mice against *Mycobacterium tuberculosis* infection with a soluble mixture of secreted mycobacterial proteins. Infect Immun 62:2536–2544

Baldwin S, D'Souza C, Roberts A et al 1998 Evaluation of neew vaccines in the mouse and guinea pig model of tuberculosis. Infect Immun, in press

Barry MA, Lai WC, Johnston SA 1995 Protection against mycoplasma infection using expression-library immunization. Nature 377:632–635

Denis O, Tanghe A, Palfliet K et al 1998 Vaccination with plasmid DNA encoding Ag85A stimulates a broader $CD4^+$ and $CD8^+$ T cell epitopic repertoire than infection with *M. tuberculosis* H37Rv. Infect Immun 66:1527–1533

Donnelly JJ, Ulmer JB, Shiver JW, Liu MA 1997 DNA vaccines. Annu Rev Immunol 15:617–648

Guleria I, Teitelbaum R, McAdam RA, Kalpana G, Jacobs WR, Bloom BR 1996 Auxotrophic vaccines for tuberculosis. Nat Med 2:334–336

Huygen K, Content J, Denis O et al 1996 Immunogenicity and protective efficacy of a tuberculosis DNA vaccine. Nat Med 2:893–898

Lim EM, Rauzier J, Timm J et al 1995 Identification of mycobacterial tuberculosis DNA sequences encoding exported proteins by using phoA gene fusions. J Bacteriol 177:59–65
Montgomery DL, Shiver JW, Leander KR et al 1993 Heterologous and homologous protection against influenza A by DNA vaccination: optimization of DNA vectors. DNA Cell Biol 12:777–783
Montgomery DL, Huygen K, Yawman AM et al 1997 Induction of humoral and cellular immune responses by vaccination with *M. tuberculosis* antigen 85 DNA. Cell Mol Biol 43:285–292
Murray PJ, Aldovini A, Young RA 1996 Manipulation and potentiation of antimycobacterial immunity using recombinant Bacille Calmette–Guérin strains that secrete cytokines. Proc Natl Acad Sci USA 93:934–939
Orme IM, Andersen P, Boom WH 1993 T cell response to *Mycobacterium tuberculosis*. J Infect Dis 67:1481–1497
Sieling, PA, Chatterjee D, Porcelli SA et al 1995 CD1-restricted T cell recognition of microbial lipoglycan antigens. Science 269:227–230
Shiver JW, Perry HC, Davies ME, Liu MA 1995 Immune responses to HIV gp120 elicited by DNA vaccination. In: Chanock RM, Brown F, Ginsberg HS, Norrby E (eds) Vaccines 95. Cold Spring Harbor Laboratory Press, Cold Spring Harbor, NY, p 95–98
Tascon RE, Colston MJ, Ragno S, Stavropoulos E, Gregory D, Lowrie DB 1996 Vaccination against tuberculosis by DNA vaccination. Nat Med 2:888–892
Ulmer JB, Donnelly JJ, Parker SE et al 1993 Heterologous protection against influenza by injection of DNA encoding a viral protein. Science 259:1745–1749
Zhu X, Venkataprasad N, Thangaraj H et al 1997 Functions and specificity of T cells following nucleic acid vaccination of mice against *Mycobacterium tuberculosis* infection. J Immunol 158:5921–5926

DISCUSSION

Colston: You undoubtedly get a diverse antigen-specific immune response, and you undoubtedly see protective immune activity, but it's not clear to me that you have demonstrated that these are causally related, i.e. that the antigen-specific immune response is causing protective immunity. There are two reasons for this: (1) your protective immunity results are almost superimposable over results from my lab on six other antigens; and (2) I'm not sure that anyone, including my lab, has used a correct control. I would like to see a control experiment that looks at whether a mycobacterial protein that is not expressed in *Mycobacterium tuberculosis* causes protective immunity. You have shown that you get an adjuvant effect from mycobacterial DNA. Is the protection you're seeing due to that, and unrelated to the antigen-specific immune response you're observing?

Ulmer: I have to believe that at least some of the protective immunity we see is antigen specific because the control we use is the same vector without the gene insert. We have seen a DNA effect, at least in the influenza system, that can be profound on the immunogenicity of a given vector. However, in terms of protection, we've only seen slight differences between mice injected with saline and mice injected with control DNA. If the protection we saw with the Ag85

DNA was unrelated to expression of antigen we would have expected to see the same level of protection with the control DNA. We could control for any possible DNA effects in the gene itself by making a frameshift mutation, but we haven't done that. We have shown, however, that in a challenge model using a virulent human clinical isolate of *M. tuberculosis* a DNA vaccine encoding the mature form of Ag85, which has the same amino acid sequence as the secreted form, does not cause protective immunity.

Mizrahi: Have you put two genes on the same plasmid to see if there are additive effects?

Ulmer: We've made dicistronic constructs, not with two different antigens, but with an antigen and a cytokine or chemokine, and showed that certain combinations can result in an enhanced immune response. But we haven't done this in the tuberculosis system, only in the influenza system.

Mizrahi: Is it reasonable to expect that a single antigen is ever going to be as good?

Ulmer: It is likely that it won't be as good. We are already starting to see, at least in these animal models, that expression of Ag85 in isolation is not as good as Bacillus Calmette–Guérin (BCG). It might be asking too much for a single antigen to give as much protection as BCG can afford, given that BCG contains all the other antigens. We have done a lot of work on this in the influenza system. We have vaccinated mice with a combination of up to seven different DNA constructs and shown that it is possible to measure immune responses against each of these seven. They're apparently not affected by the others. Therefore, a combination vaccine for tuberculosis is feasible. We don't know whether the effects will be additive or synergistic, or whether one immunodominant protective antigen will dilute the effects of the others. Stephen Johnson has used as many as 1000–2000 discrete plasmids in combination in his expression library immunization system using *Mycoplasma pulmonis* as a model, and has demonstrated protection in a first round of screening (Barry et al 1995). This gives us some hope that we may be able to play with a combination of many antigens.

Kaplan: The better the immune response following improvements in the antigens used in the vaccine, the more likely the vaccine is to induce an immune response against muscle cells. This would in the long term result in muscle damage. Have you injected mice with muscle-expressing antigen for long periods of time and then looked at muscle damage? You talked about autoimmunity against DNA, but you didn't talk about autoimmunity against muscle cells.

Ulmer: We have two things in our favour: (1) we transfect few muscle cells, probably less than 1%, so we can afford to lose some; and (2) muscle cells are highly regenerative. We examined the histology of the muscle during the preclinical safety studies that were performed on the clinical flu DNA vaccines, and other people have looked at anti-muscle responses, but no-one has reported antibodies against muscle proteins, even up to two years following the injection.

Beyers: Is it possible to coexpress the co-stimulatory molecule B7 in muscle cells in order to enhance their function as antigen-presenting cells?

Ulmer: People have tried to do that. We modelled our bone marrow chimera studies on the work of Drew Pardoll, who showed that in implanted tumours expressing foreign antigens antigen-presenting cells (APCs) are required and tumour cells are not converted into APCs. People have co-injected DNA with DNA expressing and B7.1 or B7.2, and have seen modest rises in antibody titres in some cases (Conry et al 1996, Iwasaki et al 1997, Tsuji et al 1997). It's unlikely that muscle cells are converted into APCs under these circumstances; but it is possible that B7 is not expressed on the same cell as the antigen is being expressed.

Brennan: You suggested that since the Ag85 has an asparagine in a glycosylation-competent motif that the expressed protein was glycosylated. Perhaps you should therefore look at an *M. tuberculosis* antigen that doesn't have an asparagine in a glycosylation-competent motif. The consequences of glycosylation on the APCs could be profound because there are carbohydrate-binding proteins on the surface of these cells.

Ulmer: We hypothesize that the antigens are transferred in the form of peptides, so unless the carbohydrate interferes with the processing of the protein into peptides then it probably won't be a problem. If, however, that amino acid is in the middle of the T cell epitope, it would probably interfere with the binding of the peptide either to the chaperone or MHC class I molecule. One could change the asparagine residue to glutamine in order to prevent the addition of the carbohydrate. However, although the addition would be prevented, the amino acid change itself may have a direct effect on the potency of the T cell epitope. We should consider that protein modifications that are not normally present may interfere with processing and presentation.

Hopewell: You mentioned that you were interested in routes of administration. Have you looked at whether different routes target APCs more effectively?

Ulmer: In the influenza model we looked at various routes of administration with a syringe, and we found that an immune response could be generated by intradermal and intravenous administration in addition to intramuscular administration. However, we found that protection based on cellular immune responses was superior for intramuscular injection. It has been reported that certain cationic lipids formulated with DNA can give rise to expression throughout the respiratory epithelial tract (Stribling et al 1992), so we have recently been looking at non-aerosol intranasal administration of naked DNA or lipid-modified DNA. From a vaccine point of view, it may be simplest for now to stick with intramuscular injection because that's how most vaccines are administered. However, oral and intranasal routes are also worth considering.

Rook: In view of some of the hypotheses about how peripheral tolerance to specific antigens is maintained, it may be precisely because the muscle cells are not converted into APCs that autoimmunity is not a problem. Additional tricks to turn the muscle cells into fully professional APCs should probably be avoided.

Ryfell: What sort of safety tests are carried out on DNA vaccines? Also, in your presentation you mentioned the term 'genotoxicity'. Could you explain what you mean by this?

Ulmer: DNA vaccines will undergo the same rigorous safety tests as those that have been carried out for any other vaccine on the market, as well as those that are specific for DNA vaccines, such as integration and autoimmune responses. What I mean by the term 'genotoxicity' is the degree of physical DNA changes, such as chromosomal DNA strand breaks, which are known to be induced by certain viral vaccines. This is a standard assay that is applied to the safety testing of vaccines.

Ryfell: Do you carry out any studies in non-human primates?

Ulmer: Yes. It is necessary to show in non-human primates that the construct expresses a protein and that an immune response is made against it, even if there is no challenge model.

Ryfell: In your opinion, which cytokines should be included in multiple antigenic construct vectors?

Ulmer: People have looked at many cytokines and chemokines in various models. Granulocyte macrophage colony-stimulating factor gives a modest reproducible enhancement of antibody titres, and in some cases cellular responses. This is also true for interleukin (IL)-12, IL-2 and certain chemokines, but nothing so far has been shown to give a profound response. We have found that, in general, it is difficult to improve DNA vaccines that consist of highly expressing vectors encoding protein antigens that are inherently immunogenic, whereas incremental enhancements can be made if you use a limiting dose of DNA and less than optimal vector encoding a poorly immunogenic protein.

Blackwell: In the dose–response experiments where you were testing the effect of DNA on T cell responses did you keep the total amount of DNA constant at 100 μg? I was wondering whether you need that much DNA not because you need a certain amount of specific antigen activity but because you need a certain amount of adjuvant activity provided by non-antigen-specific sequences in the vector DNA.

Ulmer: The amount of DNA in those experiments was not constant, rather we were titrating the DNA dose, and the response was antigen specific. We found that for the influenza constructs robust responses could be observed with only 1 μg administered once. Better responses were not always observed just by adding extra DNA. We haven't done these experiments with tuberculosis DNA vaccines, so I don't know whether this would give the same effect.

Blackwell: Have you tried pooling DNA vaccines for a series of unknown genes?

Ulmer: These experiments are currently in progress. There are advantages with using fewer constructs because you can keep the dose of each up in the 100 μg range, whereas if you use 1000 constructs then the dose of each would be in the nanogram range, which may not be sufficient to generate an adequate response.

Colston: Have you treated an established infection with an antigen-specific response? Also, do you give multiple injections in the protection experiments, and if so why?

Ulmer: We haven't done many dosing experiments in the tuberculosis challenge model because of limitations on how many different things can be tested at once. In the influenza model we found that we could give a smaller total DNA dose if we administered it in multiple injections. We did not get the same type of boosting response that we do with a subunit-type vaccine, but we did observe incremental increases in antibody titres. However, the titres normalize with time, irrespective of the total dose or how many doses are given. The kinetics of the responses appear different, so it is possible that the maturation of the protective immune response is more rapid with high doses and multiple injections of DNA.

Rook: I would like to ask a question that may be unfair because we're on the record. In view of the fact that it is difficult to do vaccination trials in human tuberculosis, and in view of the fact that the efficacy of the BCG vaccine is variable, so you would be superimposing a vaccine upon this variable response, how serious are your studies on tuberculosis?

Ulmer: That's a good question. I was actually going to ask Paul Fine a similar question because I become more and more discouraged each time I hear him speak. The prospect of a 20–30-year clinical study to establish the efficacy of a vaccine will deter even the most intrepid vaccine developer. What are the prospects of setting up a large-scale, short-term trial in order to establish some measure of superiority of a vaccine over BCG? This could allow licensure of the vaccine. A long-term, post-licensure follow-up study could then be set up to investigate variations in regimen and boosters, etc.

Fine: The problem is that we need protection against adult pulmonary disease, whereas we generally talk about vaccinations in childhood. Protection in adults might be shown through a long-term follow-up study starting with children, or by aiming a trial at uninfected and/or recently infected adults. We are looking at a disease where the major burden lasts into old age, so we're also going to have to face the issue of the duration of protection, and whether we need boosters and how they can be evaluated. I agree that it is discouraging to think that it will take 20 or 30 years to complete 'all' these studies, and I too am concerned that it may put a damper on efforts to develop and evaluate vaccines. Your suggestion to explore large-scale, short-term trials to show superiority over BCG is interesting, but it still raises difficult issues. If we aim at childhood tuberculosis, e.g. meningitis, we are setting a difficult target for any new vaccine to improve upon, given that BCG appears to be consistently quite effective against such disease. For this reason I might favour setting up a comparative trial against adult tuberculosis in an area where BCG has shown to provide poor protection. If we can show that a new vaccine can provide any protection under such circumstances then we can play it by ear after that.

Ulmer: It seems to me that the most expeditious thing to do is to set the bar low in the first instance, although it would have to be high enough so that you could show superiority or equivalence to BCG.

Donald: In primary childhood tuberculosis the two main manifestations that are clinically diagnosable and occur in high incidence areas are miliary tuberculosis and tuberculous meningitis. If your vaccine was effective these would be prevented, and by preventing these disseminated forms you might also prevent a certain amount of adult tuberculosis. It would only take a couple of years to find this out.

Ulmer: But you would have to do it as an adjunct to the BCG vaccine.

Donald: In the Western Cape we routinely vaccinate with BCG, and yet we have consistently high rates of tuberculosis meningitis and miliary tuberculosis. Therefore, there is certainly ample room for substituting a different vaccine for BCG in certain areas on the basis of our current figures.

Bateman: If there are ethical problems relating to administering the vaccine to paediatric age groups, it may be possible to compare the effect of the vaccine upon reactivation in adults, in both low and high prevalence areas. This could provide results within a reasonable time span.

Ulmer: We have considered whether there's a specific group within an industrialized country that we could look at. There are relatively high incidence areas within the US, but unfortunately those populations tend to be homeless and/or transient and are therefore not amenable to long-term follow-up studies. The only population that comes to mind is a prison population.

Bellamy: It seems to me that vaccines have two different aims. The first is to reduce the incidence of disease manifestations in populations that become infected. And the second is to reduce the fitness of the micro-organism so as to reduce its spread. Tuberculosis and malaria are particularly unusual diseases in that they have exerted significant effects on the human genome throughout the course of evolution. The human immune system has spent a long time attempting to develop basic mechanisms to reduce the fitness of those micro-organisms within human populations, but it hasn't been able to eradicate them. Therefore, it may be difficult to design a vaccine that will decrease the fitness of these micro-organisms, and perhaps the goal we ought to be aiming at is to reduce the manifestations of disease. In the case of malaria, perhaps what we ought to do is prevent cerebral malaria in children, and in the case of tuberculosis what we ought to do is reduce the incidence of miliary tuberculosis and tuberculous meningitis in children.

Ulmer: There is a perception that the gold standard for a vaccine is to induce immunity in the same way that an infection does, and that you can't improve upon immune responses generated by micro-organism infection. I don't entirely agree with this. If you can modulate the immune response or focus the immune response on specific antigens, so that when individuals encounter an organism they can rapidly respond to those antigens, then this can potentially result in better immunity than the micro-organism can itself induce, which as you mentioned has had plenty of time to evolve specific mechanisms to avoid detection and eradication by the immune system.

Bellamy: But malaria and tuberculosis may be different to other infectious diseases in that they are essentially parasites that sometimes cause diseases, and throughout history there has been substantial selective pressure on the human immune system to develop mechanisms of eradicating the micro-organism. The diseases for which we have developed effective vaccines may have not been around for as long, or they may not have been exposed to the same level of selective forces, because they occur in epidemics or because they have not been a cause of such high mortality rates as malaria and tuberculosis.

Fine: I question the idea of developing a new vaccine against tuberculous meningitis and miliary disease because most of the evidence suggests that BCG is effective against these forms of tuberculosis. In South Africa there seems to be a general feeling that BCG isn't protecting against tuberculosis meningitis, but BCG is administered in a rather peculiar way in South Africa, so many people question whether this is responsible for the failure of BCG here. Let us also not forget that tuberculous meningitis and miliary disease are rare, and these are not why tuberculosis is considered a global emergency.

Ryfell: What is different about BCG administration in South Africa?

Donald: It's the same procedure as that used in France, Japan and in several other countries. There are arguments about the effectiveness of the precise instrument that has been used here in the past. There may be something wrong with the method of administration; however, we can certainly document whether or not children have received their BCG injection, and we find that 90% of children who present with tuberculosis meningitis have received it.

Fine: They may have received BCG but many people have questioned whether administration with this instrument is an effective way to deliver the vaccine. To my knowledge there have been no formal studies comparing its effectiveness with other methods of delivering the vaccine.

Ress: A study to address this question is due to commence shortly.

Colston: The two most recent trials where BCG has failed to protect are the Malawi and South India trials (Karonga Prevention Trial Group 1996, Baily 1980), and these are areas where BCG has been used extensively for many years. Is it possible that there has been selection for *M. tuberculosis* strains that are unaffected by BCG?

Fine: An interesting question. The South India BCG trial commenced in 1968 in an area where BCG hadn't previously been used, and our trial in Malawi was carried out in a population where BCG had been introduced in 1975 and had not been widely used. I would therefore be surprised if there had been selection against certain strains of *M. tuberculosis*, given that most transmission is thought to be attributable to adults with pulmonary disease who, in the case of these two populations, were in general born and infected before the introduction of BCG.

References

Baily GV 1980 Tuberculosis prevention trial, Madras. Indian J Med Res (suppl) 72:1–74

Barry MA, Lai WC, Johnston SA 1995 Protection against mycoplasma infection using expression library immunization. Nature 377:632–635

Conry RM, Widera G, Lobuglio AF et al 1996 Selected strategies to augment polynucleotide immunization. Gene Ther 3: 67–74

Iwasaki A, Stiernholm BJN, Chan AK, Berinstein NL, Barber BH 1997 Enhanced cytotoxic T lymphocyte responses mediated by plasmid DNA immunogens encoding co-stimulatory molecules and cytokines. J Immunol 158:4591–4601

Karonga Prevention Trial Group 1996 Randomised controlled trial of single BCG, repeated BCG, or combined BCG and killed *Mycobacterium leprae* vaccine for prevention of leprosy and tuberculosis in Malawi. Lancet 348:17–24

Stribling R, Brunette E, Liggitt D, Gaensler K, Debs R 1992 Aerosol gene delivery *in vivo*. Proc Natl Acad Sci USA 89:11277–11281

Tsuji T, Hamajima K, Ishii N et al 1997 Immunomodulatory effects of a plasmid expressing B7–2 on human immunodeficiency virus-1-specific cell-mediated immunity induced by a plasmid encoding the viral antigen. Eur J Immunol 27:782–787

Summary

Douglas Young

Department of Medical Microbiology, Imperial College School of Medicine at St Mary's Hospital, Norfolk Place, London

From a symposium on the theme of 'Genetics and Tuberculosis', it is attractive to consider tuberculosis as a meeting of the mycobacterial genome with the human genome, and suggest that identification of the genes involved in this interaction will hold the key to understanding the disease. There are two caveats to such a conclusion, however. The first came up in the presentations from Richard Bellamy and Igor Kramnik. Variations in a large number of genes — probably at least 100 — appear to influence the host response to infection, and there may be many additional background genes that, although not having a direct role in infection, can influence those genes that are involved. Igor showed that, in a mouse model, certain genes could have a beneficial effect in some backgrounds and a detrimental effect in others, for example. Susceptibility or resistance is a complex phenotype, in which understanding of interactions between multiple genes may well prove more important than identification of individual genes.

Similarly in bacteria, genes encoding factors that are directly involved in host cell interaction are most probably influenced by other background physiology genes. The second caveat is that, whatever the genotype, factors related to nutrition, stress and endocrine status all have a profound influence on disease susceptibility. To understand tuberculosis, therefore, we need not only to identify the key host and microbial genes, but also to understand how their function is regulated by interactions with other genes and by physiological and environmental variables. Genetics is not going to provide a simple one-word answer that will solve all our problems in tuberculosis; we have to regard genetics as a powerful approach that we can add on to our existing ways of trying to dissect these problems. In summarizing our discussions over the last few days, I will divide my comments into ways that genetics are being used to answer questions firstly about the microbe, then about mechanisms of protective immunity and finally about the epidemiology of tuberculosis.

In terms of the bacteria, drug resistance is an area in which genetics has already answered important questions and generated improved tools for disease control. However, as Julian Davies and Lynn Miesel pointed out, there is more to be learned about drug resistance than simply identification of genotypic markers.

Julian encouraged us to think of what we might learn about the biology of *Mycobacterium tuberculosis* by looking more closely at the development of resistance. He referred to the horizontal transfer of genes. Although, so far, this has not played a role in resistance in *M. tuberculosis*, phage-like elements have been identified by genome sequencing, and the possibility that horizontal gene transfer has contributed to virulence, and might make a future contribution to resistance, merits careful evaluation.

Stewart Cole's presentation of the near-complete sequence of the genome of *M. tuberculosis* provided an exciting vision of future research directions. He illustrated the way in which the process of accumulation of sequence data is being transformed into functional genomics, with novel and unexpected implications that can be fed back to the 'wet' lab for experimental analysis. With rapid progress in bacterial genome sequencing, comparative genomics will revolutionize the study of bacterial physiology and pathogenesis over the next decade. The development of expertise in bioinformatics will be an important aspect in ensuring that tuberculosis research can optimally exploit progress in the field of bacterial genomics.

Paul van Helden discussed ways of studying mycobacterial virulence by looking at transmission of different strains in an epidemiological setting. He discussed questions associated with the evolution of *M. tuberculosis*, and the 'virulence' or 'fitness' of individual strains. A limitation of current studies in this area is that, although we now have useful genotypic markers for strain variation (such as IS6110), we have yet to identify variable genetic elements that have an influence on biological phenotype. Genomic analysis will provide a rich source of candidates for testing as potential phenotypic markers. Considerable benefit might be anticipated in this area from interactions with experts in the study of other bacterial diseases, in which the science of population genetics is more advanced.

The interaction of mycobacteria with macrophages is a central event in the infectious process and was discussed briefly by David Russell. He described the development of elegant approaches to study the biology and biochemistry of the mycobacterial phagosome, in combination with genetic tools to investigate the way in which the pathogen subverts the normal host response. By establishing the precise sequence of events leading to mycobacterial killing, or survival, within the macrophage, we can set a framework for understanding the significance of the other immunological events associated with infection.

We spent a lot of time discussing the immune response to infection. We agreed that macrophage activation by the γ-interferon (IFN-γ) pathway plays a central role in protection, although additional cell–cell interactions may also influence the killing of intracellular mycobacteria. There was much discussion of why, if IFN-γ is protective, we can have IFN-γ production and still have disease. Where does the immune response go wrong? What is the mechanism of disease-associated immunopathology? The effect of IFN-γ might depend on the presence or absence of other immune mediators. A combination of IFN-γ with excessive levels of inflammatory cytokines, such as tumour necrosis factor (TNF), may be detrimental, for example. Another suggestion was that TNF in the presence of T helper 2 cytokines might contribute to pathology. Ian Orme suggested that immune activation driven predominantly by cytokines might provide protection, whereas a chemokine-driven response might be associated with delayed-type hypersensitivity. Graham Rook discussed mechanisms by which the endocrine system can influence the immune response; potentially antagonizing or synergizing with the IFN-γ pathway. Finally, Peter Donald cautioned that humoral responses should not be ignored, because they may influence the initial stages of infection or the transfer of bacteria from one cell to another.

There is no simple way to disentangle this complex network of interacting immune mediators. Animal models represent one useful approach, with knockout mice providing a powerful tool to determine whether or not individual cytokines play an essential role in protection. It is more difficult to analyse the hypothesis that protection versus pathology is determined by quantitative relationships between different mediators, however. In the context of human tuberculosis, Gilla Kaplan and Graham Rook discussed the way in which immunotherapy trials might be used to evaluate the contribution of particular cytokines or T cell subsets to the outcome of the immune response. Paul Fine pointed out that the differential effect of Bacillus Calmette–Guérin vaccination, protecting some populations but not others, provides an opportunity to test the role of particular types of immune response in protection.

Richard Bellamy discussed two genetic approaches that could be used to study susceptibility in human populations. The candidate gene approach could have an important role in evaluating the contribution of particular cytokines to protection. Alternatively, screening based on a whole-genome approach has the potential to identify genes, and consequently mechanisms, that had not been seen by conventional biology. *NRAMP* was initially identified by whole-genome screen in mice, for example, with the product now being actively studied for its biological function. Recent candidate gene approaches suggest that *NRAMP* may also have some role in tuberculosis susceptibility in humans. The problem of the whole-genome approach is that its 'power' is relatively low in a situation in which multiple genes each make a relatively small individual contribution to overall susceptibility. It may be possible to increase the power of the genetic screen by focusing on susceptibility to particular types of disease—reactivation versus reinfection disease, for example. A combination of human genetics with bacterial genetics (to identify particular disease situations) might be advantageous. It is clear that we are only at the beginning of exploitation of human genetic analysis in tuberculosis research.

I have focused my comments particularly on the mechanisms underlying the biology of tuberculosis from the perspective of fundamental scientific understanding. With the urgent need for improved disease control, it is clear that we cannot wait for a complete scientific understanding before proceeding with attempts to develop new drugs and vaccines for tuberculosis. Progress in combating tuberculosis will ultimately depend on a synergy between the 'push' of basic science and the 'pull' of the public health priority. Our discussions over the last few days have highlighted the exciting potential that novel genetic approaches provide in facing the formidable problems of this most complex and persistent of human infections. I'm sure that if René Dubos had been here, he would wish once more to be a young researcher setting out on the fight against tuberculosis, armed with today's powerful new genetic tools.

Epilogue

Solomon R. Benatar

Department of Medicine, Medical School, University of Cape Town, Observatory 7925, Cape Town, South Africa

A quarter of a century ago there was hope that tuberculosis could be eradicated from the world, even though it was anticipated that this would take many decades to achieve. The causative organism was well known, and effective drugs were available. These drugs had been well tested in clinical trials, and rifampicin, a new and powerful drug that allowed treatment time to be reduced by two-thirds, was coming into widespread use. Today, the possibility of eradicating tuberculosis is not only more remote but also seems almost impossible to contemplate given the development of multidrug resistance and the profoundly adverse influence of HIV infection (Benatar 1995).

The lost opportunity to eradicate a disease responsible for vast human suffering can be attributed to the failure to appreciate the extent to which tuberculosis and other diseases cannot be addressed merely from a biomedical perspective. There is a desperate need also to address widening economic disparities associated with exploitative economic and military policies and wars that are killing more civilians than ever, displacing whole populations from their homes and countries, and rendering them vulnerable to old and new infectious diseases. Ecological degradation, due to both population growth in poor countries and unsustainable consumption patterns in wealthy countries, pose additional threats to the health of whole populations (McMichael 1993, Benatar 1997).

Scientific advances must continue to be made as these will lead to the development of more effective drugs and vaccines that could shorten the duration of treatment. However, this will not be sufficient, and scientists cannot complacently confine themselves to the comfort of their laboratories. They need to harness the credibility associated with their scientific knowledge to contribute to those social and political processes that could reduce exploitation, reduce wastage of resources on weapons of mass destruction and improve population health through investment in development. If they fail to do so they will be seen by future generations to have been blind, selfish or stupid!

Antituberculosis treatment is one of the most cost-effective treatments available in the world. Making such drugs available to all with the disease is not beyond the financial or organizational capacity of the world. The withdrawal of the USA and the UK from payments of their subscriptions to the IUATLD (International Union against Tuberculosis and Lung Diseases) is a sad reflection on the humanitarianism of some of the most privileged people in the world.

As we look back and recognize that it took 50 years for the antislavery campaign to render slavery illegal, we can appreciate that it is indeed possible to reverse profound social ills, although the process is difficult and slow. Endeavours to reduce the production, sale and use of weapons of mass destruction and efforts to divert resources towards more productive uses needs widespread support and could have a major impact on tuberculosis and other infectious diseases (Benatar 1997).

Can we afford to be silent and complacent? What kind of moral judgements will be made about us in the future if we neglect these obvious obligations? Will scientists be seen in the same light as South Africans who were blind to the effects of apartheid? We know the answers to these questions. It is time to take action that includes but goes beyond our justifiable commitment to sustaining scientific progress.

References

Benatar SR 1995 Prospects for global health: lessons from tuberculosis. Thorax 50:487–489

Benatar SR 1997 World Health Report 1996: some millennial challenges. J Roy Coll Physicians Lond 31:456–457

McMichael AJ 1993 Planetary overload: global environmental change and the health of the human species. Cambridge University Press, Cambridge

Index of contributors

Non-participating co-authors are indicated by asterisks. Entries in bold indicate papers; other entries refer to discussion contributions

M

O

Q

R

S

T

U

V

W

Y

Z

Subject index

D

E

K

L

M

N

O

P

Other Novartis Foundation Symposia:

No. 197 **Variation in the human genome**
Chair: K. M. Weiss
1996 ISBN 0-471-96152-3

No. 206 **The rising trends in asthma**
Chair: S. T Holgate
1997 ISBN 0-471-97012-3

No. 207 **Antibiotic resistance: origins, evolution, selection and spread**
Chair: S. B. Levy
1997 ISBN 0-471-97105-7